GEOValue

T0272011

GEOValue
The Socioeconomic Value of Geospatial Information

Edited by

Jamie Brown Kruse

Joep Crompvoets

Françoise Pearlman

CRC Press
Taylor & Francis Group
Boca Raton London New York

CRC Press is an imprint of the
Taylor & Francis Group, an **informa** business

CRC Press
Taylor & Francis Group
6000 Broken Sound Parkway NW, Suite 300
Boca Raton, FL 33487-2742

First issued in paperback 2019

© 2018 by Taylor & Francis Group, LLC
CRC Press is an imprint of Taylor & Francis Group, an Informa business

No claim to original U.S. Government works

ISBN-13: 978-1-4987-7451-2 (hbk)
ISBN-13: 978-0-367-87889-4 (pbk)

This book contains information obtained from authentic and highly regarded sources. Reasonable efforts have been made to publish reliable data and information, but the author and publisher cannot assume responsibility for the validity of all materials or the consequences of their use. The authors and publishers have attempted to trace the copyright holders of all material reproduced in this publication and apologize to copyright holders if permission to publish in this form has not been obtained. If any copyright material has not been acknowledged please write and let us know so we may rectify in any future reprint.

Except as permitted under U.S. Copyright Law, no part of this book may be reprinted, reproduced, transmitted, or utilized in any form by any electronic, mechanical, or other means, now known or hereafter invented, including photocopying, microfilming, and recording, or in any information storage or retrieval system, without written permission from the publishers.

For permission to photocopy or use material electronically from this work, please access www.copyright.com (http://www.copyright.com/) or contact the Copyright Clearance Center, Inc. (CCC), 222 Rosewood Drive, Danvers, MA 01923, 978-750-8400. CCC is a not-for-profit organization that provides licenses and registration for a variety of users. For organizations that have been granted a photocopy license by the CCC, a separate system of payment has been arranged.

Trademark Notice: Product or corporate names may be trademarks or registered trademarks, and are used only for identification and explanation without intent to infringe.

Library of Congress Cataloging-in-Publication Data

Names: Kruse, Jamie B., editor. | Crompvoets, Joep, editor. | Pearlman, Françoise, editor.
Title: GEOvalue : the socioeconomic value of geospatial information / edited by Jamie Kruse, Joep Crompvoets, and Françoise Pearlman.
Description: Boca Raton, FL : Taylor & Francis, 2018. | Includes bibliographical references.
Identifiers: LCCN 2017025193 | ISBN 9781498774512 (hardback : alk. paper)
Subjects: LCSH: Geospatial data--Economic aspects. | Geospatial data--Social aspects. | Geographic information systems--Economic aspects. | Geographic information systems--Social aspects.
Classification: LCC G70.217.G46 G467 2018 | DDC 338.4/791--dc23
LC record available at https://lccn.loc.gov/2017025193

Visit the Taylor & Francis Web site at
http://www.taylorandfrancis.com

and the CRC Press Web site at
http://www.crcpress.com

This book is dedicated to Molly Macauley whose untimely death shocked us all.

Molly Macauley, the scientist, was intensely focused on how space-based resources provide a number of valuable public goods, including information from Earth observation, and on how society could do a better job to manage those resources. Molly Macauley, the pioneer, brought new ideas from economics to the space and remote-sensing communities, thereby drawing together disciplines that often barely knew of each other. Molly Macauley, the advocate, fearlessly and tirelessly mentored and sought to open the doors for other women in space and economic sciences. Molly Macauley, the person, was deeply loyal, kind, and unselfish to a fault. To know her was to be touched by her, and her death leaves a hole in the hearts of all the many people who loved her.

Dr. Mike Toman
Lead Economist and Manager
Sustainable Development, The World Bank

Molly's intellectual contributions in both space and environmental policy spanned disciplinary boundaries. She brought the rigor of her training as an economist to policy questions that turned on the value of scientific information, and her work on remote sensing broke new ground. This would be quite enough of an achievement for anyone, but Molly also cared deeply about the people and institutions with which she was associated and worked hard to ensure their sustainability. This volume speaks to the commitment of Molly's many colleagues and friends to see her standards of excellence and integrity live on.

Dr. Debra Knopman
Principal Researcher
RAND Corporation

Molly left behind a lifelong commitment to connecting science and decision-making, and she brought a deep passion for environmental economics, renewable energy, and satellites. Beyond the many roles she played and the committees she served on, it was the way that Molly engaged with people and with issues, discerning key points and offering her counsel with grace, poise, and respect.

Lawrence Friedl
Director
Applied Sciences Program, NASA

For the past decade, when we heard about Molly Macauley, we thought of a leader

in space economics and in value of information research. Molly was an icon at

Resources for the Future and in the space economics and value of information

research communities. I had the good fortune of arranging for Molly to be a plenary

speaker at each of the first three A Community of Ecosystem Services (ACES)

conferences. In 2008, Molly spoke in the closing plenary session; in 2010, she was a

plenary speaker in the opening session, and in 2012, Molly moderated the plenary

session on research advances. In each case, Molly engaged the community with a

clear, well-thought out, and thought-provoking message on advancing the science and

practice. The last time that I saw Molly was at the GEOValue Workshop in Paris

in March 2016. I was asked to provide summary remarks immediately after Molly's

at the end of the workshop. I remember thinking and saying that it was always a

challenge to speak immediately after Molly at a public event. Her thoughtful and

engaging remarks always set a high bar that was difficult if not impossible to meet.

The research community of which she was a leader now faces the challenge of building

on the rich legacy that Molly left us to continue to advance research and applications

in space economics, value of information, and other topics of national need.

Dr. Carl Shapiro
Director
Science and Decisions Center, US Geological Survey

Contents

Foreword

Connecting People, Places, and Policy; Collaborative Choices, Better Decisions—At Any Level By Anyone

Geospatially Speaking ... More than a Map

Any mention of geospatial information or geographic information system (GIS) to a cross-section of people often elicits either a blank shrug (what is a GIS?) or possibly a reference to ... *that map stuff.* GIS as an effective tool for policy-makers must be more than a map.

A geographic information system, or GIS by the acronym, is a simple way to understand complexity through place-based information—the Where of Things. People in positions of responsibility want to understand complex subjects and weigh their options before making a decision, but once a decision is reached they must be able to explain their rationale in easy to grasp terms. Give them a story to tell. Spatial data used in today's advanced technology formats enables understandable communication through visualization, the power of a picture. We are witnessing the evolution of geospatial systems from specialized and compartmentalized niche technology to broadly embedded and integrated daily use.

GIS may not be in a decision-maker's lexicon, but all decision-makers use a georeference whether they are aware or not. That is because everything we do is some*where*. Plus, every decision influences something or someone. A map only shows geography in two dimensions. Adding spatial data and time to those dimensions provides context, insight, and understanding. Geospatial analysis enables perspective for decision-makers.

The collaboration through integration of information systems can enable better decisions. But geospatial data gathered by remote-sensing devices, satellites, statistical analysis, and other methods, including personal observations are often scattered among many public and private agencies and are not always available to enable better decisions. Information Integration is not an option but a necessity, delivering tremendous benefit when users leverage each other's information and knowledge. The challenges of today span political and agency boundaries and even business enterprises. Decision-makers should understand their shared responsibilities and concurrent jurisdictions because functional areas transcend agency boundaries.

How might geospatial data and information be shared and reused? Effective integration requires a shared purpose or a common mission, mutual, not exclusive, benefit among all, and an understanding that cooperation is not capitulation and leadership that recognizes purpose over process.

Today's knowledge economy benefits from geospatial thinking that enables greater collaboration to resolve difficult or conflicting goals—in context. With ever increasing expectations by citizens and consumers, how well are we preparing the next generation to meet global competition? Perhaps our greatest threat is not our competition but our own complacency.

The government agencies do not always recognize the benefit or the need to acquire data that can be gathered by one then used by others over many issues. Deciding who owns the data is not the important factor. Agencies want to control data ostensibly to assure that it is not compromised or misused, but in reality, controlling geospatial data represents retaining power. *Information is power* is a misnomer. Power is not in the hands of the one holding the data. Power is in the hands of those who apply the information.

Information delivered through visualization based on location is increasingly popular, particularly with today's easier to use digital tools and web connectivity.

Elected officials, business executives, and even ordinary citizens want to understand the interrelationships between and among the issues that they face every day. The top issues that dominate discussions could overwhelm our policy-makers if they do not have the right information at the right time to make the right decision.

We do not just need more data. We need data that enables a decision. That is how geospatial systems will appeal to decision-makers. One powerful way to make an impression is to put a face on the issue—personalize it. If your state leaders are having trouble grasping the significance of GIS, reference one of the high priority issues of the day, such as the economy, education, or homeland security, then by example, show how geospatial data and analysis brings the issue into focus and can offer options for decisions. Impress on your leaders that they do not have to know how GIS works, only that it does work. Let the picture do the talking. Get their attention and take advantage of the teachable moment.

We are in a data-rich society with a great variety of information available through digital access—if we can discover it, if we are willing to share it. What we need is a decision-rich society.

The power of GIS is its ability to show the interrelationships of data referenced at a given place or geography. Presentations in GIS form enable the ability to spot trends or allocate resources more quickly than any other means of presentation.

For those who want to influence or inform decision-makers, consider this—If you had ten minutes at the top, what would you talk about? What do decision-makers want? They want information in a context that leads to a decision that enables action. But the first question any decision-maker should ask is *What are my options*? Geospatial data provides the where, the context and the relationship of issues and the ability to evaluate how the options optimize the intended benefits against the known or even the unintended consequences.

Jim Geringer, Wyoming Governor, 1995–2003
ESRI—Director of Policy and Public Sector Strategy

Preface

GEOValue: The Socioeconomic Value of Geospatial Information provides an overarching framework to explore the many facets involved in value creation from geospatial information. The contributors to this book have a broad range of backgrounds, scientific disciplines, and very different domains of expertise. Coming from Europe, United States, Australia, and South Africa, and affiliated with academia, public sector, or private sector, each author has brought a unique vision for a clearer understanding of the process and the vast potential for future contributions of Earth Observation (EO) and geospatial information to human well-being. Although the field has been active since the beginning of the twenty-first century, there is much work to be accomplished before valuation methods are sufficiently vetted and consistently applied to become standards of practice. This book is a first step in identifying where we are and what gaps must be bridged to get to such standards.

This book has been organized into three sections: Section I provides an overview of socioeconomic value of EO. In Chapter 1, the editors provide background and motivate the necessity for valuation of geospatial information. They explore the challenge of describing the contribution of geospatial information to twenty-first century society. In Chapter 2, Jason Gallo, Richard Bernknopf, and Steven Lev examine current policy and use of EO products, explore the characteristics of openly accessible geospatial data as a public or merit good, and outline a conceptual framework for estimating the socioeconomic value of geospatial information products.

Section II is devoted to exploring the components of the value chain that transform data to decision support systems. In Chapter 3, Joep Crompvoets Françoise Pearlman and Jay Pearlman discuss the transition from data to information and look at the sources of geospatial data and the processes used to interpret the data, thus providing value to a user. Discussions include data quantity, quality, types, resolution, and scale; an abundance of data and archives; typology of data supplied; and typology of services supplied. In Chapter 4, Pierre-Philippe Mathieu makes the case for an integrated approach to world resource management and argues that observations from satellites and other remote-sensed data form the foundation of scientific understanding of Earth's resources especially with respect to the prospect of climate change. Examples are presented of EO pilot projects performed in partnership with scientists, industry, and development practitioners to support climate science, adaptation, mitigation, and disaster risk management. In Chapter 5, in contrast to remotely sensed data, Jay Pearlman and Joep Crompvoets examine the role of spatial data that are acquired on-site or *in situ*. They consider sensors in four domains: (1) land and water, (2) climate and atmosphere, (3) oceans and coast, and (4) economy and society. These four categories are also represented in a number of international initiatives, including the 2030 Agenda for Sustainable Development and the associated Goals (SDGs). Lesley Arnold then addresses value and uncertainty of information and the requirements to transform data into an actionable knowledge for the end user with special emphasis on the supply chain in Chapter 6. A supply chain that comprises Semantic Web technologies is discussed as a way for the spatial information sector to move beyond the business of data to become a knowledge-based industry that has enhanced value for end users. Chapter 7 on business models for geographic information (GI) by Glenn Vancauwenberghe, Frederika Welle Donker, and Bastiaan van Loenen, examines the process of creating value using the foundational format of business model theory. The aim of this chapter is to contribute to the understanding of the structure and diversity of dealing with GI. This is done by investigating how organizations use a broad variety of models to create, deliver, and capture value. In Chapter 8, Dawn Wright outlines the role of aggregators that herald a new era of dynamic big data and the challenges of organizing varieties of data streams of different volumes that arrive at different velocities. Recent advances in information technology and civil remote sensing have allowed us to advance beyond mere static data collection and archiving, which enables information awareness facilitating better decision-making. In Chapter 9, Robert Downs provides a brief overview of recommended practices for enabling the reuse of geospatial information such as geospatial data, maps, systems, products, and services. Stewardship activities are addressed throughout the data lifecycle, which facilitate the reuse for new purposes, such as scientific research, education, planning, and policy-making. Processes are suggested that engage both the creators of source data, potential user communities, and digital repositories, such as scientific archives. Intellectual property rights associated with potential reuse material are considered.

In Section III of this book, the lead chapter (Chapter 10) by contributors Alan Smart, Andrew Coote, Ben Miller, and Richard Bernknopf explores the meaning of socioeconomic value and the theoretical foundation of estimating valuation from an economic perspective. The authors provide a comprehensive overview of valuation methodologies including their applicability in three given contexts with illustrative case studies. They conclude by summarizing new and emerging techniques that will expand decision-oriented uses of geospatial data. In a two-chapter series on qualitative methods, Miriam Murambadoro and Julia Mambo examine participatory approaches in Chapter 11. Participatory approaches encourage communities, institutions, or individuals to actively participate in understanding a phenomenon or addressing a particular issue. Data that are collected is subjective and participants decide what knowledge and/or experiences they wish to share or withhold. Participatory research includes, for example, focus group discussions and participant observation and interviews. Advantages and disadvantages of focus group discussions, participant observations, and interviews are discussed. This discussion addresses validity, reliability, and ethical issues when conducting qualitative research. The second chapter (Chapter 12) from Serene Ho asserts that as geospatial information has become more heterogeneous, and its use and application more diverse, there is a need for practitioners and users to assess and understand the multiple ways that geospatial information creates value. This chapter outlines some of the key philosophies and associated methods from the social sciences that can help producers of spatial data construct a richer understanding and description of the benefits and impacts of geospatial data.

The following chapters are devoted to case studies that demonstrate applications that provide access to knowledge and understanding that show the unique contribution of geospatial information. Chapter 13 by Jacob Hochard and Evan Plous Kresch examines recent trends in the use of geospatial data for economic inferences, with particular focus on *data-poor* developing countries. After introducing the current remote sensing and other data collection techniques, the authors note that satellite-derived data having countrywide, regional, or global coverage are comparable across geopolitical boundaries; this avoids quality control problems that arise when synthesizing national surveys or census data from multiple sources. An example is the measurement of economic vitality based on observations of nighttime light on the Earth's surface. Richard Bernknopf (Chapter 14) summarizes the estimation of the value of information in two case studies focused on the food-water nexus in the agriculture sector. In the first example, the Landsat archive is used to evaluate the societal benefits of adapting agricultural land management to reduce nonpoint source groundwater contamination. The second case study targets mitigating drought disasters by utilizing Gravity Recovery and Climate Experiment (GRACE) data to determine eligibility for drought disaster assistance and insurance. In each case, EO data are transformed into information and is processed with other science-based indicators and socioeconomic data. In Chapter 15, Kecheng Xu, Linda Nozick, Jamie Brown Kruse, Rachel Davidson, and Joe Trainor present an analysis of proposed insurance affordability criteria and forecasted effects on uninsured hurricane catastrophic losses. A geospatially explicit loss model is coupled with a nested model of an insurance market, reinsurance, government, and homeowners to examine the proportion of homeowners and expected losses that would fall under the category of uninsured losses due to affordability thresholds. Results indicate that subsidization of insurance rates would not be a cost-effective policy for loss reduction. In Chapter 16, the value of hydrometeorological information in Austria is discussed by Nikolay Khabarov, Andrey Krasovskii, Alexander Schwartz, Ian McCallum, and Michael Obersteiner. The authors utilize benefit transfer to derive the value of weather information in Austria estimated for households, agriculture, transport, and construction sectors. They conclude that the wide range of obtained estimates unveils the uncertainties associated with assessments published for other countries. A case study of traffic safety monitoring as an application of location-enabled e-government processes is described by Danny Vandenbroucke, Glenn Vancauwenberghe, Anuja Dangol, and Francesco Pignatelli in Chapter 17. The integration of location information in e-government processes is considered a key challenge in adding value to Spatial Data Infrastructures (SDI's). A qualitative method is developed to estimate the impact of location enablement of e-government processes on system performance using three indicators: (1) time, (2) cost, and (3) quality. Results show that the performance in terms of time, costs, and quality are negatively influenced by lack of upstream data harmonization as well as the handling of the sharing agreements is an impeding factor.

In Chapter 18, Joep Crompvoets, Jamie Brown Kruse, and Françoise Pearlman provide a summing up of what we have learned and explore the potential ways forward. The beginning of the chapter addresses sustainable development goals within the context of GEOValue. Many of the sustainable development challenges are crosscutting in nature and are characterized by complex interlinkages that will benefit from using geospatial information as a common reference framework. Despite the intrinsic value of data, the ability to value the geospatial information that impacts on decisions is still maturing. The path forward is mainly derived from a body of knowledge and experiences relating to performance assessment activities and literature. The goal is to design an overall framework for geo-valuation that takes into account the different purposes, views, approaches, developmental trajectories, and value regime. Designing such an *evolutionary* framework will be a challenge for the GEOValue community as it is very likely that applications continue to be more comprehensive, realistic, and critical in which resource limitations must be addressed.

Acknowledgments

The editors recognize the following for the support that helped to make this book possible. Support included outreach to the GEOValue community and support of associated workshops on socioeconomic impact and value of geospatial information. Pearlman acknowledges support from NASA grants NNX14AO01G to J&F Enterprise, and USGS grant G14AC00303 to the University of Colorado. Kruse acknowledges support from USGS grant #G16AC00058 and NASA grant #NNX15AT69G to East Carolina University.

Editors

Dr. Jamie Brown Kruse is the HCAS distinguished professor of economics at East Carolina University, Senior Scientist at the Institute for Coastal Science and Policy, and director of the Center for Natural Hazards Research, North Carolina. She has held faculty positions at the University of Colorado, Texas Tech University, East Carolina University, a visiting position at Eidgenossische Technische Hochschule (ETH) in Zurich, Switzerland, and has served as the chief economist at U.S. National Oceanic and Atmospheric Administration (NOAA). Her work has been supported by the National Aeronautics and Space Administration, U.S. National Science Foundation, U.S. Geological Survey, U.S. Department of Energy, NOAA, the National Institute of Standards and Technology, the Federal Emergency Management Agency, the Department of Homeland Security, and U.S. Federal Deposit Insurance Corporation. She was the recipient of the 2012 Lifetime Achievement Award for Research and Creative Work at East Carolina University.

Dr. Joep Crompvoets is an associate professor at KU Leuven Public Governance Institute, Belgium, focusing on information management in the public sector and secretary-general of EuroSDR—a European spatial data research network linking national mapping agencies with research institutes and universities for the purpose of applied research in the domain of geospatial information management. He has held a faculty position at Wageningen University, the Netherlands, and was employed by CSIC IRNAS research institute in Spain. He has been involved in numerous (inter)national projects related to spatial data infrastructures, GIS, and e-governance. His work has been supported by the UN-GGIM, the World Bank, the European Commission, the BELSPO federal Public Planning Service Science Policy, Belgium, the Netherlands Organisation for Scientific Research (NWO), the Flemish government, the IWT Agency for Innovation by Science and Technology, and several national governments around the world. He is a member of the board of directors of the Global Spatial Data Infrastructures Association.

Ms. Françoise Pearlman has 30 years of experience in engineering and management, including system of systems engineering, software engineering, and software/system integration and testing. After a career in technical management for major aerospace corporations, she is currently coowner and manager of J&F Enterprise. Her focus is in communication and outreach, with particular interest in socioeconomic benefits of Earth observation. She earned her master's degree in aeronautical engineering from the University of Washington, and a master's in business administration from the University of New Mexico, Albuquerque, New Mexico. She is a senior member of IEEE.

Contributors

Lesley Arnold
Curtin University
Perth, Western Australia
and
Cooperative Research Centre for
Spatial Information
Melbourne, Australia
and
Geospatial Frameworks
Perth, Western Australia

Richard Bernknopf
Department of Economics
University of New Mexico
Albuquerque, New Mexico

Andrew Coote
ConsultingWhere
Chipperfield, United Kingdom

Joep Crompvoets
Spatial Applications Division (SADL)
KU Leuven
Leuven, Belgium

Anuja Dangol
Spatial Applications Division (SADL)
KU Leuven
Leuven, Belgium

Rachel Davidson
University of Delaware
Newark, Delaware

Robert R. Downs
Center for International Earth
Science Information Network
(CIESIN)
The Earth Institute
Columbia University
New York, New York

Jason Gallo
IDA Science and Technology Policy
Institute
Washington, DC

Serene Ho
KU Leuven
Leuven, Belgium

Jacob Hochard
Department of Economics and
Institute for Coastal Science and
Policy
East Carolina University
Greenville, North Carolina

Nikolay Khabarov
International Institute for Applied
Systems Analysis (IIASA)
Laxenburg, Austria

Andrey Krasovskii
International Institute for Applied
Systems Analysis (IIASA)
Laxenburg, Austria

Evan Plous Kresch
Department of Economics
Oberlin College
Oberlin, Ohio

Jamie Brown Kruse
Center for Natural Hazards Research
East Carolina University
Greenville, North Carolina

Steven Lev
IDA Science and Technology Policy
Institute
Washington, DC

Julia Mambo
Council for Scientific and Industrial
  Research (CSIR) South Africa
Pretoria, South Africa

Pierre-Philippe Mathieu
ESA/ESRIN
Frascati, Italy

Ian McCallum
International Institute for Applied
  Systems Analysis (IIASA)
Laxenburg, Austria

Ben Miller
RAND Corporation
Santa Monica, California

Miriam Murambadoro
Council for Scientific and Industrial
  Research (CSIR) South Africa
Pretoria, South Africa

Linda Nozick
Cornell University
Ithaca, New York

Michael Obersteiner
International Institute for Applied
  Systems Analysis (IIASA)
Laxenburg, Austria

Françoise Pearlman
J&F Enterprise
Seattle, Washington

Jay Pearlman
J&F Enterprise
Seattle, Washington

Francesco Pignatelli
European Commission, Joint
  Research Centre
Ispra, Italy

Alexander Schwartz
International Institute for Applied
  Systems Analysis (IIASA)
Laxenburg, Austria

Alan Smart
ACIL Allen Consulting
Sydney, Australia

Joe Trainor
University of Delaware
Newark, Delaware

Danny Vandenbroucke
Spatial Applications Division (SADL)
KU Leuven
Leuven, Belgium

Glenn Vancauwenberghe
TU Delft/Geographic Information
  Governance Centre
Delft, the Netherlands

Bastiaan van Loenen
Delft University of Technology
Delft, the Netherlands

Frederika Welle Donker
Delft University of Technology
Delft, the Netherlands

Dawn J. Wright
Environmental Systems Research
  Institute
Redlands, California

Kecheng Xu
Cornell University
Ithaca, New York

Section I

The Socioeconomic Value of Earth Observation

An Overview

1

Introduction

Jamie Brown Kruse, Joep Crompvoets, and Françoise Pearlman

CONTENTS

This book provides a forum to address methodologies for valuing the benefits of geospatial data and services and presents case studies, which illustrate those methodologies. Drawing from the contributions to recent workshops on socioeconomic benefits of Earth information (Pearlman et al., 2016), this book explores the meaning of socioeconomic value and reviews the existing and evolving methodologies underpinning these value assessments. Examples in this book illustrate the potential for geospatial data to contribute societal benefit over a wide and growing range of applications and uses. There are currently no standardized practices for measuring the contribution to society of geospatial information systems. In a world of increasing demands on shrinking resources, the future value of the proposed new geospatial information systems must become a component of an effective investment decision process. This book explores the convergence toward standardized practices and selection of use cases, which can further guide that process.

Although examples of the use of geospatial information exist that go back for centuries—for example, one of the earliest known maps of the world was found on a cuneiform tablet dated from the sixth century BC (Brotton, 2013)—the advancement in observing technologies and user interfaces in the past two decades make this century the geospatial information age. This advancement has resulted in decision support systems that are both useful and usable for a broad array of applications, furthering the trend that "Geospatial information is ubiquitous in everyday life ..." (Loomis et al., 2015). Earth observations (EO) are about our natural and environmental resources and our human interactions with them. The World Bank, the United Nations, and more than 375 international environmental and resource agreements label these resources as part of *the wealth of nations*. Information is beneficial when it helps us to better manage, enhance, preserve, protect, and use this *natural wealth*. Using geospatial information, we can address priority global challenges such as the Sustainable Development Goals* and responsive natural- and human-built environments at the local, national, and global levels. The intrinsic value of geospatial information is implied to such an extent that the science of measuring its importance has yet to catch up. However, as the maintenance of existing systems and the development of promising new systems become ever more expensive, it is imperative that we measure societal benefits that accrue and assign them appropriately to these information resources. In addition, the social and economic measures must be developed, refined, and standardized using the same scientific rigor and high standards that are the hallmark of our finest Earth-observing systems.

The expanding investment and use of EO by government entities and private industry has become a convention of the twenty-first century. In the United States alone, civil EO-estimated funding in 2014 that included $2.5 billion in satellite systems and more than $1 billion for airborne, terrestrial, and marine networks and surveys. Geospatial information is geographically referenced information—information associated with a location (e.g., latitude and longitude or physical address). The use of geospatial information has become so interwoven with everyday activity that it is hard to imagine a world without such access. Yet the acquisition and processing of this information require continuing and expanding investments. The production and accessibility of geospatial information are changing from the perspective of technology and human interactions. Technological advances affect the processing and delivery of geospatial information, as well as the quantity, quality, appropriateness, and timeliness of information available to individuals, the government sector, and the business sector. The increased availability of data and rapidly evolving analysis techniques enhance the impact that environmental information can have on both public and private decision-makers.

In order to analyze the full impact of the investments in data systems on the society, many disciplines must contribute to the understanding of the complex process that starts with data from geospatial information and concludes with human decisions supporting the desired user objectives. Macauley (2006) provides a theoretical foundation for establishing the value of space-derived information and a framework using economic principles. An important characteristic of geospatial information is that it can be held privately with restricted access or viewed as a public good in an open-source format. For the case of privately held information, the contribution or value of better information can be measured directly in terms of improved financial performance or payments for access to the information. For the case of geospatial information as a public good, *the whole is greater than the sum of its parts*. In this case, indirect or nonmarket methods may be more suited to ascribe value to information. Further, information may be parsed and combined with other data sources and analysis to create products and decision

---

* see un.org Sustainable Development Goals – 17 goals to transform our world.

support systems in new and innovative ways and thus spur entirely new industries. The desired objectives of users of geospatial information cover a wide range from addressing the world's grand challenges to individuals engaging in leisure activities. For example, improved and timely geospatial information is an important component of policy decisions that confront *wicked problems* (Rittell and Webber, 1973) of worldwide concern such as climate change, natural hazards, and health care to name a few.

This book explores the process that starts with observations that become the information resources for decision-making. The important issues pertaining to the use of data and information are: data type (form of data), data schema or structure, metadata (presence, quality, and completeness), provenance (source, storage, and processing), quality (content and accuracy), and currency (age relative to use). Meeks and Dasgupta (2004) proposed a scoring method, Geospatial Information Utility (GeoIU), to provide a filter to refine the data, and, in turn, assist search engines in the information discovery and retrieval stage (Nativi et al., 2013).

The Meeks and Dasgupta process model can be extended to articulate the process of turning observations into high value outcomes. Many of the criteria described to assess the utility of the geospatial *data sources* are also appropriate to assess utility in the context of informing the action taken by a *decision-maker*. The usefulness of the output for the decision-maker will depend on the form of the information (visual and textual), provenance, quality and currency, and the preferences of the decision-maker, as well as the specific fitness-for-purpose. However, in assessing the value of geospatial information, it is not the action *per se* that is of ultimate interest—it is the *outcome* of the action. Value is expressed as measurable improvement in a social context (the *outcome*) such as reduction in mortality and morbidity, reduced damage to capital assets, improved profit, improved community well-being, and other social or economic benefit measures. The *value chain* is often referred to when considering data to decision sequence of activities in the valuation process model. Figure 1.1 provides a simplified version of the process model.

Although applicable in general, the figure depicts some of the components of decisions surrounding a recent flood event. A subset of geospatial data sources include satellite imagery of an extreme weather event, on-the-ground reports from trusted informants, Light Detection and Ranging (LiDAR) data, and stream gauge information. These and other data are analyzed to produce location-specific rainfall estimates that are combined with topographical characteristics of the watershed to produce current and forecasted river flow rates and depths (courtesy of U.S. Geological Survey [USGS]). The information product is further refined to show the river depths with minor, moderate, and major flood stages along with the record depth to provide additional context for flood risk (courtesy of National Oceanic and Atmospheric Administration [NOAA]). These in turn, feed decision support systems. One such system predicts the extent of inundation coupled with building footprints and value to provide a real-time forecast of the private and public infrastructure

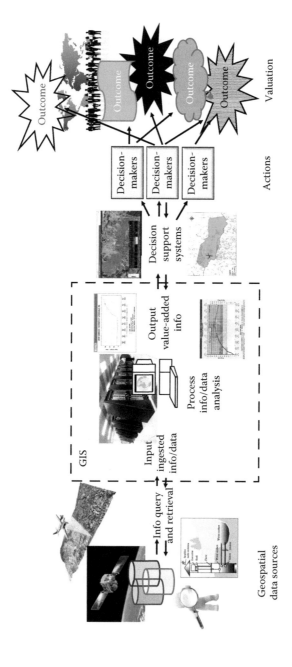

FIGURE 1.1
Data to decisions to valuation process model.

likely to be damaged, and a dollar estimate of the damage to structures at the forecasted flood depth. This system also provides an estimate of the population at risk in residential homes, and informs the second decision support product, which are the mandatory and voluntary evacuation zones and the timing of when evacuation protocols will be enforced (Flood Inundation Mapping and Alert Network Map [FIMAN] and Evacuation Map courtesy of North Carolina Department of Public Safety). The decision-makers that use the products include visitors, households, homeowners, public utilities, first responders, emergency managers, private businesses, not-for-profit organizations, insurance providers, and public institutions such as universities and hospitals. In addition, FIMAN produces a quick estimate of damages that can be used for requests for federal grants and other support. Longer time frame choices that involve home location, development and zoning decisions also rely on this suite of observations, geographic information system (GIS), and decision support systems. This description thus addresses both the relatively short time frame of emergency response, and relatively longer time frame of recovery. A more complete and even more complex figure would incorporate feedback at all stages.

New collection methods in real time make big data even bigger. In addition, processing and analysis tailored to big multidimensional data makes new approaches to meet societal needs possible. The technology of collecting observations for geospatial information has greatly expanded our ability to assess individual preferences and well-being through self-reports by social media and citizen science initiatives. This can support the formulation of bottom-up policy that meets the needs of all stakeholders as envisioned by Glynn et al. (2017).

1.1 Why GEOValue?

Assessing value and impacts of geospatial information involves a broad multidisciplinary effort across technical, social, and economic domains. The GEOValue community is a community of experts from a wide range of disciplines that form a community of practice to refine and advance standards of practice for consistent, replicable, and reliable measurement of social and economic benefits that stem from geospatial information. GEOValue fosters collaboration across specialties and helps to build collaboration across disciplines. Publications and workshops were the initial approach of the community to share information and facilitate networking. Geospatial information contributes to decisions by both societal decision-makers and individuals. Understanding and making effective use of this information is essential as issues become increasingly complex and consequences can be critical for future economic development and societal progress and equity.

This book is intended for professionals who create and use geospatial information, including decision-makers in the public or private sector. This book will provide a reference document for both its technical background, and application case studies, as a foundation for valuing Earth information. As is the case when an endeavor must span many disciplines, most professionals will find some of the topical material in their own domain foundational and straightforward, whereas other chapters present challenging new concepts. The goal is to provide a unifying framework that utilizes many disciplines to produce a fully formed picture of where observations come from, how they are transformed into usable data, how data become a GIS, how the GIS is analyzed to create value added information products, the contribution of information to knowledge ,and decision support systems that are utilized to choose actions, ultimately producing outcomes whose impact on society can be measured by quantitative and qualitative valuation methods.

References

Brotton, J. (2014) *A History of the World in Twelve Maps*. Penguin Books, New York.

Glynn, P. D., A. A. Volinov, C. D. Shapiro, and P. A. White (2017) From data to decisions: Processing information, biases, and beliefs for improved management of natural resources and environments, *Earths Future* 5. doi:10.1002/2016EF000487.

Loomis, J., S. Koontz, H. Miller, and L. Richardson (2015) Valuing geospatial information: Using the contingent valuation method to estimate the economic benefits of Landsat satellite imagery, *Photogrammetric Engineering & Remote Sensing* 81(8):647–656.

Macauley, M. K. (2006) The value of information: Measuring the contribution of space-derived earth science data to resource management, *Space Policy* 22(4):274–282.

Meeks, W. L. and S. Dasgupta (2004) Geospatial information utility: An estimation of the relevance of geospatial information to users, *Decision Support Systems* 38:47–63.

Nativi, S., M. Craglia, and J. Pearlman (2013) Earth science infrastructures interoperability: The brokering approach, *IEEE JSTARS* 6(3):118–129.

Pearlman, F., J. Pearlman, R. Bernknopf et al. (2016) Assessing the socioeconomic impact and value of open geospatial information, Open-File Report 2016-1036, https://pubs.er.usgs.gov/publication/ofr20161036.

Rittell, H. W. J. and M. M. Webber (1973) Dilemmas in a general theory of planning, *Policy Sciences* 4:155–169.

2

Measuring the Socioeconomic Value of Data and Information Products Derived from Earth Observations and Other Geospatial Data

Jason Gallo, Richard Bernknopf, and Steven Lev

CONTENTS

2.1 Introduction

The observation, monitoring, and measurement of the Earth system, its processes, and natural and built environments inform decisions that protect lives and property, support economic activity, and ensure environmental security. Data and information products derived from Earth observations (EO) and other geospatial data have improved scientific understanding of natural and human systems, and the interaction and feedback between them. This information is increasingly applied in decision-making processes across public and private sectors. Decisions are influenced by the availability of actionable information to a decision-maker. The value of that information to the decision-maker is determined, in part, by the decision-maker's ability to use that information to reduce uncertainty or to increase the probability of a successful outcome (Bernknopf and Shapiro 2015). The provision of useful information, in this case data and information products derived from EO, therefore, is a fundamental part of the value chain and needs to be accurately characterized to determine the value of the resulting information.

According to the 2016 United States *Common Framework for Earth Observation Data*, "A core principle of the U.S. Government is that Federal Earth-observation data are public goods, paid for by the American people, and that free, full, and open access to these data significantly enhances their value. The return on our annual Earth observation and other remotely sensed data investment increases in accordance with the data's widespread use in public- and private-sector decision-making" (National Science and Technology Council 2016). At the heart of this statement is the acknowledgement that the value associated with EO data and the derivative geospatial information products comes from improved public and private sector decision-making. The statement also assumes that improving access and removing barriers to Federal EO and associated geospatial data and information products will increase their value by increasing their uptake and use in decision-making.

Policies that enable free and open access to EO data and associated infor-mation products clearly have the potential to increase their use in decision-making; however, it is difficult to accurately estimate their value in a way that allows policy-makers and decision-makers to measure or project the return on investment. This is because free, publically available, and open data are not free of cost. There is a vast global architecture of EO systems, sensors, networks, surveys, and sampling activities that measure parame-ters of the Earth and Earth-system processes. There is also a complimentary networked infrastructure dedicated to capturing, calibrating, validating, processing, storing, describing, curating, and disseminating the resulting data. These are both in addition to the infrastructure needed to analyze and model EO data and develop useful information products that may combine one or more measurements. There are also myriad labor and resource costs associated with deploying, managing, using, and maintaining this global infrastructure.

To estimate value, it is necessary to account for the investment and full life-cycle costs associated with EO data and information products, including the infrastructure needed to provide them; the data management and dissemina-tion costs incurred to provide them; and end user transaction costs associated with the discovering, accessing, and using these data and information prod-ucts. In addition to costs, it is necessary to assess the value of the information to a decision-maker. One approach is to estimate the socioeconomic impacts of an investment in EO. There are currently robust methods for determining the value of information to a decision-maker; however, there are few methods for evaluating the net benefits of the infrastructure that supports the provi-sion of that information, let alone the value of those benefits.

This chapter introduces a conceptual framework to assist decision-makers in analyzing and estimating the socioeconomic value of geospatial information as a part of society's infrastructure.* The framework is supported by EO and other geospatial data that provide the foundation for decisions. The infrastructure is assumed to be publically funded (although there is an increasing use of commercial observations in both the public and private sectors). The choices that result from using the conceptual framework rely on geospatial information products and can be shown to serve the public interest, specific private interests, or both. The valuation methods discussed in this and later chapters are for evaluating the *return on investment* in EO and other geospatial data collection technology. The estimation of the value of information (VOI) must be able to describe how the decision-maker's information changes as a result of the acquisition of an additional EO and other geospatial data. The implementation of the conceptual framework is designed to provide input to specific public and private sector policies and investments that affect the provision, access, and use EO and associated geospatial data and information products.

2.2 A Changing Data Landscape

Over the past three decades, there have been rapid technological advances that have changed the way that EO, environmental data, and, more broadly, geospatial data are collected and the way information products and services are provided, accessed, and used. Many governments are committed to making public sector information (open-access data) and scientific data (open research data) more widely accessible (G8 2013; Obama 2013). These governments provide free and open access to EO data and associated information products or have reduced prices for government-produced data and information products. Private sector services, such as environmental systems research institute's (ESRI) geographic information system (GIS) products, Google Earth Engine, and Microsoft's Bing Maps Platform, are taking advantage of the proliferation and availability of EO data to provide enhanced services that allow a variety of users to perform geospatial research and analysis. Partnerships such as EarthCube, Earth Science Information Partners, and the Open Geospatial Consortium, among others, are providing the geospatial community critical cyber infrastructure and standards to enable data sharing, research and analysis, and service development.

---

* For the purposes of this chapter, Earth observations are measurements of the Earth and its processes collected either remotely or *in situ* by space-based, airborne, terrestrial, freshwater, and marine Earth observation systems, sensors, networks, and surveys. These measurements are calibrated, validated, and transformed into data and information products that contribute to decisions. Earth observations infrastructure refers to the sensing elements described earlier, as well as the cyber, physical, and human networks and systems that enable the collection, management, analysis, dissemination, and use of Earth observation data and information products in decisions.

In addition to the increasing availability of government EO and geospatial data, other forms of digital geospatial information have become useful to decision-makers. Recently, crowdsourced data has been used to support social science research and community decision-making (e.g., USGS Did You Feel It Program). Networked computing and mobile telephony have facilitated crowdsourcing and citizen science in Earth science research. These technologies allow for both the distributed collection of geospatial data and original analysis. A growing number of these initiatives are currently contributing to Earth science research and providing public services. Examples include: the U.S. Geological Survey's Did You Feel It and iCoast programs[*]; the U.S. National Oceanographic and Atmospheric Administration's Meteorological Phenomena Identification Near the Ground (mPING) program[†]; the Community Collaborative Rain, Hail, and Snow Network (CoCoRaHS)[‡]; European Environment Agency's Coordinate Information on the Environment (Corine) program using OpenStreetMap[§]; the European Union's Citizens' Observatory program[¶] (promotes the development of community-based environmental monitoring, data collection, interpretation, and information delivery systems); and Safecast[**] (a global sensor network for collecting and sharing radiation measurements, which was used following the Fukushima disaster in Japan).

The increasingly broad availability of timely and free, or inexpensive, geospatial data provided through digital infrastructure supports researchers and decision-makers by reducing uncertainty when analyzing environmental, financial, and policy choices (Coote 2010; Pearlman et al. 2014). This infrastructure supports services such as on-demand local weather forecasts, detailed mapping and transportation planning tools, and mobile applications that take advantage of geolocation. Recent developments in information processing coupled with natural and social sciences have enabled decision-makers to begin to take advantage of big data and engage with large, complex, and heterogeneous datasets to gain new insight into the Earth system and its processes.

---

[*] U.S. Geological Survey, Earthquake Hazards Program. Did You Feel It? http://earthquake.usgs.gov/earthquakes/dyfi/; U.S. Geological Survey, iCoast—Did the Coast Change? http://coastal.er.usgs.gov/icoast/index.php.

[†] National Oceanographic and Atmospheric Administration, National Severe Storms Laboratory. Meteorological Phenomena Identification near the Ground, http://mping.nssl.noaa.gov/.

[‡] Community Collaborative Rain, Hail, and Snow Network, http://www.cocorahs.org/.

[§] European Environment Agency, Coordinate Information on the Environment, http://sia.eionet.europa.eu/CLC2006.

[¶] European Union. Citizens' Observatory, http://www.citizen-obs.eu/Home.aspx.

[**] Safecast, http://blog.safecast.org/.

2.3 Geospatial Information Products
as Public and Merit Goods

Data gathered from EO serve as inputs to information products, services, and activities that inform decisions and, are therefore, considered intermediate economic goods. As an intermediate good, the same data can have many uses simultaneously. As the data are digital, the cost of supplying the data is greatest for the first user; the cost of disseminating information to additional users is much smaller than the cost of obtaining information for the first user. Economists refer to this situation as jointness of supply and are true for the great majority of information goods. For EO in particular, the fixed costs of developing systems to collect, manage, analyze, and disseminate data are large, but in most cases the variable costs of reproduction and user access are small, especially if the data are digital (Varian et al. 2004).

In this case, EO data, collected by the U.S. Government are available as open-access products on a nondiscriminatory basis. Hence the data are both nonexcludable and nonrivalrous, which are attributes of a public good. This means that potential users are not prevented from accessing and using these data and that their use by one party does not exhaust their availability to other parties. These data, and the associated information products, are often durable over long time spans and exhibit no congestion costs in which one individual's use of the data and information does not degrade their value to another (Bernknopf and Shapiro 2015). In other words, consumption of the data does not subtract from any other user's consumption, therefore the data, as a good, cannot be depleted. These data may also be considered public goods if the policies that govern their accessibility ensure their nonexcludability. These policies are often put in place because governments that invest in EO believe that the provision and use of these data are in the public interest.

Although these data are often used in a number of applications that are good for the general public, considering them public goods by nature is more complicated (Harris and Miller 2011). It is possible that these same data may be collected by entities that choose to restrict their availability to some or all potential users, meaning that these goods cannot be considered nonexcludable; however, this excludability does not impact the nonrivalrous nature of EO and geospatial data and information products. For example, some governments may provide tiered access to EO data that are collected using public funds, in which specific classes of users may be excluded from use. Other governments may either charge fees for data access or sell datasets to users to recover some or all costs associated with collecting the data. In other cases, EO data are private goods that may be collected by private entities on a contractual basis or licensed by one or more paying consumers. In this chapter, specific information refers to information that is localized and is narrow in focus.

Specific, privately collected, and access-restricted EO data and information products may also be used to further the public interest. For example, the U.S. Government purchases and uses commercial EO data such as Light Detection and Ranging—LiDAR measurements,* commercial high-resolution satellite imagery, commercial radar satellite data, and commercial lightning data, among others, to develop EO information products that are freely and openly available on a nondiscriminatory basis. The use of private data by the U.S. Government to create public good information products is governed by data acquisition and management policies, agreements, contracts, and licensing. Private intermediate goods may be inputs to end products or services that are public goods. These hybrid goods exhibit the qualities of a public good when consumed as an end product, while they are developed using, at least in part, private intermediate goods.

Harris and Miller (2011) describe a spectrum of *publicness* in which EO and other geospatial data and information products may have varying levels of nonexcludability and nonrivalry. They describe EO data and information products as *merit goods* that are nonexcludable and nonrivalrous, but specifically provide a benefit, or merit, that would not otherwise be provided by the private sector. Whether EO and geospatial data or information products exhibit the qualities of a public good or a merit good are, therefore, determined by policy decisions rather than the intrinsic nature of the observation itself or the intermediate goods that are used to develop an end product. Governments, through policy decisions, may opt to underwrite the costs of providing merit goods as a matter of public benefit.

2.4 Free and Open Data Are Not Free

The provision of EO data and information products is the result of significant public and private investments in observation systems and programs, data management and distribution, research and analysis, reanalysis, and information product development. Global governments invest in, operate, and maintain portfolios of EO systems, sensors, networks, and surveys to characterize and monitor the Earth and its processes. The resulting data can be routinely used to benefit society. They contribute to the understanding of the natural processes of the planet and closely coupled human activity.

---

* Light Detection and Ranging (LiDAR) is a remote-sensing method that uses light in the form of a pulsed laser to measure ranges (variable distances) to the Earth. These light pulses—combined with other data recorded by the airborne system—generate precise, three-dimensional information about the shape of the Earth and its surface characteristics (http://oceanservice.noaa.gov/facts/lidar.html).

EO support decisions and planning activities that address long-term challenges associated with the water–food–energy nexus and the Earth's changing climate.

Currently, the U.S. Government invests more than $3.5 billion annually in Earth-observing systems and is the largest provider of Earth-system data globally (National Science and Technology Council 2014). The U.S. *National Plan for Civil Earth Observations* prioritizes the continuity of sustained observations for public services, continuity of sustained observations for Earth-system research, and continued investment in experimental observations (National Science and Technology Council 2014). Sustained EO improves the accuracy of measurements and provides an archive that, over time, can reduce uncertainty in understanding critical environmental trends. The plan also recognizes the value of collaboration with international partners that U.S. Government agencies leverage to improve services and strengthen scientific research.

The Copernicus Programme is a multibillion Euro investment by the European Commission and European Space Agency. "Copernicus consists of a complex set of systems, which collect data from multiple sources: Earth observation satellites and *in situ* sensors such as ground stations, airborne, and seaborne sensors. The program processes these data and provides users with reliable and up-to-date information through a set of services related to environmental and security issues" (European Commission 2014). The program launched the first of six Sentinel satellites in April 2014 to provide continuous all-weather C-band synthetic aperture radar imaging for land and ocean services. Copernicus is a component of the Global Earth Observation System of Systems (GEOSS), which is an international geospatial information infrastructure network being developed by the Group on Earth Observations (GEO)* to provide geospatial information globally.

The global nature of the EO enterprise and the value of observations and geospatial data to society are highlighted by the January 2014 GEO Ministerial Declaration, "Integrating Observations to Sustain Our Planet," which stated that international collaboration in this area "has advanced the availability of long-term, global data, and information as a basis for sound decision-making for improving human welfare; encouraging innovation and growth, alleviating human suffering, including eradicating poverty; protecting the global environment; and advancing sustainable development" (Group on Earth Observations 2014). Although it intuitively follows that EO and geospatial data and information products contribute to decisions that help improve human welfare, economic prosperity, and protect the natural environment, rigorously valuing them is difficult.

---

* GEO is a voluntary, international partnership with 104 member nations and the European Commission and 109 participating organizations that ensure comprehensive and sustained Earth observations (https://www.earthobservations.org/index.php).

2.5 Value in the Use of Earth Observation Data

EO data are collected and applied to societal and market decisions across a variety of sectors, including agriculture and forestry, energy and mineral exploration, and transportation. They also inform decisions about conservation designations, the establishment of ecological preserves, and infrastructure construction and maintenance, among others. In many cases in which the benefits of EO data and information products are assumed to be sufficiently in the public interest, governments support the cost of the developing and deploying the EO infrastructure that collects measurements of the Earth and its processes, as well as associated programs, data management networks, and workforce. Global governments invest in EO to provide public benefits that are based on assumptions and calculations that investments in observations will provide a significant return on investment to both the public and private sectors and benefit national economies.

One example is the Australian Commonwealth Scientific and Industrial Research Organisation (CSIRO) commitment to a more comprehensive National Airborne Electromagnetic (AEM) survey program. The primary objectives associated with the CSIRO AEM survey programs are improved resolution and quality of shallow subsurface geological maps to inform mineral resource characterization, water resource evaluations, and food security issues associated with the agriculture–water nexus. Wide deployment of AEM surveys began in 1998 to meet the need for precompetitive data to inform mineral and water resource exploration activities. Precompetitive geoscience data acquisition in Australia refers to the collection, collation, and integration of basic geoscientific data by government agencies. These data are free and available to the public. There is currently a major initiative underway ($100 Million AU through 2020) to produce more precompetitive data on a national scale. The current estimate from CSRIO for AEM return on investment is 20:1 from the private sector for every dollar of federal investment. An independent 2014 assessment of all CSIRO activities, including AEM projects, estimated ROI at 5:1 (ACIL Allen Consulting 2014).

Estimating the public benefits and return on investment of EO and other geospatial data are made complicated due to the lack of a functioning market (Smart 2014). As intermediate goods, the value of EO data is tied to end products or services for decision-making. The most direct method for describing value in the use of geospatial information within a given domain is through a clear, well-defined, objective statement that relies on contributions from EO data to deliver a return on investment. In the public sector, there are several domains in which well-defined objectives that rely on geospatial data have been identified. Several examples are described here.

2.5.1 Value in Use for Meeting Policy Objectives

The U.S. Presidential Policy Directive-8 (PPD-8) issued on March 30, 2011 led to the development of the National Preparedness Goal: "A secure and resilient nation with the capabilities required across the whole community to prevent, protect against, mitigate, respond to, and recover from the threats and hazards that pose the greatest risk." PPD-8 identifies catastrophic natural disasters, such as hurricanes or drought, as one of the threats that pose the greatest risk to national security. To address their role in achieving the National Preparedness Goal, the U.S. National Oceanic and Atmospheric Administration (NOAA) is investing annually to provide information and services that help communities prepare for, respond to, and recover from the impacts of natural disasters. NOAA meets this objective through a range of activities and products that rely heavily on EO and other geospatial data and information products. For example, in meeting the National Preparedness Goal, the National Weather Service is responsible for providing decision support services for the Emergency Management (EM) community and widely disseminating storm tracking and forecast information to both the EM community and the public. The delivery of benefits in this domain can be evaluated against the PPD-8 policy objectives and the extent to which NOAA is meeting its designated responsibilities under that policy.

2.5.2 Value in Use for Government Operations

In the United States, the federal government is responsible for the collection and dissemination of a wide range of EO and geospatial information products. Investment in the generation of the data and information products is driven by policy objectives and operational requirements. NOAA's FY 2017 budget justification identifies two of the agency's top priorities as "Provide Information and Services to Make Communities More Resilient" and "Continue to Evolve the National Weather Service." These two priorities are prime examples of policy goals and agency mission requirements defining the objectives through which benefit can be delivered.

NOAA's mission requirements include the prediction and warning responsibilities associated with water resources. The services span information on water availability for drought prediction and management to flood prediction and warnings. To meet this mission requirement NOAA must develop and deliver water forecast products and information to local decision-makers and the public. These information products rely heavily on EO and geospatial data that are direct inputs to decisions that help avoid losses in the case of a flood warning or mitigate the impact of a drought on natural systems and communities. The delivery of benefits in this domain can be evaluated against the mission requirements (e.g. whether a flood warning is issued and whether it is accurate) and the effectiveness of data or an information product in achieving the desired outcome (e.g., whether a local official use a drought prediction to mitigate the impact on the community).

2.6 Methods for Evaluating the Value of Earth Observations

Weighting the socioeconomic benefits of EO and geospatial data to establish VOI in a decision is a function of the application of the data and information products. End use information products derived from EO and geospatial data may be either public or private goods that enable other activities and value-added services. Underlying this definition is the assumption that improved decision-making is an improvement in economic welfare. It is, therefore, necessary to both accurately characterize the *publicness* of the EO and geospatial data and information products and to demonstrate their economic value. To do so, there must be rigorous methods for evaluating the value of the information products to specific decisions. Testable methods, case studies, and use cases that demonstrate improvements to decision-making processes are critical if public and private sector EO providers are to measure the socioeconomic value of their investments (NASA 2013).

How does a decision-maker decide what EO and geospatial information product or products to use? How is the choice motivated? How do public decisions improve with the use of a public good or a private good?

The conceptual framework in Figure 2.1 is a process for evaluating the socioeconomic benefits of EO and other geospatial data. In application, the conceptual framework can be used to correlate a qualitative (experiential) or a quantitative (statistical) relationship between geospatial information and observed activities in an economic sector. This relationship can

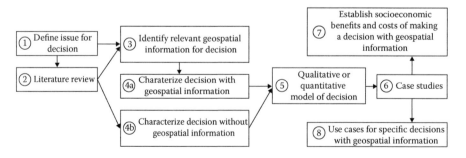

FIGURE 2.1
A conceptual framework for estimating the socioeconomic value of geospatial information products derived from Earth observations and other geospatial data.

be analyzed qualitatively with participation observation, interviews, focus groups, and historical comparison. Alternatively, the relationship can be evaluated quantitatively with an economic model or with data mining and machine learning. Both qualitative and quantitative approaches can utilize deductive and inductive methods of evaluation to assess or estimate the marginal effect of a change or a reduction in uncertainty on an economic or social sector of interest.

The conceptual framework in Figure 2.1 is a stepwise process for estimating the socioeconomic value of information products derived from EO and other geospatial data. The first step for a researcher is to establish the context in which geospatial information will be used in a decision (box 1) and to identify examples of similar decision problems, if they exist, in the scientific literature (box 2). The next step is for the researcher to identify general or specific geospatial information derived from EO and other geospatial data that are best suited for implementation in the decision (box 3). The researcher must then characterize the decision with geospatial information (box 4a) and to characterize the same decision without the geospatial information (box 4b). The researcher constructs either a qualitative or quantitative model of the decision (box 5) and further develops a case study or studies based on the model (box 6). At this point, the researcher conducts either a qualitative or quantitative analysis of the case study or studies to establish socioeconomic benefits and costs (box 7) and then develop a formal use case (box 8). The type of study is determined by whether the decision-maker needs an experiential or a statistical understanding of the impact of incorporating geospatial information into a decision process.

If a qualitative approach is appropriate, the resulting socioeconomic value is an estimate of the changes in human well-being or benefits that result from making better informed decisions that use geospatial data. Models that rely on qualitative inputs, including crowdsourcing and machine language ontologies to provide rapid response and pattern recognition, can be used to yield valuable insights into better informed decisions.

A quantitative case study can be made for the application of geospatial information products derived from EO to decisions that affect individuals, communities, and economies. Economists seek to monetize the difference between decisions made with and without geospatial information in cases studies as the impact on the economy of the application of technology.

As depicted in Figure 2.1, the analysis starts with a statement or definition of the issue for decision and a literature review of existing evaluations for similar decisions. Here the analysis would be to estimate whether a change to an economy would occur by modeling the outcome of a specific introduction of a new technology relative to a baseline or current decision process. EO represents a new technology to be included in a decision, whereas the baseline case that assumes the new technology is not available.

In a quantitative VOI analysis EO data, whether from public or private sources, is transformed into information products when coupled with natural and anthropogenic process models and other data to create new EO and geospatial information products. These products enable value-added services and activities that inform decisions. It is difficult to accurately estimate the information's value in a way that allows policy and decision-makers to measure or project the return on investment. It is also necessary to assess the transaction costs to users for discovering, accessing, and using *free* data in the transformation to useful information. Assessing the socioeconomic impacts of the investment is about demonstrating the value of how the information is used.

Decision models and analysis can be based on a theoretical foundation or on an empirical approach. Evaluations can utilize applications of Bayesian decision or machine learning frameworks to quantify the role and value of the information. One example of the Bayesian decision model is VOI analysis. In the evaluation, EO provide an input to other activities and are considered an intermediate economic good. Determining VOI is based on evaluating the change in the availability of information to a decision-maker and the implications of that change on the decision-maker's objectives and constraints. In box 7 of the conceptual framework, the results of the model developed in the case study describe how the decision-maker's information changes as the result of the acquisition of additional geospatial information. It is important to note that additional information may be valuable even if it does not lead to a different choice on the part of a decision-maker. This is especially true when decision-makers are risk averse; they will have an *ex ante* willingness to pay for information to reduce the uncertainty associated with the decision.

The use case in box 8 is the last step in the conceptual framework. The use case consists of the interplay between public and private sector providers of EO and other geospatial data and users who make operational decisions. To transition to the operational use of the geospatial information, a use case includes the interaction of various stakeholders within a decision process. The objective is to design implementation experiments of a system from the user's and contributor's perspective, and to communicate system behavior in their terms. A use case requires communication of system requirements, how the system operates and may be used, the roles that all participants play, and what value the user will receive from the system. The use case is broader than a technical demonstration of capability and involves scientific experts, decision-makers, and their representatives.

The implementation of the conceptual framework assumes that the private sector relies on public goods to provide value-added products and services. A public good is provided when there is a market failure in which privately produced goods cannot meet demand. However, the public sector is increasingly reliant on private sector data and information products to provide public services (i.e., merit goods). Given this interplay, from a Federal government perspective, it is important to look at the origin of the data.

That is, did it originate in a Federal agency or in the private sector? Who provides the data has a nontrivial bearing on who can use it and how it is intended to be used. Therefore, it is important to determine what information is essential to the public interest and whose collection and dissemination should remain an inherently governmental function. When working with the private sector, what provisions can or should be made for ensuring data are free and open? What sort of agreements or contracts can facilitate cost-effective implementation?

2.7 Justifying the Provision of the Information Good: Benefits and Costs

Accurately quantifying the full range of benefits requires that the scope of the potential benefits be defined. This can be challenging given the common *off-label* use of geospatial data and data products to derive benefit. The same data or data product may be more (or less) critical for a given decision-maker. Identifying the VOI of an information product in a decision type can be achieved through a Value Tree Analysis (VTA) approach. A VTA is a tool used to support decision-making processes (Keeney et al. 1990; Sylak-Glassman et al. 2016). A VTA is useful for defining concepts, identifying objectives, and creating a hierarchical framework of objectives. A VTA hierarchy establishes the connection from societal benefits to the set of observation inputs that contribute to delivering those benefits. Through a series of discussions with end users of geospatial data and data products, the relative reliance on specific types of information can be established.

The methods for expert elicitation that may be used in VTA have been developed by the Stanford Research Institute (SRI) and Sandia National Laboratories/U.S. Nuclear Regulatory Commission (Bonano et al. 1990; DeWispelare et al. 1994; Kotra et al. 1996), among others, notably by Morgan et al. (1992). These methods are a useful starting point for data collection. More recently, these methods have been used in conjunction with structured data collection tools and a mathematical model to derive impact scores by the NOAA Observing System Integrated Analysis project (National Oceanic and Atmospheric Administration 2015), the U.S. Geological Survey Requirements Capabilities and Analysis for Earth Observations,[*] and the U.S. Group on Earth Observation Assessment in 2016.

Deductive evaluation approaches use microeconomic (cost-benefit and cost-effectiveness analysis, Bayesian decision models, and stated preference models)

---

[*] U.S. Geological Survey. Requirements Capabilities & Analysis for Earth Observations. https://remotesensing.usgs.gov/rca-eo/.

and macroeconomic analyses (computable general equilibrium models). These approaches have shown that quantification of VOI is possible. Microeconomic models focus on quantitative evaluations of individual decisions. Quantitative evaluation requires a specific application of geospatial information from EO and involves comparing the benefits in a scenario in which data are used (i.e., introduction of new technology) to one in which the observation data are not available (i.e., baseline). Macroeconomic models are used to evaluate policies and regulations. The output of the analysis is to forecast VOI in a local, regional, or national economy. Inductive empirical approaches that utilize big data and machine learning are being tested for use in decisions also.

2.8 Summary

What is the value of an Earth-observing system? Can its full value—including societal benefits beyond its designed uses—be accurately estimated? Is it possible to evaluate the effect of the geospatial information derived from EO data are having on decisions that affect the protection of life, property, and the environment?

EO and geospatial data are collectively an intermediate economic good that enables other activities and value-added services. If EO and geospatial data and information products are openly accessible they can be used by many decision-makers simultaneously and can possess characteristics of merit or public goods. Furthermore, the cost of supplying the digital data is greatest to the first user and decreases with each additional user. This property is true for the great majority of information goods giving rise to the need for studies to demonstrate the societal and economic benefits of the information. For examples, the benefits of EO and geospatial data and information products are best demonstrated when an analysis explains how they can be applied and used in a decision. Likewise, estimation of the value of this information depends on clearly defined cases in which the information is or could be used to quantify the impact, but such evidence is scarce.

To accurately assess the value of EO and geospatial data and information products more examples are needed in making scientific, policy, economic, and social decisions. Further research is needed to establish broadly applicable methods for estimating full lifecycle costs, including externalities and costs associated with data collection and administration. As methods are further developed and refined, the research community should consider the following questions: What information is valuable to which decision-makers and in what context? How do we know what information is valuable for a specific decision (an *ex ante* question)? How do we measure the effectiveness of applying information to a decision or to reducing uncertainty (could be *ex ante* or *ex post*)? What is the full range of costs, including externalities, of collecting and using EO data? What is the full range of benefits associated with the use of EO, including from *off label* uses?

The societal value of scientific research is directly tied to improved decisions and outcomes. Earth science research and the availability of geospatial data help reduce uncertainty and provide additional measurable context to decisions. Decisions rarely affect all actors equally, so it is important to engage the social, economic, and political sciences in understanding and estimating the value based on EO and geospatial information products. The perspectives from these disciplines may help explain how public/private sector partnerships, crowdsourcing and citizen science, and public outreach activities influence outcomes, value, and impact. If EO and geospatial information products are to be used to improve decision-making, the research community must accurately capture all impacts of a decision, consider the range of affected actors, and communicate the analytical results clearly and efficiently.

Acknowledgment

The authors would like to acknowledge the contribution of Eoin McCarron to an earlier draft of this chapter.

References

ACIL Allen Consulting. 2014. *CSIRO's Impact and Value*, Canberra, Australia: ACIL Allen Consulting.

Bernknopf, R. and C. Shapiro. 2015. Economic assessment of the use value of geospatial information. *ISPRS International Journal of Geo-Information* 4 (3): 1142–1165. doi:10.3390/ijgi4031142.

Bonano, E.J., S.C. Hora, R.L. Keeney, and D. Von Winterfeldt. 1990. Elicitation and use of expert judgment in performance assessment for high-level radioactive waste repositories. Nuclear Regulatory Commission, Washington, DC. Division of High-Level Waste Management, Sandia National Laboratories, Albuquerque, NM.

Coote, A. 2010. The value of geospatial information in local public service delivery. *2012 Socio-Economic Benefits Workshop: Defining, Measuring, and Communicating the Socio-Economic Benefits of Geospatial Information*, Boulder, CO, June 12–14.

DeWispelare, A.R., L.T. Herren, E.J. Bonano, and R.T. Clemen. 1994. *Background Report on the Use and Elicitation of Expert Judgment*. San Antonio, TX: Center for Nuclear Waste Regulatory Analyses.

European Commission. 2014. *Earth Observation: First Copernicus Satellite Sentinel 1A*. Brussels, Belgium: European Commission.

G8. 2013. G8 open data charter and technical annex. Accessed February 20, 2017. https://www.gov.uk/government/publications/open-data-charter/g8-open-data-charter-and-technical-annex.

Group on Earth Observations. 2014. The group on earth observations (GEO) Geneva declaration: Integrating observations to sustain our planet Geneva.

Harris, R. and L. Miller. 2011. Earth observation and the public good. *Space Policy* 27 (4): 194–201.

Keeney, R.L., D. Von Winterfeldt, and T. Eppel. 1990. Eliciting public values for complex policy decisions. *Management Science* 36 (9): 1011–1030.

Kotra, J.P., M.P. Lee, N.A. Eisenberg, and A.R. DeWispelare. 1996. Branch technical position on the use of expert elicitation in the high-level radioactive waste program. *NUREG-1563*.

Morgan, M.G., M. Henrion, and M. Small. 1992. *Uncertainty: A Guide to Dealing with Uncertainty in Quantitative Risk and Policy Analysis*. Cambridge: Cambridge University Press.

NASA. 2013. *Measuring the Socioeconomic Impacts of Earth Observation—A Primer*. Washington, DC: NASA.

National Oceanic and Atmospheric Administration, National Environmental Satellite, Data, and Information Service. 2015. NOAA observing system integrated analysis (NOSIA-II) methodology report. Edited by Department of Commerce. Washington, DC.

National Science and Technology Council, Committee on Environment, Natural Resources, and Sustainability. 2014. National plan for civil earth observations.

National Science and Technology Council, Committee on Environment, Natural Resources, and Sustainability. 2016. Common framework for earth-observation data. Washington, DC.

Obama, B. 2013. *Executive Order 13642, Making Open and Machine Readable the New Default for Government Information*. Washington, DC: Federal Register.

Pearlman, F., R. Bernknopf, M.A. Stewart, and J.S. Pearlman. 2014. Impacts of geospatial information for decision making. In *New Trends in Earth-Science Outreach and Engagement*, pp. 137–152. Springer.

Smart, A. 2014. Position paper: Evaluation methods and techniques. *Assessing the Socioeconomic Impacts and Value of 'Open' Geospatial Information Workshop*, Washington, DC, October 28.

Sylak-Glassman, E., S. Lev, and J. Gallo. 2016. The development and evaluation of a method to weight value tree elements in the United States' national earth observation assessment. *Data to Decisions: Valuing the Societal Benefits of Geospatial Information*, OECD Paris, March 10–11.

Varian, H.R., J. Farrell, and C. Shapiro. 2004. *The Economics of Information Technology: An Introduction*. Cambridge: Cambridge University Press.

Section II

From Data to Decisions

3

From Data to Knowledge—An Introduction

Joep Crompvoets, Françoise Pearlman, and Jay Pearlman

If we consider the value chain, which takes us from data to decision, the first step of the path takes us from data to information. In Chapter 3, we look at the sources of geospatial data (e.g., satellite images, remotely sensed data, and *in situ* data), and the processes used to interpret those data, giving them a meaning for a user. In the context of the well-known DIKW* pyramid, also known variously as the DIKW hierarchy, wisdom hierarchy, knowledge hierarchy, information hierarchy, and the knowledge pyramid (Rowley 2007), geospatial data becomes information when it is applied to some purpose and adds value for the recipient. The resulting geospatial information will contribute to providing a foundation for geospatially oriented decisions.

In this chapter, we will focus on the transition from data to information. Numerous information systems, including geographic information system (GIS) and Earth Observation (EO) systems exist that differ in form and application, but essentially they serve a common purpose, which is to convert data into meaningful information, which in turn enables an individual or organization to generate knowledge (Whitney 2007).

Data can be raw (Belinger et al. 2004) facts, observations, and figures without any added interpretation or analysis value (Henry 1974; Rowley and Hartley 2006). Data can also be processed to include corrections for instrument artifacts. For example, Landsat data come in the form of digital numbers, which are the light received by the sensor. However, the sensor has a calibration to turn these numbers into light intensity. The results are the calibrated response of the sensor. It is these data that can be the base for creating *information* characterizing what has been observed on the ground. Data may be considered as being objective (Davenport and Prusak 2000; Rowley 2007). "The price of crude oil is U.S. $80 per barrel." Data can be something simple or something complex. When it is put in context (organized or processed), it can then become information. However, data comes with certain embedded errors (due to the sensor or the environment) and thus quality and veracity checks are needed to have a solid foundation for information and knowledge creation. The word *Data* comes from a singular Latin word, *datum*, which

---

* DIKW stands for Data, Information, Knowledge, and Wisdom.

27

originally meant *something given*. Its early usage dates back to the 1600s. Over time *data* has become the plural of *datum* (Diffen 2013).

During the past 50 years, data have increasingly been processed by computers and has been formatted into digital representations. When the processing is done on a local computer, a scientist can create it in a format that is best adapted for the analyses to be used. When data are shared between research scientists or across communities, then it is best to format data according to certain rules that help interoperability and exchange. Those rules broadly adopted by the community are usually referred to as *standards*. For example, the Open Geospatial Consortium or OGC has developed a rich set of standards related to sensors and the geospatial data that they produce (OGC 2017). Among those, the OpenGIS® Sensor Model Language Encoding Standard (SensorML) specifies models and encoding that provide a framework within which the geometric, dynamic, and observational characteristics of sensors and sensor systems can be defined; the OGC® Sensor Web Enablement (SWE) framework standard defines low level data models for exchanging sensor-related data between nodes. But even with standards, the wide variety of data may result in the digital string of numbers representing, for example, temperature, to be in a different location in the string. Thus there is a need to have a file, which describes the way the data has been formatted and what is included. The data about data are called Metadata. Metadata provide an implementation agnostic description of the data, and a variety of standards codify the detailed implementation. Use of such tools (e.g., OGC CSW—Catalogue Service for the Web), although essential to the successful interpretation of the data, may generally be transparent to the end user. To simplify the use of data, most major systems such as satellites provide both data and metadata and indicate what are the adopted standards and to what level the data have been calibrated.

Information is data that has been organized, processed, and interpreted in a given context, so that it has a meaning for the user. If we consider the radar images of weather, there is a big step between digital radar returns and the images available to the public. The radar intensity and timing are turned into storm intensity and then integrated over time to show trends and movement of weather. Information may or may not be useful. If the radar weather is shown for a faraway country, it is many times less valuable than the local weather for making decisions on whether to schedule an outdoor event. Thus information can be useful, but does not have to be (Belinger et al. 2004). For example, "The price of crude oil has risen from U.S. $70 to $80 per barrel" gives meaning to the data, and so is said to be informative to someone who tracks oil prices. Ackoff (1989) as well as Rowley and Hartley (2006) infer information from data in the process of answering interrogative questions (e.g., *Who, What, Where, How many, When*), thereby making the data useful (Belinger et al. 2004) for decisions and/or action (Liew 2007). *Information* is an older word that dates back to the 1300s and has Old French and Middle English origins. It has always been referred to *the act of informing,* usually in regard to education, instruction, or other knowledge communication (Diffen 2013).

Knowledge is a combination of information, experience, and insight that may benefit the individual or the organization. For example, when crude oil prices go up by $10 per barrel, it is likely that petrol prices will rise and the impact will be few car sales. This is more representative of knowledge because it has not only the information, but has implications of impacts that can be acted on ("I will wait to buy a car because I can be in a stronger negotiating position and get a lower price").

The boundaries between the three terms are not always clear. What is data to one person is information to someone else. To a commodities trader, for example, slight changes in the sea of number on a computer screen convey messages that act as information and lead to a trader taking some action. To almost anyone else they would look similar to raw data of random numbers. What matters are the concept and the ability to use data to build meaningful information and knowledge.

Collecting data—in particular, geospatial—can be expensive and there needs to be a clear benefit to merit the effort, one of the main reasons that individuals or organizations collect data is to monitor and improve (policy) performance. If you are to have the information you need for policy decisions for performance improvement, you need to define and collect data on the indicators that really affect performance. These data must be collected reliably and regularly and then converted into the information needed (Whitney 2007). Indicators can vary in their level of complexity. A recent complex example is that of the Sustainability Development Goals, or SDGs, associated targets and indicators. The High-level Political Forum on Sustainable Development is the central UN platform for the follow-up and review of the 2030 Agenda for Sustainable Development adopted at the United Nations Sustainable Development Summit on September 25, 2015. Seventeen aspirational goals have been identified, which are addressed through 169 targets and monitored through 230 indicators. Reporting on progress toward a goal is done by using the applicable indicators (UN Department of Economic and Social Affairs 2016).

In order to generate useful information, data must fulfill a set of conditions such as relevance to a specific purpose, completeness, accuracy, timeliness, proper formatting, and its availability at a suitable price. In this chapter, we will repeatedly see the importance of getting the right information and getting the information right. A manager investigating poor punctuality of trains on a particular line needs information showing all the arrival data on that line. Data on other lines may be irrelevant, unless late connections elsewhere are causing the problem. Just as important. The manager must use the data correctly. One day of engineering works will have a major impact on a week's schedule of arrivals. Wrongly interpreting the results could identify a problem: where no problem actually exists.

In addition to the introduction, this section includes three chapters. The first contribution (Chapter 4) focus on the characteristics of geospatial data acquired through satellite. These types of data are called remotely sensed data and mainly differ in the way the geospatial data are acquired through a variety of space-based devices such as optical instruments operating in visible and infrared lights; radar and passive radiometry; and gravity measurement devices. This chapter, written by Pierre-Philippe Mathieu from the European Space Agency (ESA), describes how EO data—in particular, from satellites, can support science and societal applications by delivering global, repetitive, consistent, and timely information on the state of the environment and its evolution. A particular focus is on climate science, monitoring, and services.

Chapter 5, written by Jay Pearlman of IEEE and Joep Crompvoets from KU Leuven, addresses data from *in situ* sensors. This sensor-type privileges the short-range contact with the object to be measured. There are many types of *in situ* sensors and sensor networks measuring physical, biogeochemical, and biology elements of the earth environment, and this chapter contains examples in a number of domains such as land and water, climate and atmosphere, oceans and coasts, and economy and society. Due to advances in technology, particularly in miniaturization, resulted in an exponential growth in the amount of *in-situ* data being generated and recorded over the past decade.

The third and last contribution (Chapter 6) provided by Lesley Arnold from Geospatial framework, Australia, considers spatial data supply chain challenges from the perspective of producers determining what activities they need to perform to deliver the right knowledge to the end user, at the right time. It also considers uncertainty from the user's perspective and their need for believability in information for it to be valued as a new knowledge.

References

Ackoff, R., 1989. From data to wisdom. *Journal of Applied Systems Analysis*, 16: 3–9.

Belinger, G., D. Castro, and A. Mills, 2004. Data, information, knowledge, and wisdom. Last accessed: February 15, 2017, http://www.systems-thinking.org/dikw/dikw.htm.

Davenport, T.H. and L. Prusak, 2000. *Working Knowledge: How Organizations Manage What They Know*. Harvard Business School Press, Boston, MA.

Diffen, 2013. Data vs. information. Last accessed: February 15, 2017, http://www.diffen.com/difference/Data_vs_Information.

Henry, N.L., 1974. Knowledge management: A new concern for public management. *Public Administration Review*, 34(3): 189.

Liew, A., 2007. Understanding data, information, knowledge and their interrelationships. *Journal of Knowledge Management Practice*, 8(2).

Open Geospatial Consortium, 2017. OGC standards. http://www.opengeospatial.org/docs/is.

Rowley, J., 2007. The wisdom hierarchy: Representations of the DIKW hierarchy. *Journal of Information and Communication Science*, 33(2): 163–180.

Rowley, J. and R. Hartley, 2006. *Organizing Knowledge: An Introduction to Managing Access to Information.* Ashgate Publishing, Aldershot, pp. 5–6.

UN Department of Economic and Social Affairs, 2016. SDG indicators. Official list of SDG indicators. Last accessed: February 21, 2016, https://unstats.un.org/sdgs/indicators/indicators-list/.

Whitney, H., 2007. How to define data, information and knowledge. TechTarget. Last accessed: February 14, 2017, http://searchdatamanagement.techtarget.com/feature/Defining-data-information-and-knowledge.

4

Satellite and Remote-Sensing Data

Pierre-Philippe Mathieu

CONTENTS

4.1 The Climate Challenge – An Application of Remote Sensing Data

With nearly three billion people entering the global middle class during the next two decades and the overall population expanding rapidly, there will be an exploding demand for food–water–energy commodities. The result will be an enormous and unsustainable pressure on natural resources such as water and arable land. Demands for energy and improved health create further impacts. Climate change induced by human activities is causing additional significant stresses on our access to ecosystems services. Humans are now the dominant driver of large-scale changes on the global environment, moving our planet into an entirely new geological epoch, the *Anthropocene* (Cruzen 2002, Kolbert 2014). Climate change, from ocean acidification and rising temperatures to sea level rise and ice melting, impacts ecosystem services and the food–water–energy nexus resources. The recent Fifth Assessment Report of the Intergovernmental Panel for Climate Change (IPCC AR5 2014) reiterated the urgent need for collective actions. The World Bank Group states in its reports (Turn Down the Heat 2012, 2013) that under current Greenhouse Gas emissions pledges, mankind is moving toward a 4°C warmer world by the end of this century, with significant impacts on agriculture, water resources, ecosystems, and human health.

Climate change has now become central to our sustainable development challenge, as it can undermine development gains and put billions of people at risk, particularly, for the most vulnerable in poor countries. The links between development and climate change have now become clear and unavoidable. Extreme climatic events could reverse years of developmental success in developing countries.

Enhancing resilience* of our society to natural hazards and climate change, while developing a sustainable low-carbon economy is now at the heart of Sustainable Development Goals (SDGs) for 2030, which recognizes our life-support ecosystem services as a prerequisite for any development (The Road to Dignity by 2030).[†] Such challenge must be addressed on a global scale with information that reaches across national and regional boundaries. Space observations provide such opportunities for information that can address these global and national scale issues and challenges.

4.1.1 The Need for a Science-Based Management Approach

A truly science-based integrated risk management approach is needed to better manage the risks to achieve sustainable development and address the challenges related to both human-induced climate changes and natural climate variability. This must be done with a quantitative approach or methodology to carry weight in the political tradeoffs that will need to be addressed. It is also necessary to consider the complete value-chain fro data to information/knowledge so that decision makers have confidence in the quality of the information.

Such an integrated approach requires accurate quantification of hazard intensity and vulnerability of populations. The uncertainty in the information plays an important, yet many time unspoken, role in willingness to take risks as part of complex decision processes. A new type of "interlinked" thinking would acknowledge and incorporate other aspects of risk contributing to the overall global change risk. As such, considerations need to include the direct impact of human activities and the cascading risks resulting from the inherent relationships between climate and the water–food–energy resource. For example, coastal flooding risk depends on both sea level rise and subsidence from water pumping.

The idea of an integrated risk management approach is not new but is gaining more momentum within the scientific community. Recent reports of the IPCC, such as SREX (2011), are advocating for a science-based evidence-based

---

* Capacity of an organization, community or country to continually evolve and adapt to gradual changes and sudden shocks while remaining able to fulfill its core function (World Forum on Global Risks).
† http://www.un.org/apps/news/story.asp?NewsID=49509#.WQH_4xQdhiM

risk approach in order to be able to make informed decisions under uncertainties, design robust early warning systems, and ultimately objectively define an optimal *safety net* to transfer risks of climate-sensitive sectors. The most recent report of Working Group II of the IPCC features the Water–Energy–Food/Feed/Fiber Nexus as linked to climate change as a cross-chapter theme. The nexus also features in the SDGs through an integrated approach.

Adopting an integrated risk management approach should hopefully lead to climate-smarter decisions at various levels for a wide variety of end users, ranging from businesses, public sector up to citizens, bringing their information needs into a unified risk framework. It would help decision-makers to better handle synergy and trade-offs between different interconnected risks related to climate, water, energy, and food security and codesign solutions to address them. Recognizing that climate change is at the heart of development, multilateral development banks, such as the World Bank Group, are now developing tools to support climate risk screening enabling them to better identify risk *hot spots* and prioritize their areas of actions and investment.

4.1.2 Earth-Observing Systems as a Key to Managing Risk

Earth Observations provide one basis for any scientific understanding, from the testing of hypotheses and the development/validation of models to the attribution and prediction of ecosystem changes.

Decision-makers require accurate, consistent, and timely information about the state of our changing environment and its evolution to form and inform their decisions regarding adaptation and mitigation strategies.

Building a full picture of a rapidly changing and interconnected environment of our planet over a wide range of scales in space and time—from minutes to decades—requires a comprehensive Integrated Global Earth Observation System of Systems (GEOSS) fusing data from multiple sources including *in situ* networks, drones, web of sensors, and satellite imagery.

The need for a *data revolution* combining all types of open data with high tech to achieve the post-2015 SDGs has been called for recently by the UN Secretary-General's Independent Expert Advisory Group on a Data Revolution for Sustainable Development (A World that counts: Mobilizing the Data Revolution for Sustainable Development 2014).[*]

Earth observations (EO) from space plays a key role in this endeavor as satellites are uniquely placed to deliver the comprehensive, global, and consistent datasets needed to support climate observation, research, and services. Space agencies around the world have been developing a set of missions (e.g., Landsat, Aura, Terra, ALOS, Copernicus, and Sentinels) covering the whole electromagnetic spectrum.

By remotely sensing radiation, EO satellites are able to derive information on some of the key climate parameters. This capability goes well beyond

---

[*] http://www.undatarevolution.org/report/

simple pretty pictures (being already very useful) to become a powerful quantitative tool. Using the unique vantage point of space, satellites deliver global data, covering even in the most remote places where no survey data exist or are possible to obtain. The ability to retrieve historical data from the satellite archive is also a key advantage to allow users to detect changes in the environment. These unique characteristics of wide-area mapping of EO data make them particularly useful to complement—but not supplement—traditional *in situ* measurements, which are typically point based, sparsely distributed, or simply completely missing in remote or difficult to access areas (e.g., mountains and polar regions).

Rapid advances in the capability to observe our planet over the past decades, have led to enormous scientific insight into how our climate and ecosystem work as part of a complex coupled system.

Today, more than hundred EO satellites, carrying multiple radar and optical instruments, are continuously monitoring the state of our planet, providing scientists with a continuous stream of data on the state of the ocean, atmosphere, ice sheets, and vegetation.

With the launch of Sentinel-1 in 2014—the first in the series of operational satellites—and Sentinel-2 in 2015, and the advent of the Copernicus initiative, Europe has entered a new era for the development and exploitation of an open EO data. Copernicus will provide a unique global observational capability to Europe across the whole electromagnetic spectrum and across a wide range of applications. These sustained observations will be complemented by exploratory measurements derived from a series of research explorer missions dedicated to Earth system science. For more information on the satellite missions worldwide and their applications, see the recent handbook of the Committee on Earth Observation Satellites (CEOS) prepared for Rio+20.

4.2 Importance of Earth-Observation Data for Monitoring Climate

The importance of *global* and *sustained* observations for monitoring climate and its changes has long been recognized by the United Nations Framework Convention on Climate Change (UNFCCC) (article 4.1g). In this context, the Global Climate Observing System (GCOS) was established in 1992 to better quantify the information needs of the UNFCCC. In particular, GCOS has defined a set of Essential Climate Variables (ECVs) required to quantify the state of our climate and related forcing.

GCOS also calls for a systematic generation of long-term, homogeneous, and continuous Climate Data Records (CDRs) of the ECVs together with a measure of their uncertainty and a documentation of the process. One of the most convincing examples of the value of CDRs, which now epitomizes the issue of climate change and has significantly driven its political agenda, is the continuous record of atmospheric composition of CO_2 compiled by Keeling at the station of Mauna Loa (Hawaii) since 1958. This simple curve revealing a continuous rise of CO_2 has forever changed our vision on the global impact of human-induced activities by unambiguously establishing the connection between the burning fossil fuel and the increase of atmospheric CO_2.

4.3 Demand for Climate Services and Associated Challenges

The demand for Climate Services[*] is rapidly growing. In order to guide their strategic plans, investments, and policy decisions, many organizations within the public and private sectors, including insurance, agriculture, health, energy, and transportation, are increasingly needing information on the specific climate risks they face.

The Global Framework for Climate Services (GFCS) (World Meteorological Organization 2011) established in 2009 during the third World Climate Conference recognizes observations and monitoring as essential pillars of Climate Services, along with research, modeling, prediction, service information system, user interface, and capacity building.

One of the key challenges of Climate Services is to convert raw climate data into *actionable* information and knowledge of sufficient quality to be of use to the end users. Such data-to-information conversion process is not an easy task and inherently depends on the type of applications and targeted sectors. Many barriers and challenges still exist in the transition from data to information and then knowledge. Delivering the right information often requires integration of EO data with other types of data, for example, *in situ* sensor observations are needed to validate satellite data. Other types of data include models (e.g., reanalysis and climate projections), socioeconomic indicators, and local knowledge. In the energy sector, energy companies would need to integrate climate information with energy demand projections to accurately forecast the load demand and improve planning of energy supply. Another challenge is to address the end to end process sequence from the initial observation to the ultimate decision. This often requires seamless integration of climate information into the user software and decision-making system, and possibly new ways of delivery such as mobile devices (cell phones or drones).

Climate Services also present very challenging requirements inherent to their climate nature, related to characterization of uncertainty, traceability

---

[*] The European research and innovation Roadmap for Climate Services defines climate services as covering *the transformation of climate-related data—together with other relevant information—into customised products such as projections, forecasts, information, trends, economic analysis, assessments (including technology assessment), counselling on best practices, development and evaluation of solutions and any other service in relation to climate that may be of use for the society at large.*

(e.g., of documentation and processing), qualification of accuracy and integrity of information, and complexity of ingesting probabilistic risk information.

Climate Services are also about prediction (i.e., forecast) and projection (i.e., scenarios). Today, thanks to rapid progress in observing technology, coupled climate modeling, data assimilation techniques, and computing power, climate scientists are able to predict seasonal-to-decadal variations of the climate with some level of skill. They also use long-term climate simulations to develop projections about the future, such as those used for the coupled model intercomparison project (CMIP). In this context, observation become increasingly important to initialize, validate, and constrain model simulations, becoming more complex (e.g., representing/parameterizing more processes) and operating at higher resolution (e.g., eddy-resolving ocean models) and regional scale (e.g., CORDEX).

Europe is now very active in the development of Climate Services with the recent advent of the European Copernicus Climate Change Service (C3S) supported by the European Commission within the framework of the Copernicus initiative. The C3S managed by the European Center for Medium-Range Weather Forecasts (ECMWF) includes elements such as the generation of CDRs, the development of long-term climate reanalyzes, the assessment of the state of climate and the prediction of its changes, and the wide distribution of data to the community through a Climate Data Store to stimulate research and the service market.

The value of Climate Service information is several times larger than the investment in the observing system. For example, the current annual cost of responding to natural disasters is about USD $6 billion, and that projection suggests that the cost could increase to up to USD $1 trillion a year by the year 2050 (Hallegatte et al. 2013).

Quantifying these risks will benefit a variety of key industrial sectors affected by the climate change, including health, water supply and sanitation, energy, transport, industry, insurance, mining, construction, trade, tourism, agriculture, forestry, and fishery, thereby resulting in major socioeconomic benefits. These benefits will likely drive a huge demand for services and open a broad market to provide customized high-added-value services to a variety of users.

In the following paragraph, we discuss some limited examples of EO-based information services supporting adaptation and mitigation of climate change, with focus on key issues such as food security, energy, and urban development. These are precursor elements of a wider and more complex integrated risk approach to climate change.

4.3.1 An EO-Based Example for Information Services - Climate-Smart Agriculture

Farmers are now facing unprecedented challenges and pressures to meet the growing global demand for food (e.g., crops for food and livestock) and energy (e.g., biofuel demand), whereas most of the limited fertile land available on the planet is already used for intensive agricultural productivity. Meeting these needs leads to huge impacts on the water–energy food nexus, as global food production is expected to increase by 35%, water by 40%, and energy by 50% by 2030 (U.S. National Intelligence 2012). In the meantime, 1 billion people remain chronically hungry today.

On the top of that, climate change induced by human activities is causing additional stress on agricultural production and water resources (e.g., 70% of freshwater is used for agriculture), due to increased frequency and magnitude of extreme events (e.g., floods/droughts) or changes in environmental conditions (e.g., early flowering in certain plants, spread of pathogens, and pests). Such extreme events are more likely to happen in a warmer world, and people in already vulnerable living conditions will be hit hardest.

Another impact of climate is its effect on the market food price and recurring spikes. In particular, climatic extreme events such as floods and droughts can induce significant volatility in the crop market, thereby suddenly putting several million people into poverty. As a result, the risk of social unrest is growing. For example, in areas that do not have sufficient domestic grain production, and which are thus dependent on imports, the problem is quickly exacerbated as grain prices rise. For example, the 2007–2008 food protests in Tunisia and Egypt, among others, are considered an important factor in the Arab Spring. This issue of volatility of the crop market has led the G20 to initiate the Agricultural Market Information System (AMIS) interagency platform, which aims to enhance international market transparency by improving the quality, timeliness, and reliability of food market information.

All these interconnected challenges of food security lie at the heart of Climate-Smart Agriculture (CSA) aiming to develop precision farming strategies for maximizing agricultural production, whereas minimizing the impact on the environment and irrigation (e.g., less drops per crops) under a climate stress.

EO can play an important role here. The Global Agricultural Monitoring (GEO-GLAM) initiative aims to coordinate satellite monitoring observation systems in different regions of the world in order to enhance crop production projections and weather forecasting data. Within this framework, GEO-GLAM has developed a unique global crop monitoring capability, providing users with a multisources consensus assessment of four primary crop types (wheat, maize, rice, and soy) in support of the AMIS market monitoring activities. One example is the monitoring of rice in Asia (Figure 4.1), where Sentinel-1 imaging radar enables the rapid and continuous assessment of the state of rice production and associated flooding. Global monitoring of crop by different sources of EO is a key building block of a wider integrated risk management system for CSA and food security. In this context, Sentinel-2 will provide a unique global and systematic view of our land vegetation at an unprecedented resolution of 10 m.

FIGURE 4.1
Land cover map in the Mekong River Delta (Vietnam) for the 2015 spring–summer rice season derived from Sentinel-1A data. The land cover map at 20 m resolution features rice production (red), nonrice (green) (e.g., other land use type such as trees, orchards, and other crops), water bodies (blue) (including sea, river, and aquaculture), and land outside the Mekong delta (gray). The total rice planted area during this rice season is estimated to be 2.14 million hectares by July 21, 2015. Copernicus Sentinel data (2015)/ESA/CESBIO/STAC/VAST/Asia-RICE/DUE.

The convergence of open climate data with rapid advances in digital technologies (e.g., cloud, Internet, and mobiles) is likely to drive new innovation in climate services for agriculture and other sectors.

4.3.2 Application to Renewable Energy Resources

Wind energy is experiencing one of the fastest growths across the whole renewable energy industry. Wind power is capital intensive. Thus, the financial success of wind farms is strongly bound to the wind resources available over the plant lifetime (hence the revenue) and to other factors affecting the initial investment such as the impact on environment, access to turbines for maintenance, regulatory considerations "h", and connection to the grid network for distribution (hence the cost). Quantifying these factors is critical to perform technical and financial feasibility studies of prospective sites in order to secure long-term investment.

The traditional way to assess the potential energy yield of a prospective wind farm is by using data from a meteorological mast, which is very expensive in terms of installation and maintenance. Although this approach is very accurate, it can only provide point-measurement data for a short period of time (typically one year), whereas the wind field is generally highly variable in space and time over periods of many years. This issue is further compounded for offshore farms, as the amount of wind offshore is sometimes estimated from onshore measurements. Using local data can therefore be an issue to assess effectively the financial viability of prospective farms.

Satellites measure wind in a synoptic manner, but only over sea through the use of active sensors, such as scatterometers, altimeters, and Synthetic Aperture Radar (SAR). EO data thereby provides a more comprehensive and spatially resolved view of the ocean wind climatology and the entire probability distribution (Figure 4.2). The new generation of algorithms is now able to extract more information on wind magnitude and direction from the radar Doppler signal.

EO radar missions such as Sentinel-1 can provide users with information on coastal wind availability, the state of the ocean (e.g., wind/wave metocean conditions), and land (e.g., roughness and vegetation cover). These missions now ensure continuity of such data streams for decades, with high revisit time of a few days over Europe and polar regions. This information will be critical to assist decision-making regarding economic viability of wind turbines which operate in the coastal environment.

Other operational user-driven missions, such as the Meteosat family in geostationary orbit, are delivering a wealth of information on the availability of natural resources, including a variety of renewable energy resources such as solar power. For example, Meteosat Second Generation (MSG) satellites deliver global maps of irradiance up to 1 km resolution every 15 min. By combining EO-based irradiance maps with other EO products, such as Digital Elevation Model and cloud cover and aerosols maps, it is possible to estimate the solar energy yield expected from a solar energy power plant. The ability to go back in time in the archive of Meteosat data—spanning several decades—provides the long-term time series and statistics of direct/diffuse solar irradiance together with cloud conditions necessary to quantify solar resources. This information forms the basis of a portfolio of Climate Services supporting solar energy managers in siting, designing, and assessing performances of solar plants, such as the Copernicus downstream services developed. The future generation of meteorological missions such as the Meteosat Third Generation (MTG) to be launched within this decade will further improve these services by providing decision-makers with enhanced sampling capability in space, time, and spectral range.

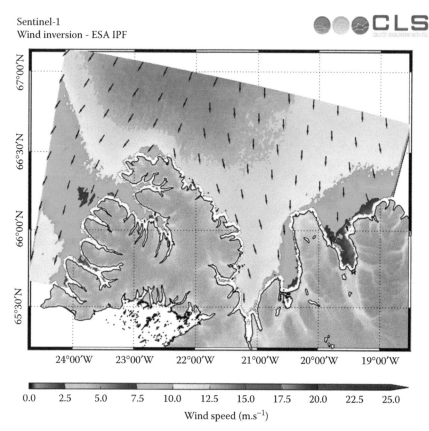

FIGURE 4.2
Coastal wind product around Denmark derived from 3.2 Satellite and remote-sensed data.

4.4 Conclusions

The ability of our society to manage climate and other natural and anthropogenic risks, adapt and become more resilient, depends to a large extent on our capacity to monitor, understand, and predict the state of our environment and the impact of and on human activities. Although our impact on climate and water, food, and energy resources has often been addressed separately in terms of risk management, they, in fact, represent diverse but inextricably interconnected facets of risk for our society, particularly under expanding human population and pressures. A new type of integrated *nexus* thinking is therefore needed to address climate problems in order to integrate multiple interrelated risks into a unified risk management framework.

Observations provide the foundation for such an integrated risk management approach. In particular, EO satellites orbiting hundreds of kilometers above the planet can play a key role here, as they *enlarge* our view of the climate system from the local to the global scale, and deliver the global, uniform (i.e., with one consistent set of instruments), and repetitive (e.g., with a rather high revisit rate) measurement of the ECVs needed to better quantify the risks.

This unique role of EO is likely to grow significantly with the new generation of satellite missions providing enhanced spectral, temporal, and spatial capabilities, delivering open, global, quality controlled, and multivariate datasets (together with a measure of their uncertainty) needed to address some of the societal risks and decision making. In particular, the availability of operational Copernicus missions and services will be improved through the Sentinel missions under the European Copernicus initiative and guided by a full and open data policy. The advent of the Copernicus services as well as systems from the Americas, Asia and elsewhere are also likely to herald a new era of global and timely environmental information services, thereby creating new science and new applications in a broad market.

References

Adams, S., Baarsch, F., Bondeau, A., Coumou, D., Donner, R., Frieler, K., Hare, B. et al. Turn down the heat: Climate extremes, regional impacts, and the case for resilience - Full report. Turn down the heat. Washington, DC: World Bank. http://documents.worldbank.org/curated/en/975911468163736818/Turn-down-the-heat-climate-extremes-regional-impacts-and-the-case-for-resilience-full-report.

Crutzen, P. J. Geology of mankind: The Anthropocene. *Nature*, 2002, 415:23.

Data Revolution Group. A world that counts: Mobilising the data revolution for sustainable development. Secretary-General's Independent Expert Advisory Group on a Data Revolution for Sustainable Development (IEAG), 2014. www.undatarevolution.org.

Food Early Solutions for Africa. *FESA Micro-Insurance: Crop Insurance Reaching Every Farmer in Africa*. Delft, the Netherlands: EARS Earth Environment Monitoring, 2014.

Global food security: Key drivers-A conference report. National Intelligence Council Report NICR 2012-05, February 1, 2012.

Hallegatte, S., Green, C., Nicholls, R. J., Corfee-Morlot, J. Future flood losses in major coastal cities. *Nature Climate Change*, 2013, 3:802–806. doi:10.1038/nclimate1979.

IPCC. Climate change 2014: Synthesis report. *Contribution of Working Groups I, II and III to the Fifth Assessment Report of the Intergovernmental Panel on Climate Change* [Core Writing Team, R. K. Pachauri and L. A. Meyer (Eds.)]. Geneva, Switzerland: IPCC, 2014, 151 pp.

Kolbert, E. *The Sixth Extinction*. New York: Henry Holt & Company, 2014.

The road to dignity by 2030: Ending poverty, transforming all lives and protecting the planet, synthesis report of the secretary-general on the post-2015 agenda. 2014.

World Bank. Turn down the heat: Why a 4°C warmer world must be avoided. Turn down the heat. Washington, DC: World Bank, 2012. http://documents.world bank.org/curated/en/865571468149107611/Turn-down-the-heat-why-a-4-C-warmer-world-must-be-avoided.

World Meteorological Organization. Climate knowledge for action: A global framework for climate services: Empowering the most vulnerable. Geneva, Switzerland: World Meteorological Organization, 2011.

5

In situ *Data*

Jay Pearlman and Joep Crompvoets

CONTENTS

5.1 Introduction

In situ is a Latin term that translates literally to *on-site* or *in position*. It has come to mean *locally, on-site, on the premises,* or *in place* to describe an event where it takes place, and is used in many different contexts. For example, in fields such as physics, chemistry, biology, as well as earth and atmospheric sciences, *in situ* may describe the way a measurement is taken, that is, in the same place the phenomenon is occurring without isolating or altering the conditions of the test.[*] In the context of geospatial information science, it refers to the spatial data acquisition on-site. This is in contrast with remotely sensed data that observe from afar or outside the media being observed as space-based satellite observations.

    In situ measurements require that the instrumentation should be located in contact with the subject of interest, for example, sensors that produce a response to a change in a physical condition, such as temperature or thermal conductivity, or to a change in chemical concentration. In contrast, remote

---

[*] Mainly derived from Merriam-Webster, Lewis & Short Latin, and Collins Latin dictionaries.

sensors are located some distances away from the subject of interest. Both *in situ* and remote sensors include passive systems (instruments that receive information naturally emitted by the media) and active systems (instruments that emit either acoustic or electromagnetic energy and record the characteristics of this energy after it reflects off an object or surface and returns back to the sensor).

All is not black or white. A sensor can be both *in situ* and remote sensing. A simple example is a man standing outside in the rain. He feels the rain on his head and is an *in situ* sensor monitoring rainfall. He also looks out to the horizon and sees the Sun shining and the look up to see the clouds moving away overhead. Using remote sensing, he predicts that the rain will stop. Although this is an example of a *human observation system* using multiple sensors, other observation systems may be configured with sensor packages having multiple sensors, as shown in Table 5.1 below.

This section considers the different types of *in situ* data, and how they confer value to users. Table 5.1 gives a sample of the *in situ* sensors and their applications. Generally, *in situ* data has a number of attributes: quantitative value of a medium's character (e.g., temperature, moisture level for soil, or salinity for ocean water), location of measurement, time of the measurement, and identity of sensor. In addition, almost all measurements have some level of uncertainty or error and this should be specified and recorded. *In situ* data are increasingly reliant on satellite-based GPS along with geodetic reference frameworks to accurately determine where the measurement is made. So, monitoring stream water, point *in situ* measurements provide a time history of stream height at the measured location. Combining information from different *in situ* sensors can allow analysis of high water propagating downstream.

TABLE 5.1

Examples of *in situ* Sensors and Their Applications

Sensor	Platform	Application
Radiosonde	Ballon	Atmospheric profiles
Local weather station	Ground-based	Local atmosphere and precipitation
Tide gauge	Shore-based platform	Monitoring tide cycles
Atmospheric gas monitoring	Long-term ecological research towers	Climate and atmospheric monitoring trends
Atmospheric contaminants monitors	Buildings and towers	Air quality assessments
Soil moisture monitors	Ground-based	Soil moisture assessment
Pressure at ocean bottom	Ocean bottom fixed mount	Tsunami warning
Seismic sensors	Ground-based	Earthquake or volcanic eruption monitor
Precision GPS	Multiple platforms	Crust deformation monitor

The aim of this section is to indicate the roles that *in situ* data play in generating valuable information for society. In what follows, four domains are presented to provide perspectives on different characteristics of *in situ* monitoring: (1) land and water, (2) climate and atmosphere, (3) oceans and coast, and (4) economy and society. The last of these is not monitoring physical domains as it is addressing the dynamics of economic and social interactions. These four categories are represented in a number of international initiatives, including the 2030 objectives and associated Sustainable Development Goals (SDGs) of the United Nations* and the intergovernmental Group on Earth Observation (GEO) (https://www.earthobservations.org/index.php). The guidance provided by these initiatives creates structures for *in situ* monitoring objectives based on socioeconomic needs and requirements. GEO, for example, is focused on environmental information that can support policy and decision. These requirements flow down to selection of sensors, observations systems, and data management processes. This is a *top-down approach* as monitoring system attributes flow down from socioeconomic requirements. There is also a bottom's up approach in which citizen contributions support an overall objective, but the volunteer contributions are more informal (Goodchild 2007). In this section, both modalities will be examined. By examining both, a range of processes for *in situ* data are considered, from a formal structure for predefined *in situ* data types to a more inclusive environment of both formal monitoring and volunteer data. In this context, a path from survey data that adheres to rigorous detailed government imposed standards, to commercial data collected by vehicles such as Google's Streetview project, and finally ventures toward the voluminous data that are being personally and continuously produced through smart technologies and voluntary or automatic citizen contributions. In addition, other developments (e.g., Internet of Things/Big Data) are briefly introduced.

Although the representations of *in situ* data provided here are by no means exhaustive, the exploration of these recognizable examples of *in situ* geospatial and socioeconomic data serves to illustrate the indicative role that such data continues to play in enabling us to derive meaningful information about our environment.

5.2 *In situ* Monitoring Characteristics

In comparison with remote-sensed data, which are averaged over a large footprint of terrain governed by the sensor and system characteristics, and which requires significant computation (Section 5.1), *in situ* data are

---

primarily localized data, relevant to a single location, or set of locations. It is often difficult to compare *in situ* and remotely sensed data directly, because a point measurement may not represent the average character of the remotely sensed footprint. These comparisons generally rely on models for such comparisons (Yoder et al. 2015). There are other challenges because some processes or features, even of topographic nature, could dominate the signal received by the remote sensor and thus the signal is not a *uniform* indication of the cell attributes. This is sometimes referred to as cross-channel contamination, which is often noted in remote sensing of the oceans in the coastal or near shore observations. Such *cross-channel contamination* is observed more often for satellites with large spatial footprints. For airborne remote sensing, the footprint can be much smaller and cross-channel artifacts may be less significant with airborne measurements depending on the scale of the phenomena being observed. However, cross-contamination between pixels may still occur in high contrast scenes.

The sampling rate of *in situ* measurements is important to monitor trends in the environment and for supporting modeling. The sampling may be continuous or at a specified interval. The data may be transmitted to support quasi real-time monitoring or it may be recorded for future analyses. Which of these is used depends on the application. Weather radar data, for example, is handled real time, whereas for subsurface ocean monitoring with glider platforms, the data are stored until the glider surfaces and relays information by shore links or satellite.

The geographic distribution of *in situ* sensors is again dependent on the application. Coastal sea level and tide gauge monitoring sensors are placed at key places along the coast. In the open ocean, monitoring is done through surface floats or Argo profiling floats, which drift with the ocean currents. Automobiles that provide traffic flows through their GPS may have random spacing on the highways, but have sufficient frequency to allow active monitoring of highway congestion. *In situ* road sensors also monitor traffic flows. These two together are a good example of multiple sensors supporting a common goal of informing drivers of traffic congestion.

Finally, *in situ* data may not need the corrections for atmospheric and irradiance effects that are typically applied to remote-sensed data. Although this is an advantage, the reproducibility and stability of *in situ* instruments may introduce uncertainties as data and information are integrated over large areas. The combination of remote-sensed data and *in situ* data, aforementioned, is a subject of current research as satellites have large footprints (from meters to kilometers), whereas the *in situ* sensors are typically point measurements (Yoder et al. 2015). Despite this difference, *in situ* data are used for calibration of instruments performing remote-sensed measurements on either satellite or on airborne platforms (Pe'eri et al. 2013).

5.3 Domains

This section focuses on the information that comes from local measurements. Many times, this is thought of as a sensor that looks at a particular local characteristic such as temperature or air contaminants. *In situ* data may also come from people when addressing more than the natural environment. For example, information about the *as built* environment of cities can come from people as well as built sensors. Closures of city streets come from an announcement or from people commenting on difficulties with traffic. It can also come from monitoring cameras and *in situ* road sensors.

As exemplars, we consider *in situ* sensors and data from four domains: (1) land and water, (2) climate and atmosphere, (3) oceans and coast, and (4) economy and society, where the last encompasses the interactions in the as-built environment and also the interactions of humans with the natural environment.

At a high level, a good perspective of Earth observations (EO) and their impacts can be gleaned from the intergovernmental GEO mentioned earlier. As referenced in the GEO strategic plan "the Societal Benefit Areas (SBAs) are the areas in which Earth observations are translated to support for decision-making" (GEO 2016). Eight SBAs are distinguished: (1) disaster Resilience, (2) food security and sustainable agriculture, (3) water management resources, (4) energy and natural resources, (5) health surveillance, (6) biodiversity and ecosystem conservation, (7) urban resilience, and (8) infrastructure and transportation management. Underlying these are knowledge bases from, for example, weather, climate, and oceans. The development of solutions to societal challenges within these SBAs is facilitated by mobilizing resources, including observations, science, modeling, and applications, to enable end-to-end system solutions and deliver services for users. Within each SBA, *in situ* data are collected, combined with other data (including remote-sensed data), and analyzed to produce information. This results in information and eventually knowledge to support the decision-makers and policy implementation. In this context, relevant stakeholders are public authorities at different administrative levels, scientists, communities, national mapping agencies, commercial users, and so on. In addition to the SBAs, GEO has Flagship Programs and initiatives that focus on disciplines and environmental types. For example, the Global Forest Observations Initiative (GFOI) address forest environment and sustainability, and the Blue Planet Initiative focus on oceans and their coasts. GEO is one example of an intergovernmental cooperation. There are many international cooperation initiatives/organizations (e.g., World Meteorological Organization [WMO],[*] Integrated

---

[*] https://www.wmo.int/pages/index_en.html

Carbon Observation System [ICOS]*, and International Ocean Commission [IOC]† in observing area applications such as drought, air quality, ocean conditions, and others).

5.3.1 Land and Water

Land cover represents an important set of Earth surface characterizations due to its crosscutting nature. It can be used to examine land use, climate-induced changes, monitoring of invasive species, flood channels, and so on. Despite many technological advances, there are still significant variations in the way land cover is analyzed using data from different sensors. Synoptic views of land cover usually come from space-based remote sensing using vegetation indices as land cover measures. *In situ* monitoring of land cover does not provide the same synoptic view. For agricultural areas, farmers will report their crop types to the government and this can be included in the land cover assessments. Physical monitoring for soil moisture or chemical constituents can be done with *in situ* sensors. "Reducing inconsistencies between land cover products, nesting of finer-scale information within broader schemes, and standardized accuracy assessments still remain major challenges" (GEO work program 2016). In the context of SBA Food Security and Sustainable Agriculture, relevant examples of applications for forestry include identification of tree species, tree heights, tree crown density, foliar chemistry, and trees fixing carbon dioxides (Pause et al. 2016); and for agriculture include crop type, fertilizer impact, and precision agriculture. A recent study of the impacts of fertilizer applications on ground water quality using both remote-sensed and *in situ* data showed the potential impact on well water (Forney et al. 2012). In this context, relevant instruments are electrical conductivity (ECa) sensors, gamma-radiometric soil sensors, and soil moisture devices, among others.

Another area of Earth monitoring is observing events that are below the surface or movement of the surface. Earthquakes, landslides, and tsunamis are well known for the damage they inflict. Monitoring of the Earth motions and their downstream effects are done with *in situ* seismic sensors (Lee et al. 2002), water pressure sensors‡ precision GPS (Langley 2010), and other modern measurement systems. The seismic sensors measure vibrations and then correlate data from multiple sensors to give the location and size of Earth movements. Precision GPS has sufficient accuracy, typically on the order of a millimeter, to observe changes in the shape of volcanoes and thus anticipate potential eruptions. Soil moisture measurements for certain types of soils and surface topography can be used to anticipate where landslides may occur. The ability to accurately forecast the location and timing of

---

* https://www.icos-ri.eu
† http://www.ioc-unesco.org
‡ http://www.bom.gov.au/tsunami/about/detection_buoys.shtml

earthquakes, eruptions, and landslides does not yet exist, but tsunami monitoring with *in situ* pressure ocean sensors exists and is expanding. When the time for tsunami propagation is long enough (at distant locations from the earthquake source), real-time forecasts of arrival are now practical. This could save a significant number of lives and costs on the economy, for example, the 2011 Tōhoku earthquake and tsunami (Japan) resulted in 16,000 lives lost and $300 billion in cost; in the 2008 Sichuan earthquake (China), more than 69,000 people lost their lives and had a cost of $148 billion; and for the 1995 *Great Hanshin earthquake* near Kobe, 6,000 *lives* and $100 billion were lost.

Water is one of the key elements in economic cycles, for agriculture, human consumption, and industry. Numerous public and commercial research institutes in different application areas (agriculture, emergency management, water/energy utilities, and transportation) are involved in *in situ* measurements of land and water. As an example, the amount of water in rivers, lakes, and smaller flows is monitored with *in situ* gauges—stream gauges are an example. There are also the tide gauges in coastal regions. These support drought as well as flood predictions. As important, they allow for allocation of water resources for production and human use. Stream gauges are not pervasive, but there is expansion of their applications to developing countries (Esri 2016). The drought monitor is created through a process that synthesizes multiple indices, outlooks, and local impacts, into an assessment that best represents current drought conditions. The final outcome of each Drought Monitor is a consensus of federal, state, and academic scientists.[*] In addition to stream gauge monitors, soil and water processes can be monitored through autonomous *in situ* sensors for ground water through soil sensor networks that may include unmanned, mobile sensor platforms for water quality and quantity. In order to improve the irrigation and water distribution over agricultural land, information from high resolution soil moisture sensors is combined with a soil hydrology model to guide irrigation, resulting in lower water use and higher yields. The sensors allow farmers to water crops where and as needed and eliminate the tendency to overwater.

The value of land and water data covers a broad range of activities, including agriculture, the impacts of precision farming, water allocations, and sensors allowing more area to be cultivated in irrigated production, often referenced to as *more crop per drop*.[†] As another example, the PrecisionAg Institute published a survey of soybean growers indicating "an average savings of about 15% on several crop inputs such as seed, fertilizer, and chemicals."[‡]

The impacts of improved earthquake monitoring and tsunamis are substantial as indicated by a U.S. National Academy of Sciences study on the benefits from improved earthquake hazard assessments (Committee on the Economic Benefits of Improved Seismic Monitoring 2005). The ability

---

[*] https://www.ncdc.noaa.gov/temp-and-precip/drought/nadm/index.php
[†] http://www.fao.org/english/newsroom/focus/2003/water.htm
[‡] http://www.precisionag.com/institute/precision-agriculture-higher-profit-lower-cost/

to manage drought impacts are economical and socially significant as well "Since 1980, major droughts and heat waves within the United States alone have resulted in costs exceeding 100 billion dollars, easily becoming one of the most costly weather-related disasters on the continent during that time" (Lott and Ross 2000).

5.3.2 Climate and Atmosphere

In atmospheric sciences, *in situ* refers to measurements obtained through direct contact with the media such as balloon-mounted instruments as opposed to remote-sensing observations with weather radar or satellites. *In situ* measurements for temperature and wind direction have been taken for centuries. They were made more quantitative by the invention of the thermometer in the seventeenth century.[*] However, it was the ability to integrate many point measurements along with satellite-based and ground-based remote observations through models that allowed the quantification of forecasts with broad applicability; improved weather modeling and prediction of the atmosphere incorporates observations from key *in situ* components such as radiosondes measuring the vertical profile of a parcel of air, anemometer measuring wind velocity, small weather stations with thermometers and hygrometer, rainfall monitors, aircraft sensors which are used to monitor the local (enroute) conditions of temperature, and moisture and wind vectors also combined with satellite measurements to produce weather forecasts. Lightning occurrences are monitored by using ground-based detectors. Such systems measure time, location, flash polarity, and stroke count of lightning strikes. When the observations from systems at different locations are combined, distribution maps of lightning strikes, and hence thunderstorm occurrences, can be made. In the United States, there are groups of volunteer observers that report local conditions to the weather service on a daily basis.

As mentioned earlier, an important capability is integration of the earlier information into models for forecasts. Two major weather models in the United States and Europe are the National Centers for Environmental Prediction (NCEP) ensemble model and the European Center for Medium-Range Weather Forecasts (ECMWF) forecast model, respectively. The use of *in situ* and satellite data in these has improved forecasts over the past two decades. This allows industry and individuals to plan future activities and events with greater certainty.

Weather forecasting has important societal impacts. Monitoring of air quality is another area that impacts health and lifestyle decisions. The impacts of air quality on health have been known for many years. In some countries such as the United States, air quality monitoring is mandated by law and EPA calculates the Air Quality Index for five major air pollutants regulated by the Clean Air Act: ground-level ozone, particle pollution (also known as

---

[*] http://galileo.rice.edu/sci/instruments/thermometer.html

particulate matter), carbon monoxide, sulfur dioxide, and nitrogen dioxide.[*] The measurements focus on airborne particles and chemicals that are known to create health issues. An international effort for air quality monitoring is also in place and expanding.[†]

There are several value of information studies conducted for atmosphere *in situ* spatial measurements (Frei 2009; Oxera 2003), which refer to cost savings and emission reductions. A more recent addition is addressed in Chapter 16, Socioeconomic Value of Hydro-Meteorological Information in Austria. The work is based on the transfer of foreign willingness-to-pay assessments and adjustments of foreign sectoral-based benefit estimates. The obtained value of weather information in Austria is in the range of several hundred million Euros per year.

5.3.3 Ocean and Coasts

The state of the ocean impacts everything from climate to fisheries to transportation to recreation. Many countries with ocean coasts treat the ocean as an essential resource. The Europeans, for example, have developed the Marine Strategy Framework Directive[‡] (MSFD) to support a sustainable and integrated approach to the monitoring of the ocean. There are multiple techniques for such monitoring. Satellite-based optical systems and radars measure surface temperature, ocean color, and wave conditions. Optical sensors using the blue–green spectrum can observe up to 10 m in depth, but most electromagnetic signals observed by optical and radar sensors do not penetrate the surface. The dynamics of the ocean below the surface are important for marine life cycles, carbon uptake, and physical attributes such as mixing and upwelling of waters rich in nutrients. Thus, the majority of the subsurface ocean observations use *in situ* sensors. These monitor large-scale dynamics such as El Niño oscillations, the upwelling of coastal currents whose nutrients support ocean populations, and so on.

In situ observations of the ocean have been important because ships have traveled the ocean. Early measurements included temperature and currents along with the tides.[§] These were done with mechanical sensors and the information was used locally. During the past century, with the expansion of electronics, more pervasive quantitative *in situ* measurements became practical. One of the earlier electromechanical *in situ* measurements in the marine environment was the determination of salinity on the basis of measurements of conductivity (C) and temperature (T). This was reported in 1964, and is considered to represent a turning point in marine observations (Wangersky 2005). The evolution of technology has made conductivity temperature and depth

---

[*] https://www.airnow.gov/index.cfm?action=airnow.pointmaps
[†] https://www.airnow.gov/index.cfm?action=airnow.international
[‡] Directive 2008/56/EC
[§] http://www.divediscover.whoi.edu/history-ocean

(pressure) measuring instruments (CTDs) as standard sensors on many *in situ* platforms. These sensors for ocean physical properties are relatively mature as compared with ocean chemistry and biology. *In situ* chemistry monitoring is traditionally being done with sample collection and then postcollection analyses on ships or in land-based laboratories. New techniques such as laboratories on a chip leveraging new solid-state technologies (Mowlem 2016) and automated reagent analyses (Tercier-Waeber 2015) are significantly advancing the state of the art. There are trends to put multiple sensors on a platform to provide simultaneous observations of multiple variables. The trend is seen, for example, in Argo floats that originally measured physical ocean characteristics and have been expanded to include ocean chemistry (Roemmich et al. 2009). Another example of this is the system of Smart Buoys operated in UK waters.[*] These devices house a range of measuring instruments (e.g., monitoring salinity, temperature, turbidity, chlorophyll fluorescence, oxygen saturation, and nitrate concentration) and transmit data telemetrically. There are fewer sensors in use for monitoring concentrations of heavy metals, organic pollutants, and algal toxins (Miles and Fones 2012). A new sensor generation has emerged from the European Oceans of Tomorrow projects and others which can measure chemical and toxicity in marine waters (Pearlman and Zielinski 2017; Zielinski et al. 2011, Wollschläger et al. 2016).

For biology observations, *in situ* observations have used acoustic sensors that can be active (emitting and receiving sound) and passive (listening only). These observe fish and marine mammals and are collecting profiles and identifying mobility patterns of fish and biological species. Recent developments in passive acoustic sensors (Delory et al. 2014) are changing the way the marine ecosystem is observed and managed. Although larger fish and marine mammals are most commonly thought of in terms of ocean biology, the population of living organisms in the oceans is dominated by microscopic creatures called plankton. New video and imaging techniques are providing broader monitoring capabilities of plankton than previously available. The application of genetic sequencing for the oceans observation over more than 10 years (Karl and Church 2014) has created new knowledge of types and relations of marine organisms. As important is the field of eDNA in which DNA left in the water from living organisms can be observed days after a fish has swum by. This introduces the potential for dynamic understanding of the environment in ways previously unavailable.

5.3.4 Economy and Society

In the context of Smart Cities, the application of *in situ* sensors is playing a significant role in facilitating services and helping citizens make informed decisions. The application of *in situ* devices and associated technologies are making intelligent sensing and machine learning widely available through

---

[*] http://www.cefas.co.uk/publications/posters/29756.pdf

information sharing, collaboration and the intelligent use of large datasets. The easy access and interaction with a wide range of devices finds application in different domains, such as home and industrial automation, medical aids, mobile health care, intelligent energy management, automotive, traffic management, to name only a few. Prominent examples can be found in the energy sector where the Smart Meter is already being widely adopted as a tool to enhance user experience—remote access to household appliances, reduced utility bills—and drive a reduction on total energy consumption, supporting the goal of sustainability. It is very likely to see an increase in the interconnectivity of new energy-related systems such as electric vehicles, storage devices, or small-scale renewable energy systems at the household level. The use of sensors in roads identifies locations of slowed or stopped traffic and helps to identify alternative routes. Cameras are well known to help with security. The challenges for *in situ* sensor implementation are the potential of being overwhelmed by data (leading to studies on how to handle big data) and the issues that arise about privacy. Video face recognitions and other software tracking tools, along with low-cost ubiquitous storage and parallel computing have the potential for pervasive monitoring.

5.4 Developments in *in situ* Monitoring and Policy

During the past years, there has been an exponential growth in the amount of *in situ* data being generated and recorded. During the past 10 years numerous devices have come into use that have Global Positioning Satellite System (GPS) functionality and an Internet connection and that, as a result, both use and create *in situ* data. The proliferation of low-cost, low-tech, and network-enabled mean that a previously unimaginable amount of *in situ* data are created and used (UN-GGIM 2013). Nevertheless, the significant growth in deployment of geospatially enabled devices and the increasing use of geospatial data in everyday life bring with them a need for stronger policy and legal framework to manage privacy concerns and protect the interests of those who are providing the data as noted earlier.

GPS technology is now mainstream, and the receiver technology is small enough to allow GPS wristwatches. This enables fast data collection in very challenging environments, with high accuracy and great integrity. User devices and environmental sensors along with greater integration with other technologies produce a more complete and ubiquitous positioning solution and precision environmental monitoring. The United Nations General Assembly adopted the United Nations resolution on Global Geodetic Reference Frame (GGRF) for Sustainable Development. The resolution is the first of its kind to be agreed by the UN and recognizes the global importance of location and positioning for many different areas of development (UN-GGIM 2015). The GGRF is a necessary tool to help address the growing demand for more precise positioning frameworks on which the *in situ* measurement devices can be linked. It is also essential for effective decision-making and a vital underpinning infrastructure, which is applied in areas of natural hazard and disaster management, climate change and sea level monitoring, mapping, and navigation by society at large every day.

The paradigm of geospatial information is changing; no longer is it used just for mapping and visualization as an end product, but also for integrating with other data sources, data analytics, modeling, and policy-making.[*] Once the geospatial data are created, it can be used over and over again to support a wide range of different applications and services. Thus, geospatial information becomes an essential component as a baseline for information in decision-making processes.

A number of uses of geospatial information rely on the provision of information that is detailed in a local region or country and beyond, is trusted, and is regularly maintained. Recognizing the increase in value of data sources, a continuing role for national mapping agencies will be to define and maintain quality standards and data currency programs for the data that is required by governments, citizens, and industry for their operation. Governments are still in a unique position to carry out this role, and to fulfill requirements for such information. Governments, in the role as authoritative suppliers of quality geospatial information, will become increasingly aware of the increased value of geospatial information for sustainable economic and social development.

5.5 National Mapping and Cadastral Applications

Traditionally, governments have had their own formal channels for collecting public sector geospatial information, normally through national mapping and cadastral agencies (NMCA). These organizations, usually publicly owned, produce topographic maps and other geographic information (GI) (e.g., elevation, hydrography, and geographical names) of a country. Some national mapping agencies also deal with cadastral[†] matters. Most countries have their own national mapping (and cadastral) agencies. Besides NMCAs, most coastal countries have separate hydrographic offices dealing with

---

[*] http://www.nytimes.com/2012/03/04/business/ibm-takes-smarter-cities-concept-to-rio-de-janeiro.html
[†] Matters relating to ownership of the real property in a district, including boundaries and tax assessments.

proper mapping, surveying, and charting the seas and navigable waters of their national territories.

NMCAs are generally responsible for and/or provide some of the following:

- *National geodetic frame*: A ratified framework on which mapping and measuring work is based upon.
- Positioning services for determining accurate, satellite positions.
- *Digital maps*: Production and management of national digital map series.
- *Printed maps*: Printing the national map series.
- *Land registry*: Land property rights registrations.
- *Property information*: Management and operation of a national registry for public property information.
- *Place names*: Administration of the national place names register.
- *Standards*: Development and maintenance of national standards for maps and GI.

The first national mapping agency was the Ordnance Survey in the United Kingdom; its roots go back to 1747, when the compilation of a map of the Scottish Highlands was proposed. This survey at a scale of 1 inch to 1,000 yards (1:36,000) was the starting point of the Principal Triangulation of Great Britain (1783–1853), and led to the creation of the Ordnance Survey itself; work was begun in earnest in 1790, when the Board of Ordnance began a national military survey at one-inch-to-the-mile (1:63,360 scale). At world level, around 110 countries have established Mapping Agencies.

In Europe, EuroGeographics is the membership association and acknowledged voice of the European National Mapping and cadastres organizations (including land registry authorities) from 46 countries. The association's main activities focus on representing members' interests, transferring knowledge, developing capacities, creating and implementing interoperability projects, and producing pan-European products based on the datasets from NMCAs.

Originally, internal resources were used, but over the past 30 years, the private sector has increasingly been involved in the collection and maintenance of data through outsourcing and partnership agreements. However, a dramatic shift in how geospatial data are sourced is unfolding through the direct involvement of citizens in crowdsourcing. Its roots lie in the increasing convergence of three phenomena: (1) the widespread use of GPS and image-based mapping technologies by professionals and expert amateurs; (2) the emerging role of Web 2.0, which allows more user involvement; and (3) interaction and the growth of social networking tools, practices, and culture. This crowdsourcing approach is also known as *citizen cyberscience, volunteered GI*, and *neogeography*.

5.6 Volunteer Geographic Information

There is a rapid growth of citizen and personal measurements that are contributing to EO—particularly for *in situ* data collection and analyses. These range from contributions to Open Street Map (OSM) to optical characterization measurements in coastal waters to contributions to noise monitoring. There are ongoing debates about the scientific value of these data, which are large in number but have greater uncertainty in individual measurements versus a limited number of precise measurements (Fischhoff and Davisa 2014). In practice, both have substantial value but require different approaches to translating data to information. The question of the value of citizen data to science research has been raised and addressed from different perspectives (UN-GGIM 2015).

Today, there exist multiple instances of *in situ* data sourced through participatory methods. These are also known as volunteered GI or citizen science (Goodchild 2007). OSM is an example to illustrate the nature of these data and their use. Starting in 2004 in the United Kingdom, OSM was a response to the dominance of proprietary maps, and seized upon the participatory enablement and social nature of Web 2.0 technologies to contribute and curate mapping data. It is predicated on the contribution of *in situ* data by users through means such as GPS and ground survey methods. OSM's wiki page (OSM 2016) provides a range of mapping techniques now used to collect *in situ* data. This includes satellite or aerial imagery, where higher resolution data made available to OSM or through appropriate licensing agreements that provide the basis for improving the quality of data in OSM.

There are, of course, varying perceptions as to the quality and value of OSM data. For example, in his analysis of OSM data for England, Haklay et al. (2010) found that although the quality of such *in situ* data has improved dramatically since its inception, quality—as measured by accuracy and coverage—is highest when there is a baseline available from aerial or satellite imagery. Nonetheless, Haklay found that significant value lay in the speed at which data was amassed; this is substantiated as well in mapping initiatives that support disaster recovery efforts, such as those in Haiti and more recently, in Nepal (Simmons 2015), where timeliness, the generation of data in hitherto unmapped regions, and the availability of the data are all critical factors for disaster response (Smith 2014).

Government entities can, and do, work in partnership with active members of the volunteer geographic information (VGI) community. There are many examples of *in situ* data of national mapping agencies being incorporated into volunteer mapping databases; changes to government licensing that can enable further collaboration between governments and VGI communities. Currently, it is perceived that there is a significant gap between authoritative and crowd-sourced data (UN-GGIM 2015). This gap will reduce as collaborations between all organizations increase—this includes VGI incorporating government-sourced data and governments exploring ways to incorporate both passively and actively created user-generated data.

5.7 Data Challenges

The generation of large amounts of *in situ* data will bring with it a requirement to address processing, visualization, and data management. The last includes a number of key capabilities such as archiving, provenance, and the use of permanent identifiers (PID). An important part of managing data is to know how to locate it uniquely over long time spans. Thus, the need to address PIDs has been recognized both for individual dataset access and for sustainability. Currently, digital object identifiers (DOI) (Gallagher et al. 2015) provide a standard naming process that has gained widespread acceptance. Another attribute of effective information management is that the data should have descriptors, called metadata. This is high-level information about the data, so that the data are easily discoverable. Standards for metadata have been developed for geospatial data.* However, the metadata needs to be created by the data originator and this is not done consistently. The other issue is that different disciplines use different terms in the metadata that is more characteristic of their community. The results are that uniform discovery and access of data are still challenging.

Large *in situ* data sets, such as other big data, offer challenges in translating data to information. This is done through applying calibration to the data and then putting the data through models, which then create the desired information. With the evolution of the cloud, storage is now less of a concern and the bottleneck for creating information is the communication overhead to move large data from the cloud to a local computer and then putting it back. There is, thus, a trend to bring the models to the data, so that the processing is done in the cloud. This significantly reduces the overhead associated with data transfer. It also allows easier reuse of data and continued processing of the information by a broader group of users. There are costs associated with using the cloud and data providers who must take this into consideration. With very large datasets, processing efficiency can also be improved through use of artificial intelligence (AI) or machine learning technologies. These not only can supplement traditional processing, but also allow new relations within the data to be discovered.

---

* (ISO19115) (https://www.iso.org/standard/53798.html).

Interoperability of data, particularly between different communities and disciplines is essential to provide a broad and reliable information base for decision and impact assessments. There are a number of standards organization that address geospatial data (e.g., international organization for standardization (ISO) and open geospatial consortium (OGC)). Although these organizations serve as a focal point for community inputs and standards creation, the recommendations have enough flexibility that the implementation of agreed standards may vary enough to reduce interoperability between disparate systems. The federation approach to improve interoperability through the creation of more rigid guidelines for data and metadata has been tried. Where there is use of data by a single discipline or for a dominant application, such as open street map, the application standards may be adopted and access to the open data is improved. In general, conforming to these guidelines has associated expenses, particularly as the federation guidelines evolve and large-scale updates are required. Experience has shown that repositories, unless specifically funded to follow guidelines, will abandon the conformity over time. A more recent alternative to improve interoperability is the use of brokering or mediation (Nativi et al. 2013). Brokering also provides transformation and semantic services to facilitate cross discipline sharing of data. Open data may be available for sharing, but in such diverse formats that accessing and understanding it in machine readable formats requires either conformance to standards or a form of brokering that allows for translation of formats across disciplines Here, brokering again allows for improved interoperability.

The network of tomorrow, built on an increasing number of sensors and thus increasing data volumes, will produce a hyperconnected environment or *Internet of Things*, with estimates of more than 20 billion things connected by 2020 (Gartner 2016). The concept of linked data is emerging and allows processing through automated connection between models and data.[*] For linked data, a machine-readable form of data identification is needed in the form of a digital object identifier or similar widely accepted standard. For research publications, author identification with ORCID is gaining broader acceptance. This is valuable if follow-up to the published information is desired. The *omnipresence* of geospatial information, whereby almost all pieces of data have some form of location reference, will continue to grow, with location providing a vital link between the sensors that will participate in the *Internet of Things*. Again a form of address or identifier is necessary. For this, a Uniform Resource Identifier (URI) is assigned to an object in order to maximize its (re)usability.

Finally, built environments are also an emerging frontier but one that still presents major challenges. Although a number of technologies exist that can be used to improve data in this area, including ultra-wideband

---

[*] http://www.w3.org/standards/semanticweb/data

communication, accelerometers, and radio frequency identification (RFID), no single source is able, as of yet, to provide the widespread coverage that may be expected in the years to come. The potential applications are broad and could include variable heating systems, greater access for disabled people, and more advanced security monitoring. Although solutions are likely to be closer to ten years than five years, we will see the greater availability and widespread use of indoor geospatial *in situ* data. It is expected that some of these new technologies will lead to new industry standards, conformant with current standard development processes.

5.8 Conclusions

In situ data are playing an increasing role in our understanding of the natural and built environments. This comes from a combination of many different advances. Sensors are becoming small and cheaper, using less power to allow a longer term sustained operations. They are also able to observe a broader range of environmental attributes. Connectivity has increased through wireless Internet access. Computers along with storage have increasing capabilities at lower cost. Software enables more detailed analyses even for big data. The ability to change data to information to knowledge is evolving. These developments should improve decision-making in areas as discussed earlier. The impact is most immediately seen on the lives and resources, whether from disaster prediction and mitigation to production and ecosystem preservation. The means to quantify these impacts is the subject of Chapter 10.

References

Committee on the Economic Benefits of Improved Seismic Monitoring, 2005. Improved seismic monitoring - Improved decision-making: Assessing the value of reduced uncertainty, Committee on Seismology and Geodynamics, National Research Council, ISBN: 0-309-55180-3, 196 p, Chapter 4, pp. 77–105. http://www.nap.edu/catalog/11327.html.

Delory, E., Corradino, L., Toma, D., Del Rio, J., Brault, P., Ruiz, P., and Fiquet, F., 2014. Developing a new generation of passive acoustics sensors for ocean observing systems. *2014 IEEE Sensor Systems for a Changing Ocean (SSCO)*, October 14 through 16 2014, Brest France.

Esri, 2016. Stream gauges and weather stations. http://www.arcgis.com/home/search.html?q=stream%20gauges&start=1&sortOrder=desc&sortField=relevance (accessed Jul 23, 2017).

Fischhoff, B., and Davisa, A., 2014. Communicating scientific uncertainty. *Proceedings of the National Academy of Sciences U S A*, 111(Suppl 4): 13664–13671.

Forney, W.M., Raunikar, R.P., Bernknopf, R.L., and Mishra, S.K., 2012. An economic value of remote-sensing information—Application to agricultural production and maintaining groundwater quality: U.S. Geological Survey Professional Paper 1796, 60 p.

Frei, T., 2009. Economic and social benefits of meteorology and climatology in Switzerland. *Meteorological Applications*, 17: 39–44.

Gallagher, J., Orcutt, J., Simpson, P., Wright, D., Pearlman, J., and Raymond, L., 2015. Facilitating open exchange of data and information. *Earth Science Informatics*, 8(4): 721–739. doi:10.1007/s12145-014-0202-2.

Gartner, 2016. Gartner sys 6.4 billion connected "thing" will be in use in 2016, up to 30 percent from 2015. Accessed on July 21, 2017. http://www.gartner.com/newsroom/id/3165317.

GEO, 2016. GEO strategy plan. Accessed on July 21, 2017. http://www.earthobservations.org/documents/GEO_Strategic_Plan_2016_2025_Implementing_GEOSS.pdf.

Goodchild, M., 2007. Citizens as sensors: The world of volunteered geography. *GeoJournal*, 69(4): 211–221.

Haklay, M., Basioukam, S., Antoniou, V., and Ather, A., 2010. How many volunteers does it take to map an area well? The validity of Linus' law to volunteered geographic information. *The Cartographic Journal*, 47(4): 315–322.

Karl, D., and Church, M., 2014. Microbial oceanography and the Hawaii Ocean Time-series programme. *Nature Reviews*, 12(10): 699–713.

Langley, R., 2010. Innovation: Accuracy verses precision. *GPS World*, May 1. http://gpsworld.com/gnss-systemalgorithms-methodsinnovation-accuracy-versus-precision-9889/.

Lee, W.H.K., Jennings, P., Kisslinger, C., and Kanamori, H., 2002. *International Handbook of Earthquake & Engineering Seismology*. Academic Press. pp. 283–304. ISBN 978-0-08-048922-3. https://en.wikipedia.org/wiki/Seismometer.

Lott, N., and Ross, T., 2000. *NCDC Technical Report 2000–02, A Climatology of Recent Extreme Weather and Climate Events*. Asheville, NC: National Climatic Data Center.

Miles, G., and Fones, G., 2012. A review of in situ methods and sensors for monitoring the marine environment. *Sensor Review*, 32(1): 17–28.

Mowlem, M., 2016. Lab on chip technology applied to in situ marine biogeochemical sensing. http://sites.ieee.org/oceanrcn/files/2017/02/Mowlem_Lab_on_a_Chip_OceanObs_RCN_Oct2015.pdf.

Nativi, S., Craglia, M., and Pearlman, J., 2013. Earth science infrastructures interoperability: The brokering approach. *IEEE Journal of Selected Topics in Applied Earth Observations and Remote Sensing*, 6(3): 1118–1129.

OSM, 2016. Accessed on July 23, 2017. http://wiki.openstreetmap.org/wiki/Map_Features.

Oxford Economic Research Associates (OXERA), 2003. The non-market value of generation technologies, report for British Nuclear Fuels Limited (BNFL), June 30. Oxford, UK: OXERA.

Pause, M., Schweitzer, C., Rosenthal, M., Keuck, V., Bumberger, J., Dietrich, P., Heurich, M., Jung, A., and Lausch, A., 2016. In situ/remote sensing integration to assess forest health—A review. *Remote Sensing*, 8: 471.

Pearlman, J., and Zielinski, O., 2017. A new generation of optical systems for ocean monitoring—Matrix fluorescence for multifunctional ocean sensing. *Sea Technology*, pp. 30–33.

Pe'eri, S., McLeod, A., Lavoie, P., Ackerman, S., Gardner, J., and Parrish, C., 2013. Field calibration and validation of remote-sensing surveys. *International Journal of Remote Sensing*, 34(18). doi:10.1080/01431161.2013.800655.

Roemmich, D., Johnson, G.C., Riser, S., Davis, R., Gilson, J., Owens, W.B., Garzoli, S.L., Schmid, C., and Ignaszewski, M., 2009. The argo program: Observing the global ocean with profiling floats. *Oceanography*, 22: 34–43.

Simmons, A., 2015. Measuring the impact of humanitarian mapping in Nepal. "Tableau," Blog, June 1. Accessed on October 20, 2016. http://www.tableau.com/about/blog/2015/5/disaster-mapping-project-nepal-reveals-power-many-39358.

Smith, N., 2014. OpenStreetMap for disaster risk management. "Development Seed," Blog, June 30. Accessed on October 20, 2016. https://developmentseed.org/blog/2014/06/30/understanding-risk-forum/.

Tercier Waeber, M.L., Bakker, E., Nardin, C., Mongin, S., Prado, E., Cuartero Botia, M., Mazaikoff, B. et al., 2015. FP7-OCEAN-2013-SCHeMA: Integrated in situ chemical mapping probes. doi:10.1109/OCEANS-Genova.2015.7271560.

United Nations Committee of Experts on Global Geospatial Information Management, 2013. Future trends in geospatial information management: The five to ten year vision. 1st ed. Ordnance Survey.

United Nations Committee of Experts on Global Geospatial Information Management, 2015. Future trends in geospatial information management: The five to ten year vision. 2nd ed. Ordnance Survey.

Wangersky, P.J., 2005. Methods of sampling and analysis and our concepts of ocean dynamics. *Scientia Marina*, 69: 75–84.

Wollschläger, J., Voß, D., Zielinski, O., and Petersen, W., 2016. In situ observations of biological and environmental parameters by means of optics—Development of next-generation ocean sensors with special focus on an integrating cavity approach. *IEEE Journal of Oceanic Engineering*, PP(99): 1–10. doi:10.1109/JOE.2016.2557466.

Yoder, J., Davis, J., Dierssen, H., Muller-Karger, F., Mahadevan, A., Pearlman, J., and Sosik, H., 2015. Report of the ocean observation research coordination network in-situ-satellite observation working group - "A modern coastal ocean observing system using data from advanced satellite and in situ sensors—An example." http://hdl.handle.net/1912/7351.

Zielinski, O., Voss, D., Saworski, B., Fiedler, B., and Koertzinger, A., 2011. Computation of nitrate concentrations in turbid coastal waters using an in situ ultraviolet spectrophotometer. *Journal of Sea Research* 65: 456–460.

6

Value and Uncertainty of Information—Supply Chain Challenges for Producers and End Users

Lesley Arnold

CONTENTS

6.1 Introduction

Providing value to end users of spatial data products will become a reality when industry is able to provide access to knowledge, and not just data, in a way that allows end users to achieve their decision-making goals; be their understanding of the relationship between phenomena, the likelihood of an event, or simply navigating to a location.

This *value proposition* focus requires a fundamental shift in thinking—both in terms of how knowledge can be made more discoverable and how the trustworthiness of the information is communicated to the end user.

Governments, businesses, and individuals have information needs that are a mix of vague, precise, and implicit requirements (Mass et al. 2011). End users, such as insurers (Valas 2016), want to query data *at will* to gain knowledge. Anecdotally, what they do not want are the costly and time-consuming overheads associated with the finding, downloading, reworking, and analyzing data; nor uncertainties about information quality and reliability. This suggests that the value of spatial data to consumers is not the actual data itself, but rather the knowledge that is extracted from the data and its quality for decision-making.

Current spatial data supply chains are geared toward data provision and not the discoverability of *knowledge* for planning and decision-making in the user's context. A common problem for organizations is the lack of information about the reason for which data will be used and the knowledge end users are seeking. Conversely, end users (businesses and individuals) know what knowledge they require, but find it increasingly difficult to obtain the information they need, when they need it, and in a format that is of use to them.

End users not only want to make informed decisions, they also want to know that their conclusions are based on the most reliable information. With the continuing expansion of the Internet, end users are faced with an excess of information causing increased levels of uncertainty about the reliability of information, the factual nature of the knowledge they derive, and the appropriateness of the decisions they make (Voinov et al. 2016). Although current metadata are adequate for communicating information quality; for end users, there still remains uncertainty around the level of truth in the information they are presented with. Therefore, ranking query responses according to the applicability of purpose and rating them on trustworthiness is a new requirement.

The consumer *value proposition* statement is, not so much about understanding *how* consumers use information, such as for in-car navigation; but rather *why* they are using it, that is, to evacuate from wildfire. The answer to the *why* question provides far more insight to what the end user is likely to value in terms of knowledge and the reliability of the information they are seeking.

However, for organizations that collect, process, and analyze data, trying to preempt the goals, and thus the information needs of all potential end users are unrealistic given the plethora of *why* possibilities. In addition to the need for basic road network data, the wildfire evacuation scenario infers the need for knowledge about temporal safe zones, road closures and alternative routes, weather conditions, and the likely trajectory of the wildfire itself.

The breadth of information required is often fragmented across a number of agencies that may or may not participate in the same spatial data supply chain (Dessers 2013, p. 40). The *knowledge* value proposition will therefore, stem from the richness of spatial analytics implemented across the broader spectrum of available datasets, rather than a single dataset from a single source.

So how do organizations determine what information to collect and what quality criteria to apply when the knowledge end users want is multidimensional and requires cooperation with several supply chain partners?

The answer is likely to stem from the use of semantic web technologies that enable meaning to be drawn from data automatically. Looking to the future, the value of spatial information will be amplified when spatial analytics can be conducted on-the-fly in response to an end user query using fundamental location data in combination with social, economic, and environmental geographies from multiple sources; and when end users are able to make an informed assessment as to the reliability of the query response as a source of knowledge.

This chapter considers a new Semantic Web-based approach to spatial data supply chains; one that interfaces the data producer's *Push* production system with a timely *pull-based* knowledge service in which end user requirements can be open-ended and the producer does not need to know these requirements. Rather than draw down data that requires further manipulation to gather the required information, the user draws down knowledge on-the-fly through an open query Web interface powered by Semantic Web technologies and processes. The aim is to create more direct value for end users by delivering knowledge as well as reducing uncertainty pertaining to this knowledge by communicating reliability at the same time.

6.2 Supply and Value Chains

Spatial data supply chains today are characteristically *push* production systems that span several organizations (supply chain nodes) that store, process, and create component parts of a product that may be progressively aggregated and combined to deliver a product or service to a user further downstream. This is particularly true of extended spatial data supply chains, such as National Supply Chains that cross several levels of government (ANZLIC 2014; CRCSI 2013).

Supply chains are often referred to as *value chains*. However, there is a subtle difference. The supply chain refers to the transfer (or conveyance) of data products between organizations and their distribution to consumers; whereas value chains refer to the value activities that each organization participating in the supply chain performs, so that a *valued* product or service can be delivered to market (Naslund and Williamson 2010).

Value to the consumer is the key differentiating factor. The supply chain focuses on the supply base and producers; whereas the value chain focuses on the consumer base and places the needs and wants of customers first (Porter 2008, p. 38). Value chains flow in the opposite direction to supply chains. The customer is the source of value, and *value* flows from the end user, in the form of demand, to the supplier (Feller et al. 2006) (Figure 6.1).

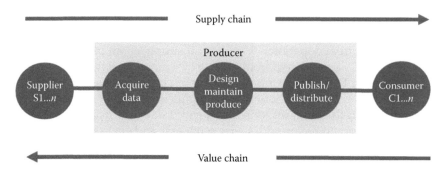

FIGURE 6.1
A spatial data supply chain represents the flow of products and services in one direction, whereas the value chain represents consumer demand in the other. (Adapted from Feller, A. et al., Value chains versus supply chains, available at http://www.bptrends.com/bpt/wp-content/publicationfiles/03–06-ART-ValueChains-SupplyChains-Feller.pdf, 2006, where the concept is applied in manufacturing. With Permission.)

6.3 Supply Chain Challenges

The terms *supply chain* and *value chain* are relatively new to the spatial industry. Historically, organizations handled everything from data collection, presentation, marketing, and distribution. In this *vertically integrated* enterprise model, the difference between supply and value chains was less apparent than today.

Gradually, spatial data supply chains have transitioned to *vertically disintegrated* models that involve multiple organizations working together to deliver a data product or service to market. This is particularly evident where outsourcing value activities and global markets have evolved, and where spatial data infrastructures (SDIs) have developed.

Vertical disintegration has created a complex network of spatial data supply chains, operating predominantly but not entirely, in a digital environment. As with manufacturing supply chains (Maxwell and Lal 2013), spatial information is typically transferred to more than one node (often simultaneously) and derived (or value added) datasets are often created. These derivatives may reenter the supply chain at any point in which the residual value is deemed to be recyclable.

A downside of this practice is that several versions of the same or similar datasets can evolve and therefore, inconsistencies (e.g., currency and accuracy) can occur between the originating dataset and the generalized or value-added datasets created further downstream.

Duplication in spatial data supply chains is not new. In the 1980–1990s, organizations found it cheaper and simpler to perform activities in-house rather than spend time and money interacting with other parties. Data and technology interoperability arose and information sharing became problematic. As a result, duplicate data management practices thrived.

Nowadays, cooperation between supply chain partners, although not mainstream, is technically possible with the use of collaborative data management and editing technologies to reduce double handling (Grech 2012). However, duplicate datasets are still a characteristic of supply chains, creating ambiguity for end users and a financial burden on the industry as a whole.

With multiple versions of the same data potentially being pushed out from different organizations (as well as the crowd), at various stages of completeness and often with similar (but not exact) information, the end user has no way of knowing which information is best suited to their purpose, and no global query tools to sort through the plethora of data that may or may not be useful to their quest for knowledge.

In addition, as data are transmitted through the web of supply chains, it undergoes various geoprocessing refinements and generalizations, and interpreted information is often a result of combining more than one dataset. Data provenance is typically not collected along the supply chain, nor is it accessible to end users. Not surprisingly, end users have a degree of uncertainty as to the reliability of the interpreted spatial information (McConnell and Weidman 2009, p. 17).

Governments globally recognize the complex and cross-functional nature of supply and value chain management issues, and are developing SDIs to facilitate spatial data sharing and coordination between mapping organizations. However, although SDIs have addressed data supply and access issues, the underlying data duplication and integration problems remain and there is insufficient coordination of value activities to meet the aspirations of end users for specific knowledge.

To be truly meaningful and relevant in today's markets, the activities in supply and value chains should specifically work together to deliver value and confidence to end users. In national and global supply chains, there are often missed opportunities to leverage upstream and downstream activities to deliver higher value knowledge-based products to end users.

The issue for the industry is that the supply chain strategies of individual organizations tend not to consider the supply chain needs beyond their direct supplier nor the value activities ahead of their immediate customers. Each organization performs value activities and sets performance and quality criteria to meet their internal business needs and those of their immediate customers. They do not necessarily have the want or need to recognize consumer needs further along the supply chain.

Yet, the knowledge value of spatial data is amplified when it is combined, contrasted, and analyzed with other datasets. This means SDIs need to move beyond current information sharing and coordination strategies between organizations, and consider the value activities within the supply chain and the partnerships necessary to maximize the benefits of spatial information more broadly.

6.4 End User Value Proposition Points

The organizations that focus on consumer value consider what benefits the end user will receive when using the spatial information products they produce. There are typically four distinct consumer value proposition points in spatial data supply chains. These are referred to here as primary, secondary, tertiary, and quaternary value propositions (Figure 6.2).

6.4.1 Primary Value Proposition

The primary value proposition is aimed at internal business users. The value activities required to get to this point involve the collection and/or sourcing of raw data and its refinement for internal business processes. Cadastral survey plan lodgment, road naming, and property street addressing are examples of business processes contributing to the business of land administration. Thus the value proposition statement for business is the ability to deliver certainty in land ownership for the land owner.

6.4.2 Secondary Value Proposition

The secondary value proposition is aimed at the external business user, often a spatial data specialist. Organizations further develop their data holdings to create an enhanced data product or service. For example, the digital cadastre, essentially a by-product of the land administration process, has supplementary value as an aggregated information product for value adders and brokers. The value proposition statement for the digital cadastre is that it provides external businesses with a seamless visualization of property boundaries and related property descriptions that are the basis for searching, planning, and analyzing land-related information.

6.4.3 Tertiary Value Proposition

The tertiary value proposition is directed at decision-makers and researchers. It refers to the point where knowledge is derived from interpreted information and has value for subsequent decision-making, policy setting, and innovation. For example, knowledge-based systems assist governments to determine the best course of action and develop policy, such as declared fire risk zones, building codes, and fuel load reduction programs. The right course of action is dependent on the reliability of information.

Knowledge-based systems respond directly to an end user query. The query is executed as a geoprocess on one or more information layers. The value of knowledge-based systems is that the analytical process is conducted automatically on behalf of the end user and the query response is in the end user context. For example, a real-estate application, underpinned by spatial information, provides property characteristics, proximity to services, and the rights, restrictions, and responsibilities pertaining to land. The tertiary value proposition statement to the end user in this instance is that the real-estate application removes the guess work when buying property and enables informed lifestyle choices to be made.

6.4.4 Quaternary Value Proposition

The quaternary value proposition is aimed at the beneficiaries of a decision or policy. It refers to the point at which benefits are realized as a result of the outcomes from an action, such as the enactment of a new policy, process, or standard. The quaternary value proposition often has a strategic focus, such as reduced poverty, affordable housing, and improved health services. The measurement of socioeconomic benefits require a benchmark of the current state (without knowledge) to calculate the benefits, and thus value, gained in the future state (with knowledge), including tangible, intangible, financial, and nonfinancial benefits.

An important point to note is that decisions, policies, and/or new processes, may have a negative impact on the community, resulting from the incorrect (Hunt 2013) or unethical use of knowledge (Figure 6.2). Ethical value, drawn from value theory (Stanford 2012) is intrinsic to decision-making and therefore an underpinning measure of the quaternary value proposition. The decision-making dynamics of an individual faced with choices involving ethical issues are complex (Bommer et al. 1987). In the case of wildfires, emergency responders make decisions on where to focus resources; governments must prioritize fire prevention works; insurance companies decide whether to insure a home or not; and people want to know if they should stay to defend their property or leave early. In each case, there is a duty of care to others that creates an ethical dilemma.

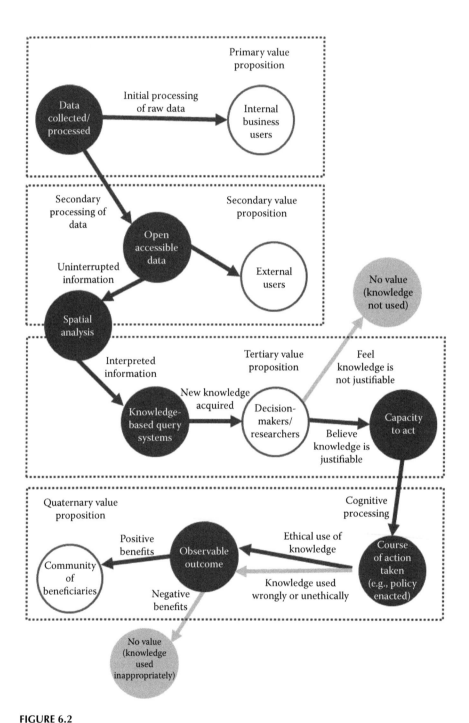

FIGURE 6.2
Primary, secondary, tertiary, and quaternary consumer value proposition points along the spatial data supply chain.

6.5 Communicating Believability to Reduce Uncertainty

The value of spatial information in the context of the user necessitates an understanding of *to what degree a person (or business) values the knowledge they receive.* To a large extent, the value of the knowledge is dependent on how much a person *believes* the information to be true.

According to Voinov et al. (2016), knowledge cannot truly be created without an explicit assertion of beliefs. Skepticism breeds doubt in the validity of information and therefore the knowledge derived may not be acted upon, and as a consequence, the information has no value to the end user (Figure 6.2).

Although spatial information providers take considerable care when collecting and processing data, the end user community is shrewd enough to know that maps are not always correct, but not sure enough to know how right or how wrong they are. The same applies to interpreted spatial information. Data integration and aggregation processes are known to generate uncertainty (Magnani and Montesi 2010), as are spatial distributions created through image analysis (Marinelli 2011).

This suggests that for spatial information to be valued by end users, data providers need to provide empirical evidence, so that the end user can make a value judgment of the truth or falsity of the information. The issue for the industry is how to communicate the reliability or believability of interpreted information to the end user to reduce uncertainty.

The spatial industry has traditionally relied on consumers' understanding of spatial metadata as a means of judging whether a data product is fit for a particular purpose or not. However, metadata are often not reliable because it is frequently out-of-date and incomplete. Nor does it lend itself to knowledge-based systems, as the lineage of analytical geoprocesses is rarely captured. In addition, human biases can negatively influence the way spatial models are constructed, and therefore how the results are interpreted (Voinov et al. 2016). Currently, an explanation *of fit for purpose* is generally through descriptive metadata that are at best a subjective interpretation from the perspective of the data custodian and not the end user.

A solution to this challenge may be found in the manufacturing industry, which assumes that consumers do not necessarily have the information to understand whether a product will suit their needs or not. Toy manufacturers include age suitability on packaging, food producers include nutrition panels, and the hotel industry has adopted a *star* quality rating system. These systems build an expectation that a product will be satisfactory for a given purpose and can be trusted. Going one step further, the food labeling industry in Australia and United Kingdom are considering including *walking time* kilojoule *burn off* to help consumers make sense of nutrition panels that are difficult to interpret for weight loss programs (Daily Mail Australia 2015; Runners World 2016; The Age 2016). There are also systems, such as TripAdvisor™, that put the quality rating process in the hands of the consumer, and for other consumers.

It is time for the spatial sector to rethink how it communicates the reliability of information to end users in an intelligible way. The spatial sector is no longer simply a *data industry* but a *knowledge service* industry. Use of these knowledge services, and thus the value of the underpinning information, will depend on end users having trust in the information and its source. Communication tools, be rating or ranking systems, need to convey the credibility of the knowledge. Methods will need and take into account the underpinning quality and lineage of the data used to derive knowledge. This is the subject of research into geodata provenance (Sadiq 2016).

6.6 Data Quality Requirements

The quality of spatial data and the way in which it is processed and made accessible has a direct impact on the effectiveness of information for decision-making. The term quality is used to indicate the superiority of a manufactured good, but it is more difficult to define for data, even though the processes are similar (Veregin 1999).

Data quality components from an organizational perspective include the accuracy (spatial, temporal, and thematic), resolution, consistency, and completeness (Caprioli et al. 2003; Veregin 1999) and pertains to dataset features within a dataset and the thematic attributes of features.

Governments typically define spatial data quality guidelines and standards to establish best practice and consistent data management across organizations. These guidelines and standards are meaningful in terms of primary and secondary value propositions where the end user is likely to have the skills to understand data quality components and therefore able to make a determination about fitness-for-purpose.

Even so, for specialists, managing the increasing availability of multisource and heterogeneous spatial data with complex lineages is challenging, particularly when quality may be undocumented, of mixed origin, or unknown. It is worth noting that organizations today typically assume the existence of error and use a disclaimer to shift the burden of data quality problems to the end user (Veregin 2009).

6.7 Forecasting Demand and Quality

A challenge for spatial industry is what constitutes a reliable approach to forecasting the demand and quality requirements of data. Market surveys are one of the few mechanisms available to profile end users to understand their needs for data themes, feature attributes, geometric and graphic representations, data quality requisites, and service requirements such as query tools and display customization.

The traditional *push* production supply chain approach is based on forecast demand—previous sales, service requests, data download, and/or views. Aerial imagery is classically supplied using the push model according to a prescheduled program based on anticipated needs. The maturation of the market means that forecasting is plausible and, unlike the manufacturing industry that experiences excess inventory problems (Buchmeister et al. 2013), imagery has a long life span and its accumulation over time contributes to new products for climate change monitoring and planning, and so on.

However, spatial data acquisition is costly and time consuming and vector data in particular are expensive to update. As such, focusing effort where it is needed and at the right time is critical. The approach will be different depending on the spatial data product in question and its likely usage. In a push production system, there are four approaches typically used by the spatial industry to meet consumer demands. These are Just-in-case, Just enough, Just-in-time, and Just-in-real-time approaches (Figure 6.3):

Just-in-case is used when it is difficult for an organization (supplier or producer) to forecast the likely need, and therefore the organization collects spatial data to meet an unexpected spike in demand. An example is the collection of data in case of a natural disaster and an emergency response.

Just enough is where an organization provides a range of products to meet 80 percent of the needs of 80 percent of end users, 80 percent of the time. An example is the Foundation Spatial Data Framework Themes, which provide fundamental spatial data coverage in order to serve the widest possible variety of users (ANZLIC 2014).

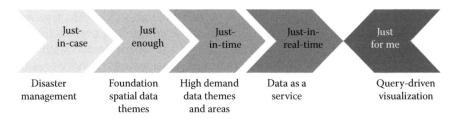

FIGURE 6.3
Four approaches to forecasting the end user demand in spatial data supply chains, and next generation Just-for-me approach.

Just-in-time is where each part of the supply chain is synchronized to deliver products downstream just before they are required. The approach requires sophisticated planning and cooperation between supply chain participants, and a common understanding of consumer needs. The lodgment and processing of cadastral survey plans for the land registration and titling are an example of Just-in-time processing.

Just-in-real-time applies to data-as-a-service, where supply chain processes are responsive to well-defined end user needs and a mature application market. This approach is mainly used in image-based products and services, and for applications that consume readings from sensors, such as water level gauges for flood monitoring. As soon as data are acquired, it is processed automatically and made available online where it is consumed routinely into business applications.

However, none of the above methods cater for instances where demand for knowledge is intrinsically unforeseeable. Consumers today want to query (pull) data to meet their immediate need for knowledge. This is referred to here as the *Just-for-me* approach (Figure 6.3), where demand is a condition of the end user.

A new approach to spatial data supply chains is required to alleviate the problems associated with demand forecasting; one that interfaces the data producer's *Push* production system with the end users' requirements for knowledge. Achieving this capability is the subject of new research (CRCSI 2013) and is explained in the Section 6.8. The premise of the research is to remove the decoupling point in supply chains by automating the query process. The decoupling point, also referred to as the push/pull interface, is the point where demand push and demand pull meet. The term decoupling point is drawn from the manufacturing industry (Olhager 2012). It is where a product moves from being a generic (subassembly) product to a customized (made to order or query driven) product.

The current challenge for mapping organizations is to balance high-demand variability requirements of end users, such as open queries and customized products, with the low demand variability products and services, such as standard street maps, which are more cost-effective and timely to produce. The balance between these competing perspectives is the point up to which demand is certain and can be forecasted with relatively high accuracy, and where operational efficiencies are more feasible due to larger economies of scale, and lower setup costs, and lead times (Figure 6.4).

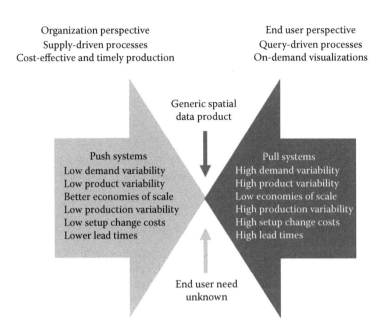

Organization perspective
Supply-driven processes
Cost-effective and timely production

End user perspective
Query-driven processes
On-demand visualizations

Generic spatial
data product

Push systems
Low demand variability
Low product variability
Better economies of scale
Low production variability
Low setup change costs
Lower lead times

Pull systems
High demand variability
High product variability
Low economies of scale
High production variability
High setup change costs
High lead times

End user need
unknown

FIGURE 6.4
Characteristics of the Push/Pull interface for spatial data supply chains. (Adapted from Sehgal, V., *Enterprise Supply Chain Management: Integrating Best in Class Processes*, John Wiley and Sons, Hoboken, NJ, p. 201, 2009, where it is applied to manufacturing. With Permission.)

6.8 Leveraging Semantic Web Technologies

The Semantic Web is a web of data also referred to as a global database containing data that anyone has produced and made accessible (W3C 2016). This includes spatial data collected by the government, business, academic, and community sectors. As a global database it is therefore possible to query the web using SPARQL, the recognized Semantic Web Query Language.

The Semantic Web has potential to remove the decoupling point in spatial data supply chains and thus, the problems associated with forecasting demand and quality requirements. With the semantic web, the end user has the potential to mobilize a broad range of spatial resources through a query to extract knowledge from data available on the web. This can be done without actually having to configure systems specifically for the end user.

Leveraging Semantic Web capabilities will require organizations to adopt the Semantic Web Resource Description Framework (RDF) format for spatial data and create ontologies that enable relationships between data to be encoded and interpreted computationally. Ontologies provide a specification to explicitly specify the semantics for particular geographic domains, so that knowledge can be extracted meaningfully. Ontologies are built using the Ontology Web Language (OWL) and are a way to enrich data with additional meaning, so that people (and machines) can do more with data (W3C 2016).

In the Semantic Web environment, pushing data out to end users will remain a central role for data producers and delivering quality information will continue to be an important element of supply chain value activities. However, it is the ability to extract knowledge from this web of data that will characterize the future SDIs, referred to here as the Spatial Knowledge Infrastructure (SKI).

6.8.1 The Future Spatial Knowledge Infrastructure

Similarly to a SDI, a SKI is the physical and organizational aspects of managing and exchanging spatial information, which comprises the people, policies, and technologies necessary to enable the use of spatially referenced data for decision-making.

The key difference is that the SKI is based on the premise of providing access to knowledge rather than data, and that this knowledge is constantly in motion as information is renewed and enhanced.

Spatial data provided by organizations are no longer the only avenue to knowledge. Community data, social media, online encyclopedias, and other data on the web also provide a rich source of diverse knowledge. When combined with government data, this web of accessible data provides potential for real-time knowledge production.

With this potential in mind, the SKI of the future is envisaged as embracing the web of collective knowledge in a way that delivers new insights to consumers in a way not possible today. New insights from data queries can then be attributed back to the web of data as new information or data layers (Figure 6.5).

In essence, the SKI of the future is envisaged here as a network of information and scientific processes that commoditize the production of knowledge for decision-making. Semantic Web technologies, Linked Data, spatial analytics, and new search and query capabilities, are the key to this new form of knowledge production.

FIGURE 6.5
Conceptual diagram of the spatial knowledge infrastructure of the future. (From Woodgate, P. et al., *Geospatial Inform. Sci.*, 20, 109–125, 2017.)

6.8.2 A Pull-Based Query Model

Research is examining Pull-based Query Models in an endeavor to realize the future SKI (Arnold 2016). Semantic web technologies in combination with domain ontologies, orchestrated web services (process ontologies), and natural query languages are being studied with the aim of automating the acquisition of knowledge from the web of spatial data (CRCSI 2013; Varadharajulu et al. 2015). The premise of the Pull-based Query Model is that suppliers make their data accessible as web services and expose semantic attributes using RDF, so that the meaning of the data can be interpreted automatically.

In the Pull-based Query Model, RDF structures and ontologies work together to interpret the relationships inherent in data. Natural language query processing is then used to decipher an end user query and locate appropriate data. Web service orchestration is used to link and execute geoprocesses in a specific way (i.e., according to a process ontology) to answer a query. Geoprocessing is the transformation of input elements into output elements with specific properties, parameters, and constraints, such as those required for flood modeling. A geoprocess ontology is defined as the workflow and sequence of events inherent in query processing, such as enquiring on *flood prediction in a given area* (Bing Tan et al. 2016).

Query processing has six steps that constitute the *automation of services* at the existing decoupling point in today's supply chains (6.6). Research is addressing each of these steps. They are

Interpret: Natural language query decomposition is used to interpret the end user query and thus what data and domain ontologies are required to answer the query (Reed et al. 2016).

Retrieve: Semantic search and filtering techniques are used to find and retrieve the data and domain ontologies for processing the query (Gulland et al. 2015; Reed et al. 2015).

Process: A process ontology is used to automatically orchestrate the linking and running of geoprocesses in the correct order to answer a query for a particular domain of knowledge (Bing Tan et al. 2016). Depending on the query, linked processes may include calculate, buffer, and overlay.

Portray: The visualization of a query response may take the form of a map, image, chart, text, table, and so on. Understanding the most appropriate visualization for a response requires further research.

Rank: A rank is assigned to the results based on quality criteria (accuracy, completeness, timeliness, cost, lineage, etc.) and a rating assigned to indicate the trustworthiness of the derived knowledge to the user.

Deliver knowledge: A list of ranked results is provided from which end users may select a response (answer to their query).

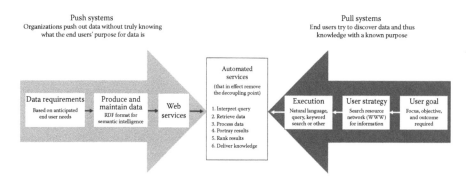

FIGURE 6.6
Next-generation supply chains with incorporate a Pull-based Query Model enabled through Semantic Web technologies.

Figure 6.6 illustrates where the automation of these services occurs at the push/pull interface. The aim is to remove the decoupling point in spatial data supply chains. With these automated services, organizations can continue to push out data to end users; making information discoverable (RDF formatted) without having to preempt the variability of end user needs. Instead, end users drive the request for information in the context of their own needs through the execution of a query. The success of the approach relies on the ability to successfully automate the query processing services at the push/pull interface. This research is continuing and there are several geoprocessing challenges that need to be overcome.

6.8.3 Geoprocessing Challenges

The Pull-based Query Model assumes that geoprocesses exist to manage the data integration, and aggregation of heterogeneous geographic datasets. This is not always the case and research continues to examine methods using Semantic Web technologies to reduce uncertainty in data conflation in which duplicate datasets exist (Yu et al. 2016) and federation processes, in which datasets require aggregating on-the-fly (Fa et al. 2016).

Process ontologies also require further investigation. Workflows are likely to stem from environmental and socioeconomic case studies where geoprocesses and spatial analytics have successfully been applied by specialists to analyze information and derive knowledge. Although these processes and spatial analytics are often manual, case study workflows and the data linkages and relationships developed, form the basis for process ontologies that provide the rules to automatically chain together resources, such as web services and data from multiple disparate sources.

There are significant advantages to be had from reusing/sharing ontologies and from the rules and data relationships devised. However, there is likely to be a growing challenge to achieve richly interconnected domain ontologies with the Semantic Web, as similar or overlapping domain ontologies are likely to increase in use and become more macroscopic but be logically inconsistent (Ibrahim et al. 2012). This is because domain ontologies may use different terms to describe the same features and different semantics can be associated with the same terms (Hong and Kuoa 2016). This will impact on how an end user query is responded to, and different results may arise. This calls for domain-ontology standards, such as a flood ontology standard, and a *domain-ontology resource library* from which spatial scientists can reuse or recast existing ontological efforts.

Although domain-ontology developers should not have to worry about developing their own upper ontologies, there is a need for a metaontology that maps all domain concepts to an upper ontology/ies, so that potential inconsistencies can be managed. The challenge for the spatial data sector will be to manage inconsistencies between overlapping domain ontologies that are developed for specific spatial analysis applications, such as flood risk analysis.

6.9 Conclusion

Within the spatial industry, it is commonly understood that spatial information can be used to analyze complex data and visualize answers to multifaceted questions. However, this is not necessarily the case for end users. The consumer has no way of knowing which information is best suited to their purpose and no centralized query tools to sort through the plethora of data and/or services that may or may not be useful in answering their question definitively.

A new approach is required. However, benefits have already been wrung out of existing supply chain processes and revolutionary methods are needed to take the industry to the next level of automation to deliver knowledge-based services to end users.

Semantic Web technologies today have the potential to transform the spatial industry into more than just a distribution channel for data. Spatial analytics, combined with real-time data feeds and agile visualizations will put spatial data and tools well and truly in the hands of the user. The end user will be able to pose complex queries and make real-time decisions on-demand.

The real value for end users as well as financial benefits to organizations will be realized when three factors (a) cooperation, (b) end user needs and uncertainty, and (c) value activities are considered together along the entire supply chain to deliver knowledge services and not just data to end users.

Acknowledgments

This work has been supported by the Cooperative Research Centre for Spatial Information, whose activities are funded by the Business Cooperative Research Centres Programme.

References

ANZLIC (2014) *ANZ Foundation Data Framework.* Available at http://spatial.gov.au/anzlic/anz-foundation-spatial-data-framework, accessed May 2016.

Arnold, L. (2016) Improving spatial data supply chains: Lessons from the manufacturing industry, *GEOProcessing 2016, The Eighth International Conference on Advanced Geographic Information Systems, Applications, and Services*, Venice, Italy.

Bing Tan, C., West, G., McMeekin, D., and Moncrieff, S. (2016) CIAO-WPSS automatic and intelligent orchestration of geospatial web services using semantic web ontologies, *2nd International Conference on Geographical Theory, Applications and Management (GISTAM)*, April 26–27, Rome, Italy.

Bommer, M., Gratto, C., Gravander, J., and Tuttle, M. (1987) A behavioral model of ethical and unethical decision making, *Journal of Business Ethics* 6: 265–280.

Buchmeister, B., Friscic, D., and Palcic, I. (2013) Impact of demand changes and supply chain's level constraints on bullwhip effect, *Advances in Production Engineering & Management* 8(4): 199.

Caprioli, M., Scognamiglio, A., Strisciuglio, G., and Tarantino, E. (2003) Rules and standards for spatial data quality in GIS environments, *Proceedings of the 21st International Cartographic Conference (ICC)*, August 10–16, Durban, South Africa. Cartographic Renaissance Hosted by The International Cartographic Association (ICA).

CRCSI (2013) *Spatial Infrastructures Research Strategy 2013.* Available at http://www.crcsi.com.au/library/resource/spatial-infrastructures-research-strategy, accessed June 2016.

Daily Mail Australia (2015) Popular snacks may soon be labelled to show how many minutes WALKING will be needed to burn off the calories if you eat them. Available at http://www.dailymail.co.uk/news/article-3193903/Australian-food-soon-labelled-minutes-WALKING-needed-burn-calories.html, accessed August 2015.

Dessers, E. (2013) *Spatial Data Infrastructures at Work: Analysing the Spatial Enablement of Sector Processes.* Leuven University Press, Leuven, Belgium, pp. 40, 255.

Fa, J.S.H., West, G., McMeekin, D., and Moncrieff, S. (2016) Brokered approach to federating data using semantic web techniques, *The Eighth International Conference on Advanced Geographic Information Systems, Applications, and Services*, pp. 146–153, Venice, Italy.

Feller, A., Shunk, D., and Callarman, T. (2006) Value chains versus supply chains. Available at http://www.bptrends.com/bpt/wp-content/publicationfiles/03–06-ART-ValueChains-SupplyChains-Feller.pdf, accessed October 2016.

Grech, P. (2012) Landgate and DEC partner to consolidate data, spatial source. Available at http://www.spatialsource.com.au/2012/09/landgate-and-dec-partner-to-consolidate-data/, Accessed July 2016.

Gulland, E.-K., Moncrieff, S., and West, G. (2015) Distributed agents for contextual online searches, *Proceedings of the Free Open Source Software for Geospatial: FOSS4G*, September, Seoul, South Korea.

Hong, J.-H., and Kuoa, C.-L. (2015) A semi-automatic lightweight ontology bridging for the semantic integration of cross-domain geospatial information, *International Journal of Geographical Information Science* 29(12): 2223–2247.

Hunt, P.D. (2003) The concept of knowledge and how to measure it, *Journal of Intellectual Capital* 4(1): 100–113.

Ibrahim, N.Y., Mokhtar, S.A., and Harb, H.M. (2012) Towards an ontology based integrated framework for semantic web, *International Journal of Computer Science and Information Security* 10(9): 91–99.

Magnani, M., and Montesi, D. (2010) A survey on uncertainty management in data integration, *Journal of Data and Information Quality* (1): 1–33.

Marinelli, M.A. (2011) Modelling and communicating the effects of spatial data uncertainty on spatially based decision-making, PhD Thesis, Curtin University of Technology, Perth, Western Australia.

Mass, Y., Ramanath, M., Sagiv, Y., and Weikum, G. (2011) IQ: The case for iterative querying for knowledge, *CIDR 2011. 5th Biennial Conference on Innovative Data Systems Research (CIDR'11)*, January 9–12, Asilomar, CA.

Maxwell, A., and Lal, S. (2013) Technological innovations in managing challenges of supply chain management, *Universal Journal of Industrial and Business Management* 1(2): 62–69.

McConnell, M., and Weidman, S. (2009) *Uncertainty Management in Remote Sensing of Climate Change Data: Summary of Workshop*. The National Academies Press, Washington, DC, p. 64.

Naslund, D., and Williamson, S. (2010) What is management in supply chain management?—A critical review of definitions, frameworks and terminology, *Journal of Management Policy and Practice* 11(4): 11–28.

Olhager, J. (2012) The role of decoupling points in value chain management, in H. Jodlbauer et al. (Eds.), *Modelling Value, Contributions to Management Science*, pp. 37–47. Springer-Verlag, Berlin, Germany.

Porter, M. (2008) *Competitive Advantage: Creating and Sustaining Superior Performance*. Simon and Schuster, Business and Economics, p. 592.

Reed, T., Gulland, E.-K., West, G.A.W., McMeekin, D., Moncrieff, S., and Cox, S. (2015) GeoMeta: A modern geospatial search engine, *Proceedings of the 11th International Conference on Innovations in Information Technology*, November, Dubai, United Arab Emirates.

Runners World (2016) Proposed food labels show how much exercise needed to burn off kilojoules. Available at http://www.runnersworldonline.com.au/food-labels-show-how-much-exercise-needed-to-burn-off-kiljojoules/, Accessed July 2016.

Sadiq, M.A., McMeekin, D., and Arnold, L. (2016) Spatial data supply chain provenance modelling for next generation spatial infrastructures using semantic web technologies, *The Eighth International Conference on Advanced Geographic Information Systems, Applications, and Services*, Venice, Italy, pp. 146–153.

Sehgal, V. (2009) *Enterprise Supply Chain Management: Integrating Best in Class Processes.*
John Wiley & Sons, Hoboken, NJ, p. 201.

Stanford (2012) Value theory, Stanford encyclopaedia of philosophy. Available at
http://plato.stanford.edu/entries/value-theory/, First published February 5,
2008; substantive revision May 29, 2012, Accessed 2016.

The Age (2016) Victorian fast food outlets to show kilojoule content under new law.
Available at http://www.theage.com.au/victoria/victorian-fast-food-outlets-
to-show-kilojoule-content-under-new-law-20160406-go0bl1.html, accessed
July 2016.

Valas, C. (2016) Data to information to knowledge: Making the transformation.
Available at http://data-informed.com/data-to-information-to-knowledge-
making-the-transformation/, accessed October 31, 2016.

Varadharajulu, P., Saqiq, M., Yu, F., McMeekin, D., West, G., Arnold, L., and
Moncrieff, S. (2015) Spatial data supply chains, *Proceedings of the 4th ISPRS
International Workshop on Web Mapping and Geoprocessing Services,* July 1–3,
pp. 41–45. ISPRS, Sardinia, Italy.

Veregin, H. (1999) Data quality parameters, geographical information systems,
chapter 12. Available at http://www.geos.ed.ac.uk/~gisteac/gis_book_abridged/
files/ch12.pdf, accessed February 2015.

Voinov, A., Kolagani, N., McCall, M., Glynn, P.D., Kragt, M.E., Ostermann, F.O.,
Pierce, S.A., and Ramu, P. (2016) Modelling with stakeholders - Next generation, *Environmental Modelling & Software* 77: 196–220.

W3C (2016) Semantic web. Available at http://www.w3.org/standards/semanticweb/,
accessed October 2016.

Woodgate, P., Coppa, I., Choy, S., Phinn, S., Arnold, L., and Duckham, M. (2017) The
Australian approach to geospatial capabilities; positioning, earth observa-
tion, infrastructure and analytics: Issues, trends and perspectives, *Geospatial
Information Science* 20(2): 109–125. doi:10.1080/10095020.2017.1325612.

Yu, F., West, G., Arnold, L., McMeekin, D.A., and Moncrieff, S. (2016) Automatic
geospatial data conflation using semantic web technologies, *Proceedings of the
Australasian Computer Science Week (ACSW).* doi:10.1145/2843043.2843375.

7

Business Models for Geographic Information

Glenn Vancauwenberghe, Frederika Welle Donker,
and Bastiaan van Loenen

CONTENTS

7.1 Introduction

In today's information society, geographic information (GI) also referred to as spatial information is extremely important to the operation of society, and thus very valuable. Citizens, businesses, and government use GI for many of their day-to-day decisions. Most of the societal, economic, and environmental challenges they face require spatial thinking and understanding.

Creating and capturing value from geographic data requires raw data to be turned into GI services and products that are consumed by end users. Although GI will only be of value once it is used (Onsrud and Rushton 1995), this value will be created through a number of stages during each of which a new value is added to the original input by various activities (Longhorn and Blakemore 2008; van Loenen and Zevenbergen 2010). Organizations can fulfill many different roles in the process of generating value from geographic data, and many different GI business models exist.

Although business models determine how organizations create, deliver, and capture value, they should not be regarded as permanent and invariable structures or settings. Business models are shaped by both internal and external forces, and will only be successful if they are able to adapt to a changing environment. In the GI domain, several technological, regulatory, and societal developments have challenged the existing business models and opened up opportunities for new business models. Among these developments are the establishment of spatial data infrastructures (SDIs) worldwide, the democratization of geographic knowledge, and the move toward open source, open standards, and open data. After the commercialization of GI technologies and the introduction of these technologies in different kind of organizations, these trends have led to the introduction of new actors into the GI domain and new ways of using and dealing with GI.

Since the development and implementation of SDIs in different parts of the world, much attention has been paid to the need to find appropriate business models for GI, and in particular, for geographic data providers in the public sector. Traditional business models in which public data providers were selling their data to customers in the private industry and other public agencies were questioned, because they restricted the opportunity for data sharing (Onsrud 1992). The concept of SDI is about moving to new business models, where partnerships between GI organizations are promoted to allow access to a much wider scope of geographic data and services (Warnest et al. 2003). A key challenge in the development of these SDIs was the alignment of different existing business models of the actors in the GI domain (Georgiadou and Stoter 2008). Moreover, the development and implementation of SDIs also led to the emergence of new business models, which was even more the case with the more recent move toward open geographic data. Open data provided an additional resource for existing business models in the GI domain, focused on the analysis, management, visualization, and integration of data (Parycek et al. 2014). In addition, new infomediary business models connecting data providers and end users aroused, focusing on creating and delivering value through the development of services on top of the geographic data made available by governments and other data providers (Janssen and Zuiderwijk 2014).

The aim of this chapter is to contribute to the understanding of the diversity of business models of organizations dealing with GI. This chapter is structured as follows. After this introductory section, the next section reviews the relevant literature on business models. Several existing business model frameworks are presented and the importance of the value concept in business models is explained. In Section 7.3, a comprehensive literature review of works on the value of GI is provided. Building further on this literature review, Section 7.4 discusses the different elements and components of the value of GI. Section 7.5 discusses the GI value chain, including the different actors, the roles, and value-adding processes within this chain. The business models for GI are then synthesized into one table and are analyzed in Section 7.6. We provide our conclusions in Section 7.7.

7.2 Business Model Theory

Business models determine how organizations can create and deliver value, for example, through the provision or use of geographic data. Although many definitions of the concept of business model exist and the literature on business models is developing largely in silos, some clear trends can be seen in the existing literature (Zott et al. 2011; Bonina 2013). To begin, business models can be seen as a new unit of analysis, in addition to traditional units or levels of analysis such as the product, the firm, or the industry. Although the business model concept is centered on a focal organization, its boundaries are wider than those of the organization (Zott et al. 2011). Moreover, business models express a more systemic and holistic approach on how organizations act and do business. In general, business models describe and explain how an organization creates, delivers, and captures value (Osterwalder and Pigneur 2010). In that way, a business model is a conceptual tool that contains a set of interrelated elements that allow organizations to create and capture value and generate revenues (Osterwalder 2004). Business models can also be regarded as the architecture of the core elements or components for supporting business conduction. The development and implementation of an appropriate business model are considered to be a key to the success of the organization and a crucial source for value creation (Yu 2016).

7.2.1 Business Models Frameworks

Several business model frameworks have been developed and can be used by organizations as a starting point for the definition of their own business model. Examples are the business model frameworks of Osterwalder and Pigneur (2010), Shafer et al. (2005), and Hamel (2000). According to Osterwalder and Pigneur (2010), a business model describes the rationale of how an organization creates, delivers, and captures value. Their Business Model Canvas is a template that can be used for developing new or describing existing business models. The Business Model Canvas contains nine building blocks: key partnerships, key activities, key resources, value proposition, relationships with customers, customers, channels, revenue stream, and cost structure. According to Shafer et al. (2005) a business model in itself is not a business strategy, but a representation of it. In their view, a business model consists of four main components: (1) strategic choices, (2) value creation, (3) value network, and (4) capture value. Hamel (2000) sees a business model as a business concept that has been put into practice. His business model framework focuses on four main elements—customer logic, strategy, resources, and network—and three interconnecting components: (1) customer benefits, (2) configuration, and (3) company frontiers. Hamel (2000) also proposes four factors that determine the profit potential of an organization: (1) efficiency, (2) uniqueness, (3) appropriateness, and (4) profit accelerators.

7.2.2 Value Components of Business Models

Value proposition and value creation are often considered as the core construct of a business model. Several authors have proposed the development of value-centric business models (Al-Debei and Avison 2010; Yu 2016; Zeleti et al. 2016). Based on a comprehensive review of the literature, Al-Debei and Avison (2010) derived a unified business model framework of which the fundamental dimensions are value-based. In their view, there are four relevant aspects to the business model framework: (1) *value proposition*, that is, the business logic for creating value for customers by offering products and services for targeted segments; (2) *value architecture*, which refers to the architecture for the technological and organizational infrastructure used in the provisioning of products and services; (3) *value network*, which is about the collaboration and coordination with other organizations; and (4) *value finance*, which refers to the costing, pricing, and revenue breakdown associated with sustaining and improving the creation of value. In addition, Zeleti et al. (2016) distinguish six-core value-based constructs that could be used to characterize a business model: value proposition, value adding process, value in return, value capture, value management, and value network. In a similar manner, Yu (2016) highlights the importance of value proposition, value creation, value capture, and value assessment in the development of business models for open data applications.

In this chapter, we will focus in particular on three value components that are proposed by several authors:

1. *Value proposition* is about the specification of the value that is delivered and offered to different stakeholders by the organization. It includes product, services, distribution channels, information, and price.
2. *Value creation* refers to the execution of particular actions to generate the desired value.
3. *Value capture*, which is the process of retaining some part of the value produced in the value-adding process.

7.3 Literature on the Value of Geo-Information

As value should be regarded as the core concept of a business model, for the development of geo-information business models it is essential to understand the notion of *value of geo-information*. Since Didier's (1990) work on the *utility and value of GI*, several studies have been conducted to determine the value of GI (Genovese et al. 2009). The existing studies on assessing the value and impact of GI can be classified into four main categories, according to the precise topic or subject of the study. A distinction should be made between (1) studies on the value of geographic data and information, (2) studies on the value of GI technologies, (3) studies on the value of GI infrastructures and SDIs, and (4) studies on the value of the entire geo-information sector. In each of these four main categories, studies can be found focusing on or addressing the social impact of GI and related technologies, infrastructures, and sectors (Genovese et al. 2009).

7.3.1 The Value of Geographic Data and Information

Many studies aim to investigate the value of geographic data or information as a whole on a macroeconomic level (ACIL Tasman 2008; Vickery 2011). Other studies focus on a particular dataset of type of geographic data. For instance, several studies exist that deal with the measurement of the benefits and economic value of meteorological data and information (Freebairn and Zillman 2002; Fornefeld et al. 2008; Lazo et al. 2009; Pettifer 2009), topographic data (Oxera 1999; Coote and Smart 2010; Bregt et al. 2013), and geological information (Bernknopf et al. 1993, 1997; Ellison and Callow 1996; Garica-Cortés et al. 2005; Häggquist and Söderholm 2015). In 2010, a study was undertaken to investigate the benefits of open address data in Denmark. The conclusion of the study was that the direct financial benefits from opening the data for society in the period 2005–2009 was around EUR 62 million, whereas the total cost of making the data open was around EUR 2 million (DECA 2010). The value assessment only included the direct financial benefits for the parties directly getting access to the address data. Recent work in the Netherlands include studies on the benefits of opening key topographic data (Bregt et al. 2013, 2014; Grus et al. 2015), small-scale energy consumption data (Welle Donker et al. 2016), and elevation data (Bregt et al. 2016).

7.3.2 The Value of Geographic Information Technologies

A second category of value-assessment studies especially focus on the technologies and systems used to collect, manage, distribute, and/or use geographic data and information. Typical examples of these are studies investigating the costs and benefits to implement and use a GI system within a particular organization. In 2013, Esri described the added value of GI systems on the Return on Investment (Esri 2013). The Geospatial Information and Technology Association (GITA 2007) outlined a methodology for the preparation of business cases for shared data and services for GI technologies. Other interesting work is the study conducted by Booz Allen Hamilton in 2005 focusing on the Return on Investment of geospatial interoperability. The study, which was delivered to NASA, compared the costs, benefits, and the risks associated with two government applications of geospatial technologies: one project utilizing a high degree of open geo-interoperable standards and another project implementing few or none of these standards (Booz Allen Hamilton 2005).

7.3.3 The Value of Geographic Information Infrastructures

Much work has been done on developing and applying frameworks and methods to monitor and measure the value of SDIs, and several interesting reports and articles have been published on this topic. In the European Union, the so-called INSPIRE Directive (Directive 2007/2/EC establishing an Infrastructure for Spatial Information in the European Community [INSPIRE]) aims to create a European Union spatial data infrastructure for the purposes of EU environmental policies and policies or activities which may have an impact on the environment. INSPIRE builds on national SDIs established by the Member States. As the overall technical coordinator of INSPIRE, the Joint Research Centre of the European Commission launched a program of activities to identify frameworks that could be used for assessing the impact of INSPIRE. Among the valuable work that has been done as part of this program are analyses of the socioeconomic impact of the SDI of Catalonia (Garcia Almirall et al. 2008) and the SDI of Lombardia (Campagna and Craglia 2012), a study on the use of spatial data for the preparation of environmental reports in Europe (Craglia et al. 2010) and a case study on e-cadastres to estimate the benefits of SDIs (Borzacchiello and Craglia 2013). In addition, several European countries have published detailed reports on the value of their national spatial data infrastructure in their country. In the Netherlands, a cost-benefit analysis of INSPIRE was conducted a first time in 2009 (Ecorys and Grontmij 2009) and repeated in 2016 (Ecorys and Geonovum 2016). In Sweden, particular effort was taken to assess and monitor the social value of the national SDI (Rydén 2013).

Prior to the implementation of INSPIRE, several scientific models and frameworks were developed to assess SDIs, and the value of these infrastructures in particular. Rodriguez Pabon (2005) created a theoretical framework for the evaluation of SDI projects through the identification and description of common success criteria across different contextual backgrounds. Crompvoets (2006) developed and implemented a procedure to assess the impacts of spatial data clearinghouses in the world. Giff and Crompvoets

(2009) proposed an approach for designing and using performance indicators to support the assessment of SDIs, aiming to evaluate both the efficiency and the short-term and long-term effectiveness of SDIs.

7.3.4 The Value of the Geographic Information Sector

In addition, the economic value of the geo-information sector has been the subject of many studies. The so-called Pira study of 2000 on the commercial exploitation of Europe's public sector information (PSI) estimated that the value adding market of GI in Europe at that time was extremely small compared to North America (Pira 2000). In this study, the U.S. geo-information industry was estimated to contribute significantly to the American economy, employing more than 3.2 million individuals and generating sales of more than €641 billion (Pira 2000, p. 50). The U.K. location market survey carried out by ConsultingWhere (2013) in the United Kingdom estimated that the market for location-related software, professional services, data, and hardware in 2012 was nearly €1.5 billion. A study of the U.S. geospatial industry suggests that it has generated at least 500,000 jobs in the United States and that geospatial services deliver efficiency gains that are many times the size of the sector itself creating a lasting source of competitive advantage for the country as a whole (Boston Consulting Group 2013). In addition, in 2013, a forecast study was delivered by Market Info Group (2013) on the market of location-based services between 2013 and 2020. An important EU-wide initiative was the smeSpire study on the geo-ICT sector in Europe, in which the market potential of INSPIRE for geo-ICT companies was investigated (Cipriano et al. 2013; Vancauwenberghe et al. 2014). ACIL Tasman analyzed the impact of geospatial information technologies on the Australian economy and estimated this impact to be between AUD 6.43 and 12.57 billion per annum (equivalent to 0.6%–1.2% of GDP) (ACIL Tasman 2008). ACIL Tasman arrived at similar results for the impact on the New Zealand economy of NZD 1.2 billion per annum or 0.6% of GDP (ACIL Tasman 2009). In Canada, Natural Resources Canada provided a detailed analysis of the Canadian geomatics industry. The study concluded that the use of geospatial information contributed CAD20.7 billion or 1.1% of GDP to the Canadian economy (Natural Resources Canada 2016).

Although all these studies show positive impacts due to geospatial information, the results cannot always be compared as many studies employ different methodologies and outcome criteria. Many of these studies cannot be validated due to poorly described methodology and a lack of the underlying data. In addition, some of these studies are based on questionable assumptions and/or selective extrapolation of only *best-case* case studies. Thus, due to extrapolating the probably too optimistic projections, the outcomes may be too optimistic and relatively small uncertainties become increasingly important (van Loenen and Welle Donker 2014).

7.4 Aspects of the Value of Geographic Information

A meta-analysis of existing studies on assessing the value and impact of the GI is provided by both Genovese et al. (2009) and Trapp et al. (2015). Genovese et al. (2009) made a classification of the literature on assessing the impact of GI based on an examination of 32 academic, business, and government studies. Trapp et al. (2015) particularly focused on cost-benefit assessments and aimed to explain the variation in returns on investments in GI. Their meta-analysis included 82 cost–benefit assessments between 1994 and 2013. The comprehensive review of the academic and professional literature on the value of GI made by Genovese et al. (2009) and Trapp et al. (2015) reveals three important insights:

1. The precise topic of the value assessment can vary, and a distinction should be made between studies dealing with the value of geographic data and the value of geographic data infrastructures. This is not only expressed in the classification of studies according to the topic by Genovese et al. (2009) but also in the meta-analysis of Trapp et al. (2015) on the return on investment of geospatial data and systems. Trapp et al. (2015) make a comparison between studies focusing on geospatial data-related investments and studies on investments in SDIs. The results of their study show that the benefit–cost ratio of investments in SDIs was slightly higher than that of investments into geospatial information. However, they also pointed out that SDI studies often were based on theoretical assumptions or few experiences, and the benefits still remain unclear.

2. Many different methods can be used for estimating and/or measuring the value. According to Genovese et al. (2009) GI assessment studies can also be classified according to the approach followed. Approach refers to the kind of methodology applied in the analysis or assessment. After a first identification of 11 different approaches, Genovese et al. (2009) regrouped these approaches into four main categories: (1) consultation and references, (2) evaluation techniques (e.g., value chain and value-benefit analysis, pricing, models, and frameworks), (3) indicators and statistical analysis, and (4) cost–benefit analyses and return on investment studies. Although Trapp et al. (2015) only addressed studies that investigated both the costs and benefits of GI investments, they also identified different methodological approaches that have been applied: cost–benefit assessments, return on investment, and other metrics.

3. Finally, it is also important to notice that the value can be defined in different ways. For instance, a distinction can be made between the financial value (monetary, exchange) and the socioeconomic value (Genovese et al. 2010), or between commercial, economic, and socioeconomic value (Longhorn and Blakemore 2008). The financial or exchange value of GI is typically reflected in the price at which the data are traded and the consumer's willingness to pay for the data offered (Longhorn and Blakemore 2008). Socioeconomic value of geo-information refers to the value of an information good and service in achieving societal goals, such as better governance, higher quality of life, or economic growth (Genovese et al. 2010). Longhorn and Blakemore (2008) consider the economic value to include the revenues and the number of people employed by the GI sector, whereas the socioeconomic value also includes noncommercial values, such as improving informed decision-making.

With regard to the third issue, several authors pointed out that information in itself has no value, and it is only of value once it is used (Onsrud and Rushton 1995; Barr and Masser 1996). Moreover, the value is rather related to the nature of the use, and less to the nature of the information. In addition, other authors argued that the value of—geographic—information is different for different users and for different applications (Longley et al. 2001). Longhorn and Blakemore (2008) add to this that the value of—geographic—information also varies with time, as information that might be of great value today, can become less or more valuable in the future. Another interesting perspective on the value of GI provided by Longhorn and Blakemore (2008) is their identification of different components of the value of GI. They distinguish:

- *The value of the location attribute*: The location attribute of GI provides special value to this information, as it provides a spatial context to the other attributes in the information package, and increase the value of data for applications where spatial awareness is essential.
- *Time dependency value*: Value of certain types of GI will be dependent on whether it is real-time data, relatively invariant data, or historical data.
- *Value determined by cost savings*: The value of SDIs and other initiatives for data sharing is not in the data itself, but in the cost savings realized by duplications in the data collection.
- *Value added through information management techniques and tools*: The value of GI will depend on the data format, the physical medium by which it is captured and stored, the availability of metadata, and the application of—open—standards.

- *Value due to legal or other mandatory use requirements*: In legal jurisdictions, certain GI is given an official or legal status for certain types of transactions.
- *Value due to network effects*: Some GI has added value simply because it is used by a large group of users (e.g., Google Maps and other online maps).
- *Value due to the quality of an information resource*: Typical quality issues such as the completeness of the data, timeliness of the information, or the accuracy of the data can add or detract from the value of the information.

Longhorn and Blakemore's identification of different components or dimensions of the value of GI also shows the complexity of measuring and assessing this value, and the need for new approaches and techniques to ease empirical analysis and gain a better understanding of the value of GI.

7.5 The Geo-Information Value Chain

7.5.1 Geo-Information Value Chain Components

To provide a better insight into the process of adding value to GI, several authors have introduced and applied the information value chain approach (Krek 2002; Longhorn and Blakemore 2008; Genovese et al. 2010; van Loenen and Zevenbergen 2010). A value chain can be defined as the set of value-adding activities that one or more organizations perform in creating and distributing goods and services (Longhorn and Blakemore 2008). The value chain concept originally was developed for the manufacturing sector, as a tool to evaluate the competitive advantage of firms (Genovese et al. 2010). More recently, the value chain concept has been applied to other sectors, including information technology where the good or service, and the benefits it provides, is less tangible in nature (Longhorn and Blakemore 2008). According to Porter (1985), a value chain involves the progress of goods from raw materials to finished products through a number of stages, during each of which a new value is added to the original input by various activities. The value chain concept was extended into the information market, with the information value chain referring to the set of activities adding value to information and turning raw data into new information products or services. Especially important in this context is the role of information and communication technologies (ICT), which have an impact on all activities in the information value chain, such as information collection, processing, dissemination, and use (Longhorn and Blakemore 2008). Some examples of information value chains are the information value chain as defined by Oelschlager (2004) and the management information value chain of Philips (2001).

In the context of GI, the value chain relates to the series of value-adding activities to transform raw geographic data into new products that are used by certain end users (Genovese et al. 2010). Although there are slightly different descriptions of the various steps of the GI value chain, in general, the essential steps in the value chain are: acquisition of raw data, the application of a data model, quality control, and integration with other sources, presentation, and distribution (cf. Van Loenen and Zevenbergen 2010; Genovese et al. 2010). In recent years, particular attention has been paid to different steps between the process of distributing data and the actual end use of an end product of GI. In addition, after the publication of the data, value can be added to the data in many different ways. Value can be added by making data from different sources easily accessible through repositories and data portals, by building and selling tailored solutions using the data to end users or by using geographic data to improve existing products and services delivered to an end user. In certain cases, this end product will be the first step of a next value chain (Welle Donker and van Loenen 2016).

Van Loenen and Zevenbergen (2010) apply the value chain approach to compare the process of adding value to two important types of GI, road center lines and parcel data sets, in Europe and in the U.S. Their research found significant differences in the GI characteristics of governments in Europe and the United States at the time of the study: European public sector GI was more accurate, more up-to-date, and more comprehensive. The geo-information value chain study indicated that the private sector datasets available in the United States were more comparable with the European public sector datasets with respect to adding value to framework datasets than the public sector data in the United States public sector. Van Loenen and Zevenbergen concluded that value adding is influenced by the different roles government and market play in the GI value chain.

Another interesting value chain approach is the approach of Attard et al. (2016) who consider the data value chain as a part of a data lifecycle, which consists of different processes of value creation. They identify six value-adding processes (Figure 7.1): (1) data creation, (2) data harmonization, (3) data publishing, (4) data interlinking, (5) data exploitation, and (6) data curation. Within each of these processes, various value creation techniques can be used. The approach of Attard et al. (2016) clearly demonstrates how different value-adding processes and techniques can be considered to be the value creation of the organizational business model.

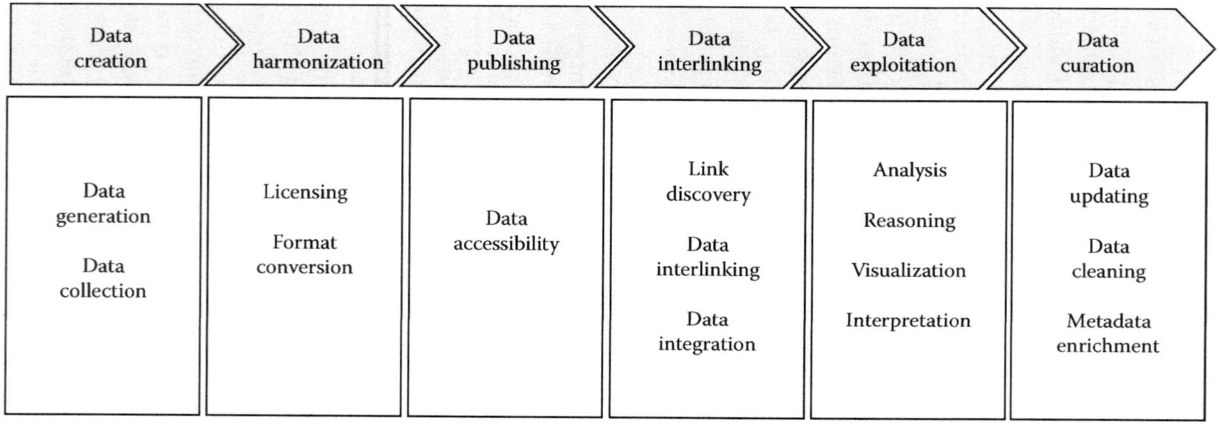

FIGURE 7.1
Value creation processes and techniques in the data value chain. (From Attard, J. et al., Value creation on open government data, in *Proceedings of the 49th Hawaii International Conference on System Sciences*, IEEE, Kauai, HI, 2016, p. 10.)

7.5.2 Roles and Activities within the Information Value Chain

Attard et al. (2016) also make a link between the value-adding activities and the different roles in the information chain. They identify four roles in which different stakeholders can contribute to creating value: *Data producers and publishers* are in charge of collecting or generating data and/ or making the data accessible to other users. *Data enhancers* create value through the actual manipulation of the data, which can be done in many different manners: through data harmonization, interlinking, curation, or exploitation. *Service creators* use available data to provide a certain service. In most cases, they exploit the data by providing additional analysis or visualization services. *Facilitators*, finally, help other stakeholders in publishing, using, or exploiting through the provision of services or technologies. Their classification compares well with the classification of the five *archetype* roles within the information value chain proposed by Deloitte LLP (2012) of (1) suppliers, (2) aggregators, (3) developers, (4) enrichers, and (5) enablers.

There are also particular identifications and descriptions of the different roles within SDIs. Hjelmager et al. (2008) identified six general roles of stakeholders in the context of SDIs: policy-makers, producers, providers, brokers, value-added resellers (VAR), and end users. For each of these stakeholders, several subtypes can be distinguished (Cooper et al. 2011):

1. *Policy-maker*: A stakeholder who sets the policy pursued by an SDI and all its stakeholders. Subtypes of policy-makers are decision-makers, legislators, promoters, and secretariat.

2. *Producer*: A stakeholder who produces SDI data or services. Subroles are captors of raw data, database administrators, passive producers and submitters, or revision notices.

3. *Provider*: A stakeholder who provides data or services, produced by others or itself, to users through an SDI. Main subtypes are data providers and service providers, whereas within both subtypes a distinction can be made between providers of their own data or services, distributors who make the data and services of other producers available, and arbiters, who test, validate, and sometimes also certify datasets or services.

4. *Broker*: A stakeholder who brings end users and providers together and assists in the negotiation of contracts between them. They are specialized publishers and can maintain metadata records on behalf of an owner of a product. Their functions include harvesting metadata from producers and providers, creating catalogs, and providing services based on these catalogs. Subtypes are facilitators, finders, harvesters, catalogers, and negotiators.

5. *Value-added reseller (VAR)*: A stakeholder who adds some new feature to an existing product or group of products, and then makes it available as a new product. Two main subtypes are publishers and aggregators/integrators.

6. *End user*: A stakeholder who uses the SDI for its intended purpose. Here, a distinction is made between naive consumers and more advanced users.

The identification and description of different roles within the—geographic—information value chain shows how many different actors are involved in this chain, and contribute in many different ways to the creation of value.

7.6 Geo-Information Business Models

7.6.1 Previous Studies

Individual organizations can be active in different parts of this value chain, and can create and offer value in many different ways. As a result, many different GI business models exist. Previous works on business models for geographic data and open data show a great variety in perspective, structures, and components used to describe business models as well as in identified business models (Yu 2016). The work of Zeleti et al. (2016) and Welle Donker and Van Loenen (2016) focused on business models for data providers, whereas Janssen and Zuiderwijk (2014), Magalhaes et al. (2014), and Yu (2016) explored business models for infomediaries and enablers.

Ferro and Osella (2013) used a *PSI*-based value chain to identify different actors and positions within the PSI-ecosystem. In their view, government organizations are mainly active in the chain and ecosystem as data providers and data users. Focusing on the roles and position of *profit-oriented actors*, they distinguish four main types of businesses: (1) core reusers, (2) enablers, (3) service advertisers, and (4) advertising factories. Focusing on the revenue models of these actors, they identified eight archetypes of business models for reusers of PSI: Premium, Freemium, Open Source Like, Infrastructural Razor and Blades, Demand-oriented Platform, Supply-Oriented Platform, Free as Branded Advertising, and White-Label development. Building further on the classification of Ferro and Osella, Zeleti et al. (2016) see five categories of business models for open data providers:

- *Indirect benefits*: Release data to support the primary goal of the business.

- *Cost savings*: Aim to reduce the costs of opening and releasing data and/or increase the quality of data through user participation.
- *Parts of tools*: Offer a first (incomplete) set of data at discount, but the complementary or dependent data at a considerably higher price.
- *Freemium*: Provide limited data for free, and apply fees for additional requests for complete higher quality data sets.
- *Premium*: Provide high quality and complete data at a certain cost.

Welle Donker and van Loenen (2016) researched the financial effects of open data policies for geographic data providers in the public sector on their business model. Business models are also about how organizations generate sufficient income to cover at least their operating costs. For government data providers, the recent political and societal evolution toward releasing data as open data requires a change in their existing business model. Welle Donker and Van Loenen (2016) demonstrated how new business models for data providers in the public sector emerged as a result of the introduction of open data.

Although Welle Donker and Van Loenen (2016) and Zeleti et al. (2016) provide insight into different business models for - geographic - data providers, other authors focused on the development of business models for users and reusers (Janssen and Zuiderwijk 2014; Magalhaes et al. 2014; Attard et al. 2015). Magalhaes et al. (2014) developed taxonomy of three open data users' business model archetypes: *Enablers* are businesses that provide customers with technologies, such as apps or software programs, mainly built on or for the use of open government data. *Facilitators* support or accelerate the access and exchange of data between the supply side and the user side by simplifying and promoting the access of the open government data. *Integrators* refer to firms that integrate open government data into their existing business models in order to further improve their existing offer. Janssen and Zuiderwijk (2014) identified six atomic business models for the so-called infomediaries: intermediaries connecting open data providers and users: single-purpose application developers, interactive application developers, information aggregators, comparison models, open data repositories, and service platforms. These atomic business models can be combined, for example, a company may develop single-purpose applications in addition to interactive applications.

In this segment, we built further on the models identified and explored in the existing literature to identify several business models in the geo-information model. In line with existing classifications, we identify and describe three main categories of organizations: (1) providers, (2) enablers, and (3) end users.

7.6.2 Geographic Data Providers

Geographic data providers are organizations that provide data or services, produced by themselves or by other data producers, to other users. Data providers can be public sector organizations, but also businesses, nonprofit organizations, and other types of organizations. For government data providers, including providers of GI, the ongoing move toward more open data has mainly caused a change in the value capture component of their business model. In the traditional cost recovery models, costs incurred in the production, maintenance, and distribution of geographic data could be recovered from retaining part of the value produced in providing the data by charging license fees. Different pricing mechanisms could be used to set license fees, such as one-off fees, subscription fees, fixed access fees, royalties, or a combination of different mechanisms (Welle Donker 2009). In current open data models, in which data are provided free of charge and without license restrictions, there is no direct revenue from supplying data for the data provider, and the data provider depends on other ways of capturing (part of the) value. Four main business models for data providers can be distinguished: (1) data providers for indirect benefits, (2) data providers for cost savings, (3) freemium data providers, and (4) premium data providers.

Data providers for indirect benefits make their geographic data available as open data for strategic reasons. Geographic data are released because opening them supports the primary goal of the organization (Zeleti et al. 2016). For instance, government organizations make their geographic data available as open data to stimulate the development of value-adding products and services on top of their data, or to improve transparency to their own activities and processes. In most cases, they will not directly benefit (financially) by making their data available. The loss of direct revenue is compensated through other funding, such as (extra) funding from budget-financing or income from supplementary fee-based products and services. In addition, public sector data providers may have access to legal instruments to, for example, levy charges for compulsory register transactions. To ensure sustainable open data business models for public sector data providers, it is essential that there are such guaranteed main sources of revenues (Welle Donker and Van Loenen 2016).

Data providers for cost savings believe opening up their geographic data will lead to significant internal efficiency gains, such as increased data quality or staff reductions. A particular example of this business model is organizations that aim to take advantage of the input from other parties to improve the quality of their data products. As such, the value creation process goes further than publishing the data, in some cases the quality of the data is improved by cleaning the data through contributions from other parties (Welle Donker and Van Loenen 2016). The capture of value is in the efficiency gains in internal processes, and in improvements to the data quality.

Both data providers for indirect benefits and data providers for cost savings do not capture value through direct revenues from making their data available. In both categories, the value created is relatively limited, and is mainly found in publishing the data. In freemium and premium business models for data providers, the value created is higher, whereas value is captured back by the provider through direct revenues from selling data and data services. *Freemium data providers* offer a limited version of the dataset free of charge, but apply fees for additional requests for complete and higher quality data (Zeleti et al. 2016). The rationale behind this business model is that opening up some datasets for free will lead to an increased demand for—and additional revenues from—high-quality data. Additional value is created through better data maintenance or through the more sophisticated data access services. *Premium data providers* do not offer any data for free, only high-quality, complete, and up-to-date data are offered at a fee. In some cases, particular data needs of users are fulfilled (e.g., 24/7 access) or additional support is guaranteed, which means additional value is created and provided. Zeteli et al. (2016) clearly show how both freemium models and premium models consists of many different subtypes. Although the freemium model is about offering limited data free of charge but applying fees for additional requests, in practice, this can be organized in several manners. For instance, fees may be charged for higher quality data, for the use of the data for commercial purposes, or for changes made to the original product. The premium business model can take many shapes; however, in all cases data are offered at a certain cost. The aim can be to meet specific customer data needs or to guarantee support and additional services on top of the dataset.

7.6.3 Geographic Data Enablers

A second category of organizations involved in creating and delivering value in the GI value chain is the category of enablers. Enablers are organizations that assist other organizations in managing, publishing, and using geographic data. While the value creation process of most geographic data providers is quite similar, that is, publishing basic and/or high-quality data, the value creation processes of enablers are much more diverse. Most of the enabling organizations fulfill an intermediary role between the providers or geographic data and the end users of these data. Three main types of enablers can be distilled: (1) supply facilitators, (2) access facilitators, and (3) service creators.

Facilitators are organizations that provide support to data providers in making their data available and to data users in getting access to available data. As such, a distinction can be made between supply facilitators and access facilitators. *Supply facilitators* facilitate the process of providing and distributing geographic data by providing technologies or services to data providers. In some cases, facilitators also facilitate the management and curation of internal data sources of the data provider, and the value creation process will not be limited to the publishing of data. Moreover, setting up data visualization and analysis tools and services can create additional value to the data publishing processes. An example of supply facilitators are companies that provide support to public organizations in meeting the requirements laid down by SDI legislation and policies, such as creating metadata, setting up web services, and harmonizing data according to predefined data specifications.

Facilitators not only provide value to data publishers through facilitating the publication of geographic data, in a similar manner, they could also support data users in getting access to geographic data from different sources. *Access facilitators* are different from supply facilitators because they deliver value to users of data. Access facilitators offer users facilitated access to available geographic data, which is realized through certain products and/or services. These access facilitators do not operate on the request of or in collaboration with data providers, but rather work together with particular users. Revenues are generated from selling products and/or services to users. In most cases, the work of access facilitators consists of structuring and classifying available geographic data and facilitating access to these data through appropriate access mechanisms. Aggregators can be regarded as a particular type of access facilitators, focusing on pooling, combining and, also often, processing geographic data from different sources. In some cases, facilitators enhance the data, which means they actually manipulate the data to create additional value for the end user. This can be achieved through data harmonization, interlinking, exploitation, and/or curation (Attard et al. 2016).

Service creators use geographic data to provide a service to certain target groups. Although in most cases facilitators cooperate closely with data providers or data users, service creators work more independently to develop their own products or services. The work of these creators strongly builds on the geographic data of data providers, whereas their products or services are used by different types of end users. It may be that service creators develop specific applications at the request of the data providers. Service creators may also develop free applications as a calling card to demonstrate their level of expertise in order to acquire fee-based orders. In some cases, even data provider themselves create additional services on top of its data to facilitate the reuse of their data by end users (Welle Donker and Van Loenen 2016). Service creators provide end users with more tailored solutions on top of geographic data from one or more data providers (Magalhaes et al. 2014). It should be noted that the category of service creators is very diverse, as many different types of services can be created on top of geographic data. The revenue models and pricing mechanisms employed for these services can also be very diverse. Examples of subtypes of the business model of service creators are developers of single-purpose apps and developers of interactive apps (Janssen and Zuiderwijk 2014).

7.6.4 Geographic Data End Users

Geographic data end users are different from enabling organizations such as service providers and facilitators, because geographic data are not at the core of their processes and activities. End users make use of geographic data products in support of their primary processes. Primary processes of government agencies are related to the main tasks and activities of government and their services to citizens and businesses. This includes services related to processes, such as maintaining registers of citizens and of businesses, or granting concessions for exploitation and management of public infrastructures. In some cases, primary processes of organizations will be decision-making processes. In the private sector, end users will make use of geographic data in combination with other types of data in order to support and augment their business capabilities (Magalhaes et al. 2014). These businesses can operate in various sectors (e.g., insurance, real estate, banking, and manufacturing) and can potentially extract extra value from geographic data in different stages of their organizational value chain (Magalhaes et al. 2014). Geographic data will be a valuable resource but will not be at the core of the value chain of the organization. The value created will mainly be in the primary activities of the organization, whereas the value will be captured through a more efficient execution of internal processes and through revenues from selling the main products and services of the organization.

An overview of the different geo-information business models and their value proposition, value creation and value capture (cf. Osterwalder and Pigneur 2010) is presented in Table 7.1. The table demonstrates the main differences between geo-information business models with regard to the three main components of a business model: the value proposition, value creation and value capture.

7.6.5 Examples

To conclude this chapter, some examples of the different business models for geographic information are shown in Boxes 7.1. through 7.3.

TABLE 7.1

Different Geo-Information Business Models and Their Value Components

Business Model	Value Proposition	Value Creation	Value Capture
Data Providers			
Data providers for indirect benefits	Open data supporting strategic business objectives	Publishing data	Improved outcomes of the organizations Lack of direct revenues compensated through other funding sources
Data providers for cost savings	Availability of higher quality data	Publishing data Cleaning data	Improved process and data Cost savings
Freemium data providers	Availability of limited data for free and high-quality data and data services at some cost	Publishing data Data maintenance More sophisticated data access services	Revenue from premium data and/or value added services
Premium data providers	High-quality data at some cost Data meeting particular user needs at some cost	Publishing data Data maintenance Data visualization services Data analysis and interlinking services	Revenue from all data and advanced data services
Enablers			
Supply facilitators	Facilitating in providing access to geographic data resources, through provision of technologies and/or services	Publishing data Harmonizing data Metadata creation (Basic) data visualization and analysis services	Revenues from selling products and/or services to data providers (different revenue and pricings models can be adopted)

(Continued)

TABLE 7.1 (*Continued*)

Different Geo-Information Business Models and Their Value Components

Business Model	Value Proposition	Value Creation	Value Capture
Access facilitators	Facilitating in access to geographic data resources, through provision of technologies and/or services access to combined and/or integrated data resources	Structuring and classifying data Aggregating data (Basic) data visualization and analysis services	Revenues from selling products and/ or services to data users (different revenue and pricing models can be adopted)
Service creators	Diversity of tailored solutions on top of geographic data	Creating applications and other solutions on top of geographic data	Revenues from selling solutions to different kind of end users Revenues from developing solutions at the request of data providers
End users			
Data users	No common value proposition, because of diversity of public and private organizations that can be considered as users	Data used within organizational processes and activities, value mainly created through use of data in key processes of the organization	Improved business processes and outcomes Revenues from main products and services delivered by the organization

BOX 7.1   DATA PROVIDER

The *Spanish National Center of Geographic Information* (*CNIG*) is an autonomous body under the National Geographic Institute (IGN), whose main mission is to produce, develop, distribute, market, and publish the data and services of the IGN. In 2014, the Center operated with a budget of approximately €6.6M, and had about 100 employees. Several geospatial data products are distributed by the CNIG: topographic database, images, land cover, raster maps, digital terrain models (DTM), LiDAR data, administrative units, gazetteer data, geodetic points, and so on.

Before 2008, users had to pay a fee to get access to the data, whereas researchers and research groups were given a 90% discount. In 2008, the business model of CNIG was changed with the adoption of the Ministerial Order FOM/956/200815 of March 31, 2008, which determines the policy for public dissemination of the GI generated by the IGN. In this Ministerial Order a distinction was made between two categories of IGN products: the digital GI included in the National Reference Geographic Equipment, which includes geodetic and leveling points, administrative units and gazetteers, and all other geographic data, such as topographic data, land cover data, LiDAR data, and so on. The first category of data, that is, the data in the National Reference Geographic Equipment, was defined as open data, and thus made available for free for any purpose. The data were licensed under a creative commons attribution license (CC-BY), which means the IGN should be mentioned as the origin and owner of the data. The rest of the geospatial data was licensed under a creative commons attribution-noncommercial-sharealike license (CC-BY-NC-SA) and thus only openly available for noncommercial purposes.

As a result of this change in their business model, only a very small part of the CNIG budget came from revenues from the sale of geospatial data products. In 2014, revenues from selling data constituted only 8% of the total budget of the CNIG. The main sources of income were government budget financing (65%) and revenues from contract-based service delivery (25%), in addition to project funding (5%) and revenues from training and consultancy (2%). In 2015, the business model of CNIG again changed with the adoption of a new Ministerial Order, under which all digital datasets and web services of the IGN were defined as open data. All data could be viewed and downloaded for free on Internet, under a license compatible with CC-BY 4.0. As a result, the CNIG now disseminates GI in five different manners, of which paper maps are the only dissemination

(Continued)

BOX 7.1 (Continued)   DATA PROVIDER

channel users have to pay for. Other ways of dissemination are files that can be downloaded, web applications, mobile applications, and web services. The most downloaded files of CNIG are the 1:25 000 National Topographic Map (3.6 million users) and the cartographic drafts (1.1 million users).

Sources: **Lopez Romero (2013); Rodriguez (2016).**

BOX 7.2   SUPPLY FACILITATORS

Between 2012 and 2014, a study was undertaken of the geo-ICT sector in Europe in the context of the SmeSpire project. The study provided an in-depth analysis of the diversity of geo-ICT companies in Europe, with a particular focus on the involvement of these companies in the establishment of national and regional SDIs. Data was collected on the activities, products and services, and customers and business models of around 300 geo-ICT companies in Europe. The results of the survey showed that the majority of these companies were users of GI and/or providers of applications and related services to support the use of GI within organizations.

Approximately one-third of all companies were actively involved in the establishment of SDIs, and could be considered as supply facilitators, that is, organizations providing products or services to data providers to support them in making their geographic data available. Key activities undertaken—or supported—by these organizations were data harmonization (26% of all companies), the development of web map services (26%), and the establishment of metadata catalogs (21%). Other products or services delivered by these supply facilitators were related to the creation and editing of metadata, the setup of web feature services and the provision of training on SDI-related topics.

Although the majority of European geo-ICT companies strongly focused on customers within their own country (mainly local and national public administrations), some examples exist of European Geo-ICT providing support to the implementation of several national SDIs. The Dutch company *GeoCat* contributes to the development of many national metadata catalog based on the open source software GeoNetwork. Several national spatial data infrastructure relies on the

(Continued)

BOX 7.2 (Continued)   SUPPLY FACILITATORS

products and services of the Finnish company *Spatineo* for monitoring and evaluating the performance of their spatial web services. The German company *wetransform* provides professional support and consulting for data harmonization with the HALE technology they have developed. The Italian SME *Epsilon Italia* has created an INSPIRE Helpdesk to support public and private organizations in the overall INSPIRE implementation process: from transformation and validation data and metadata to the publication of web services. In recent years, similar supply facilitating companies have been established throughout Europe.

Sources: **Cipriano et al. (2013); Vancauwenberghe et al. (2014).**

BOX 7.3   ACCESS FACILITATORS

Established in 2011 and located in the United Kingdom, GeoLytix is a good example of a recently created SME, which focuses on facilitating and enabling access to geographic data. GeoLytix aims to solve problems where location matters, such as location decision support and network strategies. The company offers a wide range of geospatial data products, including maps, boundary data, and points of interest. In addition to this, analysis, training, and consultancy services are provided to support customers in developing new commercial insight.

A large number of the company geodata products and services are based on open data, from data sources such as the Land Registry, the Department of Health, the Department of Education, OpenStreet Maps, and several transport agencies. GeoLytix tries to differentiate itself from other data analytics company by also creating and releasing open geodatasets themselves. For example, GeoLytix gathered and mapped data on more than 10,000 supermarkets in the United Kingdom and published it as open data—allowing supermarkets to identify market competition and opportunities for new stores, and developers to create new services for shoppers. Other open datasets released by the company include processed census data, postal sector boundaries, retail places, and workplace data.

The Open Data Institute considers GeoLytix as one of the best examples of U.K. companies using open data. Several other good examples of open geodata companies can be found in existing repositories of

(Continued)

BOX 7.3 (Continued)   ACCESS FACILITATORS

open data impact case studies and use examples such as the Open Data Impact Map and Open Data 500. The Open Data Impact Map is a public database of organizations that use open government data from around the world, and currently contains 1765 organizations from 96 companies. Around one-third of the organizations included the Open Data Impact Map are in the data/information technology and geospatial mapping sectors, for which geospatial data are the most used type of data. Many of these organizations provide data analysis and visualization services using geospatial and mapping data to provide geospatial intelligence. The Open Data 500 is a study of companies that use open government data to generate new business and develop new products and services. The study is currently undertaken in six different countries. The OD 500 U.S. focusing on open data companies in the United States identified around 30 open data companies that consider themselves as geospatial/mapping companies. Both initiatives clearly demonstrate the importance of geospatial data companies in the current open data ecosystem.

Sources: **Open Data Institute (2015); The OD500 Global Network (2017); Open Data for Development Network (2017).**

7.7 Conclusion

This chapter has provided an overview of business models for GI based on an extensive literature review. Starting from the observation that business models are all about proposing, creating, and capturing value, an overview and discussion was provided of the existing literature on the value of GI, of GI technologies and infrastructures, and of the GI sector. The notion of the GI value chain was used to provide better understanding of how value can be created and added to geographic data by turning these raw data into information products and services for end users. Public, private, and other organizations can be active in different stages of this GI value chain, and can create and provide value to actors in later stages of the chain. Although value is created through the production and creation of data, this value will be further enhanced through the harmonization, publication, interlinking, curation, and exploitation of the data. These many different ways of proposing, creating, and capturing value from geographic data and information result in different GI business models. Although data providers, data enablers, and data end users could be seen as three main categories of GI business models, each of these categories consists of many different business models, as different value propositions will exist, and value can be created and captured in several ways. In this chapter on "Business Models for Geographic Information," only a general introduction to the topic of GI business models is given, which will be further explored in the forthcoming chapters.

References

ACIL Tasman (2008). The value of spatial information. The impact of modern spatial information technologies on the Australian Economy, p. 221. http://www.acilallen.com.au/cms_files/ACIL_spatial%20information_economicimpact.pdf, Accessed January 23, 2016.

ACIL Tasman (2009). Spatial information in the New Zealand economy. Realising productivity gains, p. 153. http://www.acilallen.com.au/cms_files/ACIL_spatial%20information_NewZealand.pdf, Accessed November 21, 2013.

Al-Debei, M.M. and D. Avison (2010). Developing a unified framework of the business model concept. *European Journal of Information Systems* 19(3): 359–376. doi:10.1057/ejis.2010.21.

Attard, J., F. Orlandi, and S. Auer (2016). Value creation on open government data. In *Proceedings of the 49th Hawaii International Conference on System Sciences* (p. 10). Kauai, HI: IEEE.

Attard, J., F. Orlandi, S. Scerri, and S. Auer (2015). A systematic review of open government data initiatives. *Government Information Quarterly* 32: 399–418. doi:10.1016/j.giq.2015.07.006.

Barr, R. and I. Masser (1996). The economic nature of geographic information: A theoretical perspective. In: *Proceedings of GIS Research U.K. 1996 Conference.* University of Kent, April 1996, pp. 59–66.

Bernknopf, R.L., D.S. Brookshire, M. McKee, and D.R. Soller (1993). Societal value of geologic maps, U.S. Geological Survey, Virginia (Circular No. 1111).

Bernknopf, R.L., D.S. Brookshire, M. McKee, and D.R. Soller (1997). Estimating the social value of geologic map information: A regulatory application. *Journal of Environmental Economics and Management* 32(2): 204–218.

Bonina, C.M. (2013). New business models and the value of open data: Definitions, challenges and opportunities. London, London School of Economics and Political Science, p. 30. http://www.nemode.ac.uk/wp-content/uploads/2013/11/Bonina-Opendata-Report-FINAL.pdf, Accessed December 4, 2015.

Booz Allen Hamilton (2005). Geospatial interoperability return on investment study. National Aeronautics and Space Administration—Geospatial Interoperability Office, p. 80. http://www.ec-gis.org/sdi/ws/costbenefit2006/reference/ROI_Study.pdf, Accessed March 8, 2017.

Borzacchiello, M.T. and M. Craglia (2013). Estimating benefits of spatial data infrastructures: A case study on e-Cadastres. *Computers, Environment and Urban Systems* 41: 276–288. doi:10.1016/j.compenvurbsys.2012.05.004.

Boston Consulting Group (2012). Putting the U.S. geospatial services on the map. Mountain View, CA, Google Technical Report.

Bregt, A., L. Grus, and D. Eertink (2014). Wat zijn de effecten van een open basis-registratie topografie na twee jaar? Wageningen, the Netherlands, Wageningen University, p. 49. http://www.wageningenur.nl/upload_mm/6/0/9/059a4c74-d6ac-4f6e-9e47-df138bf9e221_Effecten%20van%20een%20open%20BRT%20analyse%202014.pdf, Accessed October 13, 2014.

Bregt, A.K., W. Castelein, L. Grus, and D. Eertink (2013). De effecten van een open basisregistratie topografie (BRT). Wageningen, the Netherlands, p. 40. http://edepot.wur.nl/278625, Accessed July 8, 2013.

Bregt, A.K., L. Grus, T. van Beuningen, and H. van Meijeren (2016). Wat zijn de effecten van een open Actueel Hoogtebestand Nederland (AHN)? Onderzoek uitgevoerd in opdracht van het Ministerie van Economische Zaken, p. 53. http://edepot.wur.nl/393158, Accessed October 31, 2016.

Campagna, M. and M. Craglia (2012). The socioeconomic impact of the spatial data infrastructure of Lombardy. *Environment and Planning B: Planning and Design* 39(6): 1069–1083. doi:10.1068/b38006.

Cipriano, P., C. Easton, E. Roglia, and G. Vancauwenberghe (2013). A European community of SMEs built on environmental digital content and languages. Final Report, p. 160. http://www.smespire.eu/wp-content/uploads/downloads/2014/03/D1.3_FinalReport_1.0.pdf, Accessed December 1, 2014.

ConsultingWhere (2013). UK location market survey 2012. Press announcement.

Cooper, A.K., S. Coetzee, and D.G. Kourie (2011). *An Assessment of Several Taxonomies of Volunteered Geographic Information* (pp. 21–36). Hershey, PA: IGI Global.

Coote, A. and A. Smart (2010). The value of geospatial information to local public service delivery in England and Wales. Final Report, Consulting Where Ltd and ACIL Tasman, p. 128. http://www.acilallen.com.au/cms_files/ACIL_Geospatial_UK.pdf, Accessed November 1, 2013.

Craglia, M., L. Pavanello, and R.S. Smith (2010). The use of spatial data for the preparation of environmental reports in Europe. Ispra, Joint Research Centre - Institute for Environment and Sustainability, p. 45. doi:10.2788/84003.

Crompvoets, J. (2006). National spatial data clearinghouses: Worldwide development and impact. Ph.D. Dissertation, Wageningen, the Netherlands, Wageningen University, p. 136. http://edepot.wur.nl/121754, Accessed March 30, 2016.

DECA [Danish Enterprise and Construction Authority] (2010). The value of Danish address data: Social benefits from the 2002 agreement on procuring address data etc. free of charge, p. 8. http://www.adresse-info.dk/Portals/2/Benefit/Value_Assessment_Danish_Address_Data_UK_2010-07-07b.pdf, Accessed August 13, 2010.

Deloitte LLP (2012). Open growth: Stimulating demand for open data in the UK. A briefing note from Deloitte Analytics. London, Deloitte Touche Tohmatsu Limited, p. 12. https://www2.deloitte.com/content/dam/Deloitte/uk/Documents/deloitte-analytics/open-growth.pdf, Accessed March 14, 2017.

Didier, M. (1990). *Utilité et valeur de l'information géographique*. Paris, France, Presses universitaires de France.

Ecorys and Geonovum (2016). Actualisatie KBA INSPIRE. Rotterdam, the Netherlands, p. 73. http://cdr.eionet.europa.eu/nl/eu/inspire/reporting/envvzwiiw/160511_Actualisatie_KBA_INSPIRE_final_versie.pdf, Accessed March 8, 2017.

Ecorys Nederland BV and Grontmij Nederland BV (2009). Kosten-batenanalyse INSPIRE. Eindrapport. Rotterdam, the Netherlands, p. 77. http://www.geonovum.nl/sites/default/files/20091123_KBA_INSPIRE_definitief.pdf, Accessed March 8, 2017.

Ellison, R.A. and R. Callow (1996). The economic evaluation of BGS geological mapping in the UK, British Geological Survey.

Esri (2013). Return on investment: Ten GIS case studies. Redlands, CA, p. 40. http://www.esri.com/library/ebooks/return-on-investment.pdf, Accessed March 8, 2017.

Ferro, E. and M. Osella (2013). Eight business model archetypes for PSI re-use. *"Open Data on the Web" Workshop Shoreditch*, London, UK, p. 13. http://www.w3.org/2013/04/odw/odw13_submission_27.pdf, Accessed November 29, 2013.

Fornefeld, M., G. Boele-Keimer, S. Recher, and M. Fanning (2008). Assessment of the re-use of public sector information (PSI) in the geographic information, meteorological information and legal information sectors—Final report. Study commissioned by EC in 2007. Dusseldorf, Germany: MICUS, p. 101. http://ec.europa.eu/newsroom/document.cfm?doc_id=1258., Accessed January 28, 2009.

Freebairn, J.W. and J.W. Zillman (2002). Economic benefits of meteorological services. *Meteorological Applications* 9(1): 33–44.

Garcia Almirall, P., M. Moix Bergadà, P. Queraltó Ros, and M. Craglia (2008). The socio-economic impact of the spatial data infrastructure of Catalonia. M. Craglia (Ed.). Ispra, JRC Joint Research Centre, p. 62. http://www.ec-gis.org/inspire/reports/Study_reports/catalonia_impact_study_report.pdf, Accessed April 4, 2008.

Garica-Cortés, A., J. Vivancos, and J. Fernández-Gianotti (2005). Evaluación económica y social del Plan MAGNA. *Boletín Geológico y Minero* 116(4): 291–305.

Genovese, E., G. Cotteret, S. Roche, C. Caron, and R. Freick (2009). Evaluating the socio-economic impact of geographic information: A classification of the literature. *International Journal of Spatial Data Infrastructures Research* 4: 218–238. doi:10.2902/1725-0463.2009.04.art12.

Genovese, E., G. Cotteret, S. Roche, C. Caron, and R. Feick (2010). The EcoGeo Cookbook for the Assessment of Geographic Information Value. *International Journal of Spatial Data Infrastructures Research* 5: 120–144 doi:10.2902/1725-0463.2010.05.art5.

Georgiadou, Y. and J. Stoter (2008). SDI for public governance—Implications for evaluation research. In *A Multi-View Framework to Assess SDIs*, J. Crompvoets, A. Rajabifard, B. Van Loenen, and T.D. Fernández (Eds.). Melbourne, Australia, Melbourne University Press, pp. 51–68.

Giff, G. and J. Crompvoets (2008). Performance indicators a tool to support spatial data infrastructures assessment. *Computers, Environment and Urban Systems* 32(5): 365–376.

GITA [Geospatial Information & Technology Association] (2007). Building a business case for shared geospatial data and services: A practitioner's guide to financial and strategic analysis for a multi-participant program, p. 69. https://www.fgdc.gov/initiatives/50states/roiworkbook.pdf, Accessed March 8, 2017.

Grus, L., A. Bregt, and D. Eertink (2015). De effecten van open data BRT na 3 jaar, Wageningen University and the Kadaster, p. 7. https://www.kadaster.nl/web/artikel/download/De-effecten-van-open-data-BRT-na-3-jaar-1.htm, Accessed January 4, 2016.

Häggquist, E. and P. Söderholm (2015). The economic value of geological informa-tion: Synthesis and directions for future research. *Resources Policy* 43: 91–100. doi:10.1016/j.resourpol.2014.11.001.

Hamel, G. (2000). *Leading the Revolution*. New York: Lille Economie and Management.

Hjelmager, J., H. Moellering, and A. Cooper (2008). An initial formal model for spa-tial data infrastructures. *International Journal of Geographical Information Science* 22: 1295–1309.

Janssen, M. and A. Zuiderwijk (2014). Infomediary business models for connecting open data providers and users. *Social Science Computer Review* 32(5): 563–576. doi:10.1177/0894439314525902.

Krek, A. (2002). An agent-based model for quantifying the economic value of geo-graphic information, GeoInfo Series Vienna 26.

Lopez Romero, E. (2014). CNIG Spain & ePSI: Perspective and Experience. *INSPIRE Conference 2014*. Aalborg, Denmark.

Lazo, J.K., R.E. Morss, and J.L. Demuth (2009). 300 billion served. Sources, perceptions, uses, and values of weather forecasts. *Bulletin of the American Meteorological Society* 90(6): 785–798. doi:10.1175/2008BAMS2604.1.

Longhorn, R.A. and M. Blakemore (2008). *Geographic Information: Value, Pricing, Production, and Consumption*. Boca Raton, FL: CRC Press, Taylor & Francis Group.

Longley, P.A., M.F. Goodchild, D.J. Maguire, and D.W. Rhind (2001). *Geographic Information Systems and Science*. Chichester, UK: John Wiley & Sons.

Magalhaes, G., C. Roseira, and L. Manley (2014). Business models for open govern-ment data. *Proceedings of the 8th International Conference on Theory and Practice of Electronic Governance*. Guimaraes, Portugal: ACM, pp. 365–370. http://dl.acm.org/citation.cfm?id=2691273.

Market Info Group (2013). Location based services—Market and technology outlook – 2013–2020. http://marketinfogroup.com/location-based-services-market/, Accessed 20 January, 2017.

Natural Resources Canada (2016). Value study findings report. Canadian geospatial data infrastructure, Information Product 48e, 148 pages. doi:10.4095/297711.

Oelschlager, F. (2004). Enterprise information integration. Enabling the information value chain to achieve business optimization. http://www.teamplay.primavera.com/partners/files/Enterprise_Information.pdf.

Onsrud, H.J. (1992). In support of cost recovery for publicly held geographic infor-mation. *GIS Law* 1(2): 1–7. http://www.spatial.maine.edu/~onsrud/pubs/Cost_recovery_for_GIS.html, Accessed August 1, 2008.

Onsrud, H.J. and G. Rushton (1995). *Sharing Geographic Information*. New Brunswick, NJ: Centre for Urban Policy Research.

Open Data Institute (2015). Open data means business: UK innovation across sec-tors and regions. http://theodi.org/open-data-means-business-uk-innovation-sectors-regions, Accessed 18 January, 2017.

Open Data for Development Network (2017). Open Data Impact Map. http://www.opendataimpactmap.org/, Accessed 15 January, 2017.

Osterwalder, A. (2004). The business model ontology: A proposition in a design science approach. Universite de Lausanne Ecole.

Osterwalder, A. and Y. Pigneur (2010). *Business Model Generation*. Hoboken, NJ: John Wiley & Sons.

OXERA [Oxford Economic Research Associates Ltd] (1999). The economic contribution of ordnance survey, p. 54. http://www.ordnancesurvey.co.uk/oswebsite/aboutus/reports/oxera/oxera.pdf, Accessed August 1, 2008.

Parycek, P., J. Höchtl, and M. Ginner (2014). Open government data implementation evaluation. *Journal of Theoretical and Applied Electronic Commerce Research* 9(2): 80–99.

Pettifer, R.E.W. (2009). PSI in European Meteorology—An unfulfilled potential. In *7th Eastern European eGov Days: eGovernment & eBusiness Ecosystem & eJustice Conference.* Prague, Czech Republic, European Projects and Management (EPMA) (CZ). http://www.primet.org/ckfinder/userfiles/files/EU%20PSI%20Working%20Groups/PSI%20in%20European%20Meteorology%20-%20an%20unfulfilled%20potential%20distribution%20copy.pdf, Accessed May 6, 2009.

Philips, R.L. (2001). The management information value chain. *Perspectives* 3.

Pira International Ltd, University of East Anglia and Knowledge View Ltd (2000). Commercial exploitation of Europe's public sector information—Final report. Pira International Ltd, European Commission Directorate General for the Information Society, p. 132. ftp://ftp.cordis.lu/pub/econtent/docs/commercial_final_report.pdf, Accessed November 20, 2006.

Porter, M.E. (1985). *Competitive Advantage, Creating, and Sustaining Superior Performance.* New York: The Free Press.

Rodriguez Pabon, O. (2005). Cadre théorique pour l'évaluation des infrastructures d'information géospatiale, Ph. D. Thesis, Département des Sciences Géomatiques, Faculté de Foresterie et de Géomatique, Laval University, Québec.

Rodriguez, A.F. (2016). A New Data & Services Policy for the IGN Spain. *INSPIRE Conference 2016.* Madrid, Spain.

Rydén, A. (2013). Assessing social benefits in Sweden. *INSPIRE Conference 2013.* Florence, Italy. PPT presentation.

Shafer, S.M., H.J. Smith, and J.C. Linder (2005). The power of business models. *Business Horizons* 48(3): 199–207. doi:10.1016/j.bushor.2004.10.014.

The OD500 Global Network (2017). Open Data 500. http://www.opendata500.com, Accessed 18 January, 2017.

Trapp, N., U.A. Schneider, I. McCallum, S. Fritz, C. Schill, M.T. Borzacchiello, C. Heumesser, and M. Craglia (2015). A meta-analysis on the return on investment of geospatial data and systems: A multi-country perspective. *Transactions in GIS* 19: 169–187.

van Loenen, B. and F. Welle Donker (2014). Open data beoordelingsraamwerk. Deel: review kosten-batenanalyses. Delft, the Netherlands, Kenniscentrum Open Data, p. 32. http://www.bk.tudelft.nl/fileadmin/Faculteit/BK/Over_de_faculteit/Afdelingen/OTB/Kenniscentrum_Open_Data/Open_data_beoordelingsraamwerk__Review_kosten-batenanalyses.pdf, Accessed December 1, 2014.

van Loenen, B. and J. Zevenbergen (2010). Assessing geographic information enhancement. *International Journal of Spatial Data Infrastructures Research* 5: 244–266. doi:10.2902/1725-0463.2010.05.art10.

Vancauwenberghe, G., P. Cipriano, M. Craglia, C. Easton, G. Martirano, and D. Vandenbroucke (2014). Exploring the market potential for geo-ICT companies in relation to INSPIRE. In *Connecting a Digital Europe through Location and Place. Proceedings of the AGILE 2014 International Conference on Geographic Information Science, AGILE Conference on Geographic Information Science,* J. Huerta, S. Schade, and C. Granell (Eds.), Castellón, Spain, June 3–6.

Vickery, G. (2011). Review of recent studies on PSI re-use and related market developments. Paris, France, Information Economics, p. 44. http://ec.europa.eu/digital-agenda/en/news/review-recent-studies-psi-reuse-and-related-market-developments, Accessed November 1, 2011.

Warnest, M., A. Rajabifard, and I. Williamson (2003). Understanding inter-organizational collaboration and partnerships in the development of national SDI. *URISA 2003*, October 11–15, Atlanta, GA. https://pdfs.semanticscholar.org/44f9/d8e26e91a044770e26f6cf848d6b43723eb8.pdf.

Welle Donker, F. (2009). Public sector geo web services: Which business model will pay for a free lunch? In *SDI Convergence: Research, Emerging Trends, and Critical Assessment*, B. van Loenen, J.W.J. Besemer, and J.A. Zevenbergen (Eds.). Delft, the Netherlands, Nederlandse Commissie voor Geodesie/Netherlands Geodetic Commission (NCG), 48, pp. 35–51.

Welle Donker, F. and B. van Loenen (2016). Sustainable business models for public sector open data providers. *JeDEM Journal of eDemocracy & Open Government* 8(1): 28–61.

Welle Donker, F., B. van Loenen, and A.K. Bregt (2016). Open data and beyond. *ISPRS International Journal of Geo-Information* 5(4). doi:10.3390/ijgi5040048.

Yu, C.-C. (2016). A value-centric business model framework for managing open data applications. *Journal of Organizational Computing & Electronic Commerce* 26(1/2): 80–115.

Zeleti, F.A., A. Ojo, and E. Curry (2016). Exploring the economic value of open government data. *Government Information Quarterly* 33(3): 535–551. doi:10.1016/j.giq.2016.01.008.

Zott, C., R. Amit, and L. Massa (2011). The business model: Recent developments and future research. *Journal of Management* 37(4): 1019–1042. doi:10.1177/0149206311406265.

8

The Role of Aggregators

Dawn J. Wright

CONTENTS

8.1 Introduction

Data (aka content) in the form of numbers or layers, and transformed into information by the way of maps, images, graphs, charts, tables, and even stories, are the very foundation of the work of geographic information system (GIS) users in academia, government organizations, nongovernmental organizations, and commercial enterprises. All these professionals need access to useful and reliable content to help them achieve a wide variety of goals, to purvey information that impacts science and policy, and oftentimes to make decisions that affect daily lives.

Recent advances in information technology (IT), as well as civil remote sensing of the Earth, have set the stage for the dawning of a new era in geospatial big data. Vast amounts of data collected by numerous sensors and sources worldwide can now be combined dynamically, becoming discoverable, accessible, and usable by global communities of interest. This trend is on a trajectory that is rapidly allowing us to advance beyond static data collection and archiving, further enabling information awareness and understanding, and leading us toward knowledge and better decision-making (Wright 2015). However, such volumes, velocities, and varieties of data streams also bring with them serious dilemmas regarding effective organization, cataloging, and easy access. This is where the role of aggregator comes in.

Citizen crowdsourcing initiatives such as Wikimapia and OpenStreetMap create free sources of map content through volunteer efforts (aka volunteered geographic information or VGI), sometimes providing the only source in areas where access to GI may be regarded as an issue of national security

(Goodchild 2007). IT giants such as Google, Facebook, IBM (with subsidiaries such as Weather Underground), and Apple are well known for aggregating just about every aspect of life in modern society, from our mapped locations to our music and even our mood swings. Aggregators need not be commercial by definition or requirement, but it is often the private sector that is best at providing the necessary sustainability and reliability of information through proven, well-engineered platforms, all with the necessary interoperability and openness as guaranteed through the adoption of established standards. This section focuses on the private sector example of the Environmental Systems Research Institute (ESRI) as a broad-scale aggregator.

Indeed, many software companies such as Esri have come to realize the inadequacy of providing just the software, recognizing the power of coupling that software with content. For example, in the geospatial world, Esri's ArcGIS has evolved from a suite of software products in the 1980s and 1990s enabling single desktop users to build their own project-centric applications or full-scale GISs, to an extension of its products in the 2000s to support enterprise-scale GIS implementation throughout entire organizations or governments, to its current state as a *platform*. The platform goes beyond a series of standalone desktop software products to leveraging the web and the accompanying Software-as-a-Service (SaaS) cloud patterns and standard business models (i.e., subscriptions to software and services in the cloud based on the number of users, as well as packages of service credits purchased in exchange for data storage and access to more advanced analytics). This also includes a place for sharing user content, as well as a large library of ready-to-use base maps, other forms of authoritative content, and geospatial services as *part* of the platform. In the platform, open data and shared services are paramount, along with the provision of many easy-to-use tools and applications, including self-service mapping and ready-to-use templates that support individuals, organizations, governments, and developers.

The aim of this transformational shift is simply and completely to make GIS easier, not only for new users but also for advanced users. Ease of use is indeed a primary goal as several usability studies (Aditya 2010, Davies and Medyckyj-Scott 1994, 1996) have shown that the single factor of ease, stemming from effective design of software functions and user interfaces, leads to greater adoption of the technology. This in turn has large implications not only for increased revenue from a commercial standpoint, but for transformations in the use and analysis of geospatial data throughout both the private and public sector, improvements in the management of information resources and the conduct of research practices, and, in some cases, to further scientific discovery (Edsall 2003). In addition, the aim is to make the platform more open, extendable, and leverageable by others, facilitating both the integration of other geospatial technologies as well as other enterprise IT and web services. Aggregators have, over the years, compiled, assembled, and produced a carefully curated library of public content, all organized into different themes such as base maps, imagery, and detailed demographics for the United States and other parts of the world. Many of these collections of content are growing and changing daily as new content is added, and the existing content is updated, sometimes as often as every few minutes.

Who are the stakeholders, end users, and user communities of aggregated geospatial content writ large? *Stakeholders* include data producers, brokers, and value-added resellers; GIS, mapping, hardware, cloud and infrastructure providers; policy-makers; and, of course, end users. Among the most common *end users* are indigenous, local, state, territorial, provincial, and national governments; climate, conservation, marine, hydrologic, ecologic, and geological scientists; commercial interests that cover shipping, fishing, oil and gas exploration, and development; land, maritime, and survey specialists; geographers; search-and-rescue officials; diplomatic and national security officials; archeologists; and national statistical/census bodies, GI specialists, social scientists, economists, communication experts, and decision-makers.

Four of the most common *user communities* include *professionals* who use the platform for mapping, scientific geographic analysis, data management, and sharing, not only of content but GIS workflows, as well as non-GIS expert systems. *Developers* may use one or more of the open application programming interfaces (APIs) supported by the platform to leverage its content and services to make their own applications and services that embed or interact with the core platform capabilities. They may also extend the system into new areas and provide new content for aggregation through the platform capabilities, as well as focused workflows and tools for both GIS and non-GIS users. *Businesses and IT organizations* may use the platform to integrate mapping and geospatial analysis into business systems (e.g., business intelligence or BI, enterprise resource planning [ERP], and customer relationship management [CRM]), as well as several business system technologies (e.g., Microsoft Office, SharePoint, and Dynamics CRM, as well as BI solutions from IBM Cognos, MicroStrategy, and SAP). In doing so, they may create even more content for aggregation into the platform, content that integrates traditional business data with other types of GIS information traditionally housed within GIS organizations. The public can use the aggregated content to understand important issues such as public safety, disaster response, and climate change and how these issues relate to their personal situation. The resulting information products may be entirely integrated into government or other websites but are driven by the content and services provided through the platform.

It is indeed the governments of the world that are charged with ensuring the continuity of key Earth observations (EO), from outer space, the atmosphere, on the ocean surface, and within the ocean water column, and deciding which data streams are the most important (e.g., weather forecasting, land surface change analysis, sea level monitoring, climate-change research; e.g., Group on Earth Observations 2005; The Royal Society 2012). Governments should also use established methodologies and/or metrics that can be used for observations collected over varying extended periods, prioritizing the relative importance of observations over these time periods, and identifying the characteristics of an extent to which data gaps and/or performance degradation are acceptable for categories of observations (National Research Council 2013, 2015). Aggregators, with their content-serving platforms that facilitate and promote wide and shared use of information resources, play an important role in helping governments to properly assess the infrastructure necessary for achieving continuity or near continuity of the data products derived from the various sensors and platforms and for plugging critical data gaps. Given the large user communities that aggregators serve, they also have an excellent idea of what users want and need (by way of e-mail surveys, interactions with users at workshops and conferences, and general feedback from users on a range of online discussion forums), which can be of value to governments in identifying which datasets and resulting information may be of highest value. For example, Esri, in working with the National Oceanic and Atmospheric Administration (NOAA), recognizes that the agency has a highly diverse set of data centers, but as an aggregator of NOAA data it has also helped the agency to identify ways in which the access to these data can be unified and prioritized.

8.2 Aggregation Approaches

The advantage of large-scale aggregation is the colocating of both the technology and the data to ensure rapid access to up-to-date data. In the alternative SaaS model mentioned earlier, data are stored in a cloud infrastructure, and a geospatial aggregator may partner with several well-known cloud hosting companies such as Amazon, Microsoft, IBM, or Cloudera, in addition to providing their own hosted services. In either scenario, the data may be in traditional relational databases (typical for vector data; Batty 1992; Fleming and Vonltalle 1989; Gebhardt et al. 2010), various gridded file formats (typical for satellite imagery and other raster datasets; Armstrong et al. 2005; Becker et al. 2015; Holroyd and Bell 1992), point clouds (typical for light detection and ranging or Light Detection and Ranging (LiDAR) data; Martinez-Rubi et al. 2016; May 2012; Rychkov et al. 2012), and other file-based formats (Heinzer et al. 2012). Data are managed using content management tools specific to the aggregator and built into the aggregator's platform. These are therefore the same tools used regardless of the storage location or storage type.

In order to make data most readily accessible, a geospatial aggregator employs a web services approach using common service interface specifications that build on international standards from the World Wide Web Consortium (W3C), the Open Geospatial Consortium (OGC), and others. These web services provide a wide range of capabilities as listed in Table 8.1.

TABLE 8.1

Common Geospatial Web Service Features and Capabilities Used by Aggregators of Content

Feature	Function
Feature access	Provides access to vector features in a map
Geocoding	Provides access to an address locator
Geodata	Provides access to the contents of a geodatabase for data query, extraction, and replication
Geoprocessing	Provides access to analytic tools and workflows allowing scientists to derive new information from the data and publish modeling methodologies
Globe	Provides access to the contents of a globe document
Imaging	Provides access to the contents of a raster dataset or mosaic dataset, including pixel values, properties, metadata, and bands
JPIP	Provides Joint Photographic Experts Group (JPEG) Joint Photographic interactive protocol (JPIP) streaming capability when using JPEG 2000 or National Imagery Transmission Format (NITF; with JPEG 2000 compression) files and configured with a JPIP Server from Esri partner Exelis VIS
KML	Uses a map document to create Keyhole Markup Language (KML) features compatible with Google and other visualization capabilities
Mapping	Provides access to the contents of a map, such as the layers and their underlying attributes
Mobile data access	Allows extraction of data from a map to a mobile device
Network analysis	Solves transportation network analysis problems using the ArcGIS Network Analyst extension
Schematics	Allows viewing, generating, updating, and editing schematic diagrams
WCS	Creates a service compliant with the OGC Web Coverage Service (WCS) specification
WFS	Creates a service compliant with the OGC Web Feature Service (WFS) specification
WMS	Creates a service compliant with the OGC Web Map Service specification
WMTS	Creates a service compliant with the OGC Web Map Tile Service (WMTS) specification
WPS	Creates a service compliant with the OGC Web Processing Service (WPS) specification

Any aggregator should ideally consider the full data value chain that includes connecting to EO, including *in situ* sensor networks, providing mechanisms for storing and hosting content (especially when hosting is not possible at the data source), making content discoverable, and enabling use of content in different media, for both online and offline use. To further increase the visibility and use of content and information products, they may be disseminated to other global or national networks also serving as aggregators of sorts, such as the Group on Earth Observation, the United States Geospatial Platform, and others.

8.3 Ultimate Aggregations? A Living Atlas of the World Example

Several years ago, the Esri Imagery Team devised a dynamic, near real-time, highly visual, cloud-based concept for EO data known as the Living Planet. Unlike traditional globes with a single image carpet layer or *skin* that is only 1 pixel deep, the Living Planet was surmised as a surface that is *alive* and deep, with dozens of EO systems contributing image streams from space in near real-time, along with community-contributed authoritative layers for oceans (bathymetry, sea surface, and water column), land terrain elevation, geology, social media, and so on. The concept included aggregating a large variety of civilian sources from the U.S. Government and commercial providers, along with community-contributed authoritative layers for land terrain, bathymetry, ocean surface, and the ocean water column. These sources were to include open and freely available data and apps along with a *marketplace* for both commercial value-added data and apps at cost, as well as free and open data and apps from nongovernmental organizations, commercial providers, and citizen geographers. The idea was for imagery and sensor data to be streamed in from all sources—air, space, ground, and sea sensors. These data are to be stored only once in the cloud. Processing and analysis algorithms are sent from desktops to the cloud and run there where the data are stored, rather than moving the data to where the processing algorithms are stored. This *brings to life* the data in near-real time, providing citizens with immediate access, and simplified understanding of environmental and climatic conditions that affect their daily lives. As they *live* in the cloud, datasets are accessible from any device, including through smartphones, tablets, web browsers, and social media, enabling community awareness and resiliency to weather and climate disasters, making the data truly *personal*.

In 2014, the Living Atlas of the World was added to the ArcGIS platform as the realization of the initial *Living Planet* concept (Wright et al. 2015). Although other aggregators focus on static content (mainly imagery), the Living Atlas of the world is fairly unique in its intent to function not only as an organized catalog of a much wider variety of content (vector, raster, point cloud, static and near-real time), along with interactive maps, map services, data analysis tools, and apps. Authoritative GI is accessible in the Living Atlas through hosted cloud services, so that users can more quickly address scientific and societal problems and decisions at spatial scales ranging from a small study area to the entire globe, while using a range of interactive map functions to tell engaging narratives along the way (aka *story maps*). The types of narratives possible within story maps are nearly infinite, but most of the data focus on guiding the viewer through stages of an analysis, describing the methodology and showing the results at varying scales along the way; or linking a set of photos or videos to transects, tours, or any sequence of places for the viewer to follow in order; or showcasing the results of spatial inventories or observations at specific sampling or survey sites in the field (Silbernagel et al. 2015; Wright 2014, 2015; Zolnai 2014). What began as a way to build trusted, authoritative, and freely available *base maps* (where a base map serves to form the background on which other data are overlain or to orient the map's location), has grown to a larger program extending far beyond this to aggregating and provisioning layers of satellite imagery, sea surface and water column data, stream hydrology, transportation networks, demographics, natural hazards, bioclimates, rock types, landforms, land cover, elevation, 3D web scenes, and much more (Environmental Systems Research Institute 2017).

Recent activities in building the Living Atlas extend beyond just the reading and serving of datasets, to the provisioning of spatial analysis on these data services in the cloud. For example, in addition to an imagery base map, the Living Atlas now offers multispectral and multitemporal image services with data sources such as Landsat 8 from the U.S. Geological Survey and National Agriculture Imagery Program (NAIP) from the U.S. Department of Agriculture. These image services enable the user to perform various types of analysis and view changes over time, as well as the cross-walking and sharing of workflows and use cases, additional apps for mobile, web, and desktop, community-building events where people gather face-to-face. As such, there are growing interlinkages with other platforms such as the Ocean Data Interoperability Platform (Paolo et al. 2017) and the National Science Foundation's EarthCube initiative and its evolving architecture (Cutcher-Gershenfeld 2016; Davis et al. 2016).

In terms of contributors, these maps and layers are built using the best available and authoritative data from thousands of organizations, including users, business partners, and aggregators, prime examples being NOAA, which has recently contributed precipitation estimates and real-time weather observations, plus commercial partners Nearmap and Hexagon, with recent submissions of high-resolution satellite imagery at resolutions of 7–30 cm. Much of this content is contributed through a Community Maps Program (Environmental Systems Research Institute 2015) in which universities, nongovernmental organizations, and even individuals submit free *contributor data*, companies submit free or premium *commercial data*, and local, state, and national government agencies contribute *open data* as mandated by government policies. All data and metadata must be OGC- or ISO-compliant and are submitted under one or more of the following categories: base map layers, imagery, land elevation and bathymetry, hydrology and stream gauges, and various *urban observatory* themes that represent transportation, population density, public utilities, demographics and lifestyle, and land use

(Environmental Systems Research Institute 2016; Matthews 2015). As shown in Figure 8.1, data and maps are not automatically or immediately accepted into the Living Atlas of the World. Minimum requirements for consideration and inclusion include the following:

- Well-maintained data and documented with the OGC or International Organization for Standardization (ISO)-compliant metadata.
- Data services that are reliable and well performing (i.e., services run on servers experiencing minimum disruption or downtime, with holdings exposed through a OGC-compliant catalog services such as Catalog Services for the Web (CSW) and as an OpenSearch endpoint, accessible through REST API, and providing GeoRSS, KML, HTML, or JSON responses; and with registered resources monitored and synchronized according to any changes in the catalog service.
- Web maps with a well-defined legend, on a well-focused topic, and with well-configured pop-ups.
- Completion of all required elements of a *home page* for the contributed item in ArcGIS Online containing an attractive thumbnail, informative item details, descriptive user profile, data or map access use constraints, credits or attribution, and search tags.

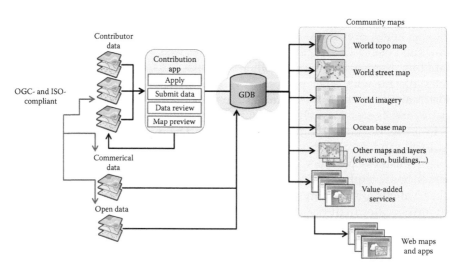

FIGURE 8.1
Architectural diagram showing the three main categories of data that are contributed to the Living Atlas, the vetting and storing of the data through cloud-hosted services, and the provisioning of the data through various divisions of the Community Maps Program, along with other maps and layers, value-added services that perform calculations of parameters in some datasets, create models, symbolize map layers, and so on, as well as web maps and web apps (such as story maps) that are created from data. OGC = Open Geospatial Consortium; ISO = International Organization for Standardization.

8.4 Conclusion

With the continuing evolution of high-volume, high-velocity data streams of varying types and formats, aggregators will continue to be important aids along the continuum from information to information-based decision support systems, and to actual decision-making. As described in this section by way of one main example of a geospatial aggregator, the provision of the necessary sustainability and reliability of information through proven, well-engineered platforms, all with the necessary interoperability and openness as guaranteed through the adoption of established standards provides an important benefit to the EO community.

Aggregators are also dynamic ecosystems, continually under construction, growing, and changing on a daily basis, as new data and maps are created and in need of curation. Over the past six years, Esri, for example, has been aggregating content, and base map requests, in particular, have doubled each year for the past few years (Matthews 2015). In addition, out of the one million items comprising more than ~200 Tb of data in the ArcGIS Online cloud-based platform (where ~160 million map requests are made by ~1.6 million users per day, 4–5 billion map tile requests are made per month, and the platform has been subject to 94 billion map views since its inception; Wright et al. 2015), the Living Atlas represents the most authoritative and highest quality of all of this content. This higher-quality data consists of imagery and base maps, as well demographic, landscape, transportation, terrain, and hazards data along with associated maps created *from* the data and suites of tools used to perform analyses *on* the data.

One significant trend that appears to be driving the need for and the use of geospatial aggregators is the quality of local content being contributed by the GIS community. As such, the services of aggregators will continue to evolve to welcome not only contributors who will publish content to be included in the aggregation platform, but curators who will assist the aggregator in reviewing, organizing, and even approving that content, thereby helping to increase and ensure its quality over the time, again, then benefiting the entire EO community.

Acknowledgment

The comments and suggestions of Françoise Pearlman and Joep Crompvoets and Jay Pearlman greatly improved this chapter. DJW is grateful to Jeanne Foust, Marten Hogeweg, Sean Breyer, Lawrie Jordan, and Gordon Plunkett for helpful discussions and ideas.

References

Aditya, T. 2010. Usability issues in applying participatory mapping for neighborhood infrastructure planning. *Transactions in GIS* 14(Supplement 1): 119–147.

Armstrong, M. P., M. K. Cowles, and S. Wang. 2005. Using a computational grid for geographic information analysis: A reconnaissance. *Professional Geographer* 57(3): 365–375.

Batty, P. 1992. Exploiting relational database technology in a GIS. *Computers & Geosciences* 18(4): 453–462.

Becker, P., L. Plesea, and T. Maurer. 2015. Cloud optimized image format and compressions. *Proceedings of the 36th International Symposium on Remote Sensing of Environment, The International Archives of the Photogrammetry, Remote Sensing and Spatial Information Sciences*, Berlin, Germany.

Cutcher-Gershenfeld, J., K. S. Baker, N. Berente, D. R. Carter, L. A. DeChurch, C. C. Flint, G. Gershenfeld et al. 2016. Built it, but will they come? A geoscience cyberinfrastructure baseline analysis. *Data Science Journal* 15(8): 1–14.

Davies, C., and D. Medyckyj-Scott. 1994. GIS usability: Recommendations based on the user's view. *International Journal of Geographical Information Systems* 8(2): 175–189.

Davies, C., and D. Medyckyj-Scott. 1996. GIS users observed. *International Journal of Geographical Information Systems* 10(4): 363–384.

Davis, R. B., J. Young, and D. Wright. 2016. Mapping the EarthCube landscape. *Paper presented at the EarthCube 2016 All Hands Meeting*, Denver, CO.

Edsall, R. M. 2003. Design and usability of an enhanced geographic information system for exploration of multivariate health statistics. *Professional Geographer* 55(2): 146–160.

Environmental Systems Research Institute. 2015. Living Atlas of the World: Community maps. Documentation page for Living Atlas of the World, Environmental Systems Research Institute. http://doc.arcgis.com/en/living-atlas/contribute/benefits.htm (accessed March 10, 2017).

Environmental Systems Research Institute. 2016. Best practices for sharing. Documentation page for ArcGIS Online, Environmental Systems Research Institute. http://esriurl.com/sharing (accessed March 10, 2017).

Environmental Systems Research Institute. 2017. Living Atlas of the World. Web site, Environmental Systems Research Institute. https://livingatlas.arcgis.com (accessed March 10, 2017).

Fleming, C. C., and B. Vonltalle. 1989. *Handbook of Relational Database Design.* New York: Addison-Wesley.

Gebhardt, S., T. Wehrmann, V. Klinger, I. Schettler, J. Huth, C. Kunzer, and S. Dech. 2010. Improving data management and dissemination in web based information systems by semantic enrichment of descriptive data aspects. *Computers & Geosciences.* doi:10.1016/j.cageo.2010.1003.1010.

Goodchild, M. F. 2007. Citizens as sensors: The world of volunteered geography. *GeoJournal* 69: 211–221.

Group on Earth Observations. 2005. Global earth observation system of systems (GEOSS) 10 year implementation. Plan Reference Document, Ed. Group on Earth Observations, 210 pp. the Netherlands: European Space Agency Publication GEO 1000R/ESA SP 1284.

Heinzer, T. J., M. D. Williams, E. C. Dogrul, T. N. Kadir, C. F. Brush, and F. I. Chung. 2012. Implementation of a feature-constraint mesh generation algorithm within a GIS. *Computers & Geosciences* 49: 46–52.

Holroyd, F., and S. M. Bell. 1992. Raster GIS: Models of raster encoding. *Computers & Geosciences* 18(4): 419–426.

Matthews, S. 2015. User community content is helping build the Living Atlas of the World. Blog post, Environmental Systems Research Institute. https://shar. es/1ZaeMS (accessed March 10, 2017).

May, D. A. 2012. Volume reconstruction of point cloud data sets derived from computational geodynamic simulations. *Geochemistry, Geophysics, Geosystems* 13(5): Q05019. doi:05010.01029/02012GC004170.

Martinez-Rubi, O., M. de Kleijn, S. Verhoeven, N. Drost, J. Attema, M. van Meersbergen, R. van Nieuwpoort, R. de Hond, E. Dias, and P. Svetachov. 2016. Using modular 3D digital earth applications based on point clouds for the study of complex sites. *International Journal of Digital Earth* 9(12): 1135–1152.

National Research Council. 2013. *Lessons Learned in Decadal Planning in Space Science: Summary of a Workshop.* Washington, DC: The National Academies Press.

National Research Council. 2015. *Sea Change: 2015–2025 Decadal Survey of Ocean Sciences.* Washington, DC: The National Academies Press.

Paolo, D., A. Leadbetter, and H. Glaves, Eds. 2017. *Oceanographic and Marine Cross-domain Data Management for Sustainable Development.* Hershey, PA: IGI Global.

The Royal Society. 2012. *Science as an Open Enterprise: Open Data for Open Science.* London, UK: The Royal Society Scientific Policy Centre.

Rychkov, I., J. Brasington, and D. Vericat. 2012. Computational and methodological aspects of terrestrial surface analysis based on point clouds. *Computers & Geosciences* 42: 64–70.

Silbernagel, J., G. Host, C. Hagley, D. Hart, R. Axler, R. Fortner, M. Axler et al. 2015. Linking place-based science to people through spatial narratives of coastal stewardship. *Journal of Coastal Conservation* 19(2): 181–198.

Wright, D. J. 2014. Speaking the "language" of spatial analysis via story maps. Blog post, Environmental Systems Research Institute. http://esriurl.com/analytical-stories (accessed March 10, 2017).

Wright, D. J. 2015. Toward a digital resilience. *Elementa Science of the Anthropocene* 4(000082). doi: 10.12952/journal.elementa.000082.

Wright, D. J., Esri Ocean Team, and Esri Living Atlas Team. 2015. Should data frameworks be inherently multiscalar? A use case of the Living Atlas of the World. *Eos, Transactions of the American Geophysical Union* 96: IN13D-03.

Zolnai, A. 2014. Map stories can provide dynamic visualizations of the Anthropocene to broaden factually based public understanding. *The Anthropocene Review* 1(3): 243–251.

9

Enabling the Reuse of Geospatial Information

Robert R. Downs

CONTENTS

9.1 Introduction

Geospatial data and related information resources often record measurements of observations that cannot be replicated and serve as cultural artifacts that have captured representations of a particular time and location. They are also valuable assets that require investments in skills and infrastructure. In addition to accomplishing the objectives for which the geospatial information was collected originally, they can also be reused for new purposes, such as scientific research, education, planning, and policy-making. By reusing geospatial information, the original investments can be leveraged further and their value can be increased as new and future communities of users obtain the benefits of reusing these information resources. Conducting stewardship activities throughout the data lifecycle can enable the reuse of geospatial information. An assessment of the needs of the designated community or communities that will reuse the geospatial information can identify the requirements for products and services to be developed and disseminated and can determine the criteria for the appraisal of any geospatial data and the related information that will be acquired to develop such products and services for reuse. Selection of candidate geospatial information can identify resources to be prepared to meet the needs of the communities that will reuse the geospatial products and services to be shared. Preparing geospatial data and related information resources for reuse also requires investments, which can be justified by appraising and selecting candidate information resources that have been identified as having potential for reuse. Geospatial information resources that have been determined to be valuable for reuse are curated to prepare data products and services for reuse today and in the future. Prior to dissemination, curated geospatial data and related information products and services are tested and evaluated to ensure that they can be reused for the purposes for which they have been curated and for reuse within diverse contexts. The value from curated geospatial information is attained when they are publicly disseminated as geospatial data products and services that are reused for new purposes. A brief overview of recommended practices for enabling the reuse of geospatial information can be informative for those involved in the collection, curation, dissemination, or reuse of geospatial data and related information on products and services.

Geospatial information resources have been useful to society for millennia, providing value for exploration, commerce, and military purposes (Brown 1979). During recent decades, the use of geospatial information resources, including geospatial data, maps, systems, products, and services, has become prominent across multiple industries and increasingly ubiquitous to include the availability of various applications on smartphones for sundry purposes. Geospatial data and related information can be useful for the investigators who are conducting the study for which the geospatial information resources were originally collected. But, geospatial information can also be useful for conducting subsequent studies that were not necessarily envisioned by the designers of the original studies that collected the data. Earth science data and geospatial information that have been previously collected and curated for subsequent reuse can serve as primary research resources for studies into phenomenon related to the original data collection study and for new studies, including studies conducted by researchers from diverse disciplines unrelated to the fields of study represented by the team that completed the original data collection study. Reuse of geospatial information can offer many opportunities for conducting a variety of scientific studies and provide value for new studies, today, and in the future. The following elements in Chapter 9 provide an overview of recommended practices for enabling the reuse of geospatial information.

9.2 Needs for and Value of Enabling Reuse of Geospatial Information

Geospatial information can be costly to collect. Often, teams of scientists invest time and incur expenses to collaborate and design data collection studies. In some cases, instruments and platforms must be developed and administered to collect geospatial information of interest. In cases where the collected geospatial observations are critical for completing an important study, such costs are justifiable. Furthermore, expected benefits exceed the costs of geospatial information by a ratio of more than three to one (Trapp et al. 2015). However, considering the cost of collecting geospatial information, enabling their reuse can increase the value of the geospatial information products beyond the value that they offer for the original study. In addition, some geospatial information, when collected to measure observations of an event may not be reproducible. In such cases, the recorded geospatial observations could be invaluable. The value of irreproducible geospatial data and related information can also be further enhanced by reusing the data after the initial data collection study has been completed. In addition, geospatial information that represents observations of unique occurrences or changing phenomenon during specific time periods also can be indispensable. Nevertheless, the value of longitudinal geospatial information can be further realized if they are reused, especially when they are combined with additional observations to extend the length of time observed or to study relationships between multiple variables.

Recognizing the opportunity to attain additional value from research data, government agencies and other funding sources have begun requiring grant recipients to ensure that the data and related information that have been collected as part of funded projects are continually managed and publicly disseminated for reuse (Holden 2013). Some publishers are also requiring authors to deposit their data in publicly accessible repositories (Gewin 2016; Michener 2015; Sturges et al. 2015). Such mandates for enabling the reuse of research data include geospatial information and other information assets that have been collected as part of funded research projects. Regardless of the motivation for data sharing behavior, norms for data sharing are increasing (Kim and Adler 2015; Kim and Burns 2016) as well as positive attitudes about data sharing (Tenopir et al. 2015). Managers of funded projects that produce geospatial information and other data products and services are depositing their data in digital repositories, such as scientific data centers and archives, to ensure that these valuable information resources are curated effectively for reuse beyond the teams of investigators that originally collected the data. For example, as part of the NASA Earth Observing System Data and Information System (EOSDIS), Distributed Active Archive Centers (DAACs) have been providing lifecycle data curation services, from archiving through dissemination and support, to enable the reuse of geospatial information that are relevant to various NASA missions (Ramapriyan et al. 2010). Investigators that collect or produce geospatial data and related information without funding should also consider depositing their geospatial information in digital repositories, so that these valuable information resources can be reused as well.

Other sources of reusable geospatial information can include projects initiated by commercial organizations, the general public, and the educational institutions that have collected data without mandates for sharing their work with others. Geospatial information that have been collected by commercial organizations could be valuable when reused for analysis or when compared or combined with other data products for reuse. Crowdsourcing, where individuals, such as the general public, contribute to the collection of geospatial information, can also produce valuable data products for reuse. For example, geolocated information about transportation routes, which have been collected by volunteers, have been compiled with satellite data and other geospatial information to develop maps of roads along with descriptions of their characteristics (Ubukawa et al. 2014). Furthermore, although crowdsourced information challenges traditional data quality assessment practices, the reuse of volunteered geographic information (GI) has the potential to contribute to research (Elwood et al. 2012). The reuse of volunteered GI also has been instrumental for preparing, managing, and responding to hazardous events, including by personnel employed in emergency rescue operations and by victims of natural disasters (Goodchild and Glennon 2010). In such cases, it may be necessary to demonstrate the benefits of sharing geospatial information, so that the data collectors and contributors will be motivated to share. It may also be necessary to provide assurances that the data collectors, contributors, and their employing organizations will be protected from misuse of proprietary information or sensitive information and from any liability or legal issues that may be associated with sharing geospatial information for reuse by others. The data collectors may also need to contribute additional resources to assist in data hygiene, curation, preparation, and description for possible reuse.

Investing in the curation and public dissemination of geospatial data and related information to facilitate their reuse can increase their value even further by enabling the reuse of geospatial information to new communities of users, today and in the future. Curation and dissemination of geospatial data and related information that enable their use across disciplinary boundaries can foster studies by individuals and groups from diverse fields of study and can provide capabilities that support interdisciplinary reuse of the geospatial information. Geospatial data curation and the development and dissemination of data products and services can also facilitate the use of geospatial information by communities representing various levels of expertise, including educators, students, planners, policy-makers, and journalists, as well as by members of the general public.

Prior to selecting geospatial information for reuse, the needs of the designated community or communities that will reuse the information resources are assessed to identify the requirements for the geospatial products and services to be developed and disseminated. More than one community may be considered for the potential reuse of geospatial information. When assessing the needs for reuse of geospatial information, representatives of the community or communities that will be served should be involved to identify the questions that their communities are trying to answer and the hypotheses that they will be investigating. For example, when developing geospatial products or services for one or more scientific communities and for planning and decision-making communities, representatives of the scientific, planning, and decision-making communities should contribute to the need assessments for their respective communities. Furthermore, representatives of the designated communities should be involved throughout the entire development process, including the appraisal and selection of source data and related information that will be used to develop geospatial products and services for reuse, and in the curation, development, and evaluation of the products and services that will be produced and disseminated.

Scientific data centers, archives, and other digital repositories often establish advisory boards, consisting of members who represent the communities that will reuse the geospatial information being developed, to guide the need assessment for data, products, and services that will serve their communities. In addition, reviews of relevant scientific research and professional and trade literature offer insight into the questions being asked within particular scientific disciplines and into the particular needs of the designated communities. The needs and requirements for reuse of geospatial information can also be identified by attending workshops, meetings, and conferences of the professional associations represented by the designated communities. Administering focus groups and surveying instruments designed to elicit knowledge on the needs of specific communities can also be valuable sources of information for determining the requirements for reuse. Based on the requirements that have been identified for developing geospatial products and services for reuse, the criteria for appraising and selecting geospatial information are determined.

Curating geospatial information for reuse by either a particular community or by multiple communities also includes appraisal and selection. Before making curation investments to support the reuse of geospatial data and related information resources for particular purposes, the candidate resources should be assessed to determine whether they can be selected for potential reuse for new purposes (Morris 2013). For example, the selection criteria established by the Long-Term Archive of the NASA Socioeconomic Data and Applications Center includes appraisal in terms of "scientific or historical value, potential usability and use, uniqueness, relevance, documentation, technological accessibility, legality and confidentiality, and nonreplicability" (Downs et al. 2007, p. 3). Furthermore, prior to curating geospatial data and related information for reuse by scientific communities, quality assessments should be completed to ensure that the information resources are of sufficient quality for reuse in scientific research. The results of such quality assessments also need to be described in the documentation and within relevant metadata elements to inform potential users of any issues that must be considered when reusing the data to conduct scientific research. Any errors, missing values, validation, and issues that the data collection team described should be evaluated independently to identify any concerns that could affect the potential for reusing the geospatial information in new scientific studies.

Similar to the examples describing the importance of appraisal and selection of geospatial data and related information for scientific research, quality and liability are also important factors when considering the potential reuse of geospatial information for other purposes. When geospatial information is being considered for potential reuse for planning and policy-making purposes, the geospatial data and related information must be determined to be of high quality, so that they can be relied on for important decisions that can affect the lives of many people. Similar to geospatial information being considered for reuse in scientific research and in planning and policy-making, geospatial information resources that are being considered for reuse within other contexts must also be properly assessed to determine whether the information resources are of sufficient quality to be eligible for reuse by other designated communities. Attaining the needed quality and documenting the quality can reduce uncertainty and the risk of liability associated with the reuse of geospatial information (Gervais et al. 2009).

Similar to other scientific data, geospatial data and related research information need to be documented properly during data collection, so that they can be assessed and prepared for reuse by others beyond the original data collection team. Exact definitions of variables are needed and descriptions of instruments must be unambiguous. In addition, in recognition of the quality of the data and documentation needed for reuse by new communities, geospatial data documentation (metadata), including quality assessments, must be evaluated to determine whether they have the potential to be reused for new purposes, so that investments in data curation and in the development of new products and services can be limited to the geospatial information that have the potential to be valuable resources for reuse. Researchers generally agree that data assessment includes a review of the documentation (Kratz and Strasser 2015).

Without an assessment of the potential usefulness and utility of the geospatial information for reuse, the decision to invest resources in data curation may have to be delayed until the resources have been determined to be of sufficient quality to be worth the investment to prepare them for reuse. When the source data and the related geospatial information have been assessed and determined to be of sufficient quality for potential reuse by one or more designated communities and purposes, the level of effort, in terms of curation and public dissemination for use by others, should be estimated. However, prior to making such determinations, the intellectual property rights[*] for reuse must be obtained to ensure that others, including the repository and the communities that will reuse the geospatial information, have been authorized to reuse the resources for any purposes, without restrictions.

In addition to quality assessments, geospatial data and related information resources that have been identified as possible candidates for reuse must also

---

[*] This chapter is not providing legal advice. Readers seeking legal advice should consult qualified legal counsel.

be assessed to determine whether the intellectual property rights that are associated with the geospatial information will allow their dissemination for reuse. The producers of the geospatial information or their legally authorized representatives should authorize the dissemination and reuse of their information resources, so that distributors, such as digital repositories, scientific data centers, and archives have the right to share the data and so that, ultimately, users will have the right to reuse the data. Such authorization should be written and included with the geospatial information resources when they are submitted by the data collectors to an archive, data center, or repository, so that the rights associated with dissemination and reuse are completely understood by all interested parties.

Prior to data collection, the team that collects the geospatial data and the related information should decide on the intellectual property rights that are going to be associated with the data. Such decisions should be agreed on by all members of the data collection team and reflect the policies of their employing organizations and sponsors, if applicable. Collectors of geospatial information may be motivated to openly share their data and related information by knowing that studies that enable the reuse of data through a publicly accessible repository are cited more often than other data (Piwowar and Vision 2013). Although several choices may be available for consideration, assigning intellectual property rights that authorize unrestricted access should be considered, so that the geospatial information will have enduring value for a variety of uses. Authorizing use by anybody for any purpose, without restrictions or expiration of such rights, allows the geospatial information to be open for any type of reuse in the future. If the data collection team can agree that their data should remain open, in this manner, the geospatial information can be designated as *open data* and their potential for reuse will not be limited by intellectual property rights.

The data collection team should consider applying an open license to the geospatial information that will allow unrestricted use for all others who may have an interest in potential reuse of the resources. In the absence of a clear declaration for the use of their geospatial information, creators may be limiting the potential for reuse. Several open licenses are available for enabling unrestricted access and use of geospatial information for any purpose. For example, in many cases, the Creative Commons Attribution 4.0 International (CC By) license can be applied to geospatial information by the copyright holders of the resources. The CC By 4.0 license allows various uses for any purpose if attribution to the source is provided (Creative Commons 2013a). Such open data licenses are consistent with current norms of science and publishing. In many countries, authors hold the copyright to original information resources that they create and can assign an open license to information, including geospatial information, which they own (Onsrud et al. 2010). Creators of geospatial information should consult their organizational policies to determine whether an open license can be applied to enable reuse.

The CC By 4.0 license, when applied to geospatial information, allows anyone to use, reproduce, adapt, or redistribute the resources for any purpose, including commercial reuse, in accordance with the terms of the license (Creative Commons 2013b). When the CC By 4.0 license is affixed to geospatial information, such as a map, those who plan to reuse the resource can easily determine their rights for reusing the map. If the creator of geospatial information applies the CC By 4.0 license to their information resources and submits those resources to a scientific data center for redistribution, the scientific data center can redistribute the resources under the same CC By license. Anyone can then reuse the geospatial information for any purpose under that license. By licensing their work with unrestrictive licenses, creators of geospatial data and related information resources can reduce the need to negotiate over intellectual property rights with the data center that will distribute their work and also reduce the need for the data center to negotiate with anyone who would like to reuse those resources or combine them with other open data products for reuse. The potential for fostering reuse of geospatial information can increase by clearly describing any licenses or restrictions associated with such resources (Campbell and Onsrud 2014). When a digital repository or scientific data center is considering the acquisition of geospatial information with clearly described intellectual property rights that allow unrestricted distribution and reuse for any purpose, understanding of the rights associated with the work can be improved, enabling the repository to archive, integrate, and disseminate the resources, if desired, and enabling the reuse of such resources. In addition, the CC By 4.0 license contains legal language providing a "Disclaimer of Warranties and Limitation of Liability" (Creative Commons 2013b). Guidelines for implementing the Principles on the Legal Interoperability of Research Data are directly applicable to enable the reuse of geospatial information (RDA-CODATA Legal Interoperability Interest Group 2016).

9.3 Acquiring and Preparing Geospatial Information to Facilitate Reuse

When the digital repository has decided to acquire geospatial information for reuse, all of the geospatial data and the related information resources that are associated with the acquisition should be obtained from the creators. For projects that involved a team of collaborators, all of the components must be received, including documentation that describe the provenance of the information resources and how they were collected and developed. In cases where different individuals are completing aspects of the project, it may be necessary to determine who is in possession of the latest version of each component, so that the current version of each component of the geospatial information can be acquired. In addition, as mentioned, previously, the geospatial information should be acquired with an open data license or a written declaration from the authors that clearly describe the rights for using the resources.

On acquiring geospatial information from the creators, a digital repository can begin preparing the information resources for reuse. Curation, including preservation and archiving of the geospatial information, along with the development and public dissemination of geospatial data products and services, facilitates the potential reuse of the data and the related information resources long after the geospatial information has been acquired. Such preparation can allow current and future users of the data to discover, explore, access, understand, analyze, integrate, and reuse the geospatial information to meet their needs. The acquiring digital repository should also involve representatives of designated communities that would benefit from the reuse of the geospatial information and should begin preparing the information resources while the creators are available, so that the resources can be reused by those communities in the near term and in the future (Downs and Chen 2017).

When preparing geospatial data and related information for reuse, the documentation that describes the geospatial information must also be compiled for dissemination. Without written descriptions of the variables that are measured and descriptions of how they were measured, reuse could be restrained and the value of the geospatial information will be limited. Descriptive information about the quality of the geospatial information and how the quality was measured or assessed is necessary for reuse and is especially important if the geospatial information resources are going to be integrated with other data to create new geospatial information products and services. In addition, as previously mentioned, descriptions of the provenance of the geospatial information resources should be included as part of the information packages that are disseminated, so that the history of the collection, processing, evaluation, and product generation is transparent to anyone who plans to reuse them. Collaborating with the creators of geospatial information may be necessary to obtain and develop sufficient documentation to ensure that the geospatial information can be reused for a variety of purposes, including purposes that were not envisioned by the creators.

Although preparing geospatial information products and services for public dissemination, summary descriptions and discovery metadata also need to be developed, so that they can be found and explored for potential reuse by those who might have an interest in reusing them. Enabling discovery and initial exploration is critical for increasing the reuse of geospatial information. The discovery metadata should be developed in compliance with standards that are compatible with catalogs and cataloging services that serve the designated communities for which the geospatial information resources are targeted. Links should also be established within the metadata to associate the data with documentation, related data resources, and relevant publications. Such links, along with complete descriptions of the data can contribute to the interoperability of geospatial information and foster additional opportunities for reuse. In addition, providing summary descriptions and discovery metadata to cataloging services that allow collection of such descriptions by automated harvesters can lead to increased exposure of available geospatial information products and services, further improving the possibility of reuse.

Once geospatial information has been discovered for potential reuse, summary information and other documentation about the data product or service should also be available to facilitate decision-making on whether reuse of the data would be appropriate for addressing an intended information need. In addition to facilitating discovery, summary information and other metadata elements can be used to determine whether a particular data product might be a candidate for potential reuse. Likewise, detailed documentation should be accessible to provide additional context that can be used to assist in the determination of the applicability of a particular data product for reuse. Furthermore, in addition to assisting in the decision regarding their potential for reuse, the availability of documentation can increase the usability of geospatial information by improving the understanding about the methods and instruments that were used during the collection, processing, and development of the data product or service. Algorithms and variables should be sufficiently described, so that the decision for potential reuse can be informed by the details that might be necessary to know prior to the reuse decision. In such cases, the availability of sufficient documentation can enable decisions on whether the methodologies that were used to collect the data are consistent with the purpose of the study for which the potential reuse of the geospatial information has been envisioned.

Facilitating discovery and exploration also includes promoting practices that enable those who reuse the data to properly cite the data within the publications that describe their work. Providing a recommended citation for each data product, visibly on the landing page for the data product, can help the authors, who have reused the data, to cite the data properly, when preparing manuscripts for publication. The recommended citation for the geospatial information should also include a persistent identifier to enable access over time, even if the location has changed (Duerr et al. 2011). By promoting the citation of geospatial information with a recommended citation that includes a persistent identifier, interested readers of published works can easily locate the geospatial information for potential reuse.

Geospatial information, including data products and related services, as well as maps and map services, should be easily accessible for reuse. Online tools and services can be developed to facilitate access to geospatial data and information resources. However, such tools should meet the needs and expectations of the anticipated user communities, without imposing additional requirements for reuse on potential users. For example, data products and maps should be available in the common formats that the user communities are experienced in using. Similarly, data services and map services should be accessible with common tools or through tools that are described and documented to foster reuse by novices as well as by experienced users of geospatial information. Providing a selection of tools that can accommodate a variety of users can also facilitate access when serving diverse user communities, enabling them to choose from the available tools to select their tool of choice.

In addition to the technical aspects of accessing geospatial information resources, providing them as fully open access, without restrictions, is necessary to facilitate reuse of the data for any purpose. The potential for reuse can be enabled if the geospatial information is released in the public domain or under a license that authorizes reuse for any purpose, as previously described. Another important aspect of making data accessible for reuse includes enabling access to the tools and other software products that can be reused to recreate, reprocess, and access the data. Releasing such software products as open source and publicly disseminating them facilitates accessibility to the capabilities needed to reproduce, integrate, and analyze geospatial information.

Prior to release, testing and evaluation of the developed geospatial products and services should also be completed to verify that the geospatial information resources are ready for public dissemination and to improve them, if necessary. Considering that reuse of geospatial information could include diverse contexts, such as scientific research, analysis, integration, learning, interpretation for planning and policy-making, and republication, and many other activities, extensive testing and evaluation should be completed to ensure the usability of the data products and services as well as the interfaces that have been developed to enable reuse of the geospatial information. After internal testing has been completed and any errors or omissions have been corrected, an external evaluation should be performed by representatives of the designated communities that have been identified as potential users of the geospatial information products and services that have been developed for reuse. The results of such external evaluation activities should identify any opportunities for further improvement prior to release and confirm that the disseminated geospatial information resources can be reused effectively by the represented communities. The results of such evaluation activities should also be documented to capture information about the quality of the geospatial information.

9.4 Assessing the Impact of Reuse

The reuse of geospatial information has the potential to benefit society in many ways. Scientific discoveries, comparisons, and longitudinal analyses can produce findings that save lives and improve the living conditions of many who otherwise would live in extreme poverty. Plans and policies developed from the reuse of geospatial information might lead to safer conditions for municipalities, counties, and even nations. However, in the absence of an assessment that describes the benefits of reusing geospatial information, the value of such reuse may go unrecognized.

Many scientific articles improperly provide attribution to the source of geospatial information that was reused to produce the results. Published reports that provide recommendations for humanitarian development, based on the reuse of geospatial information, do not always credit the source. Unpublished plans for new infrastructure and other beneficial projects, also based on geospatial information, often are proprietary and remain confidential. In such situations, the phenomenal benefits attained from the reuse of geospatial information are not recognized. Without recognition of the value of the geospatial information that was reused to attain such benefits, the efforts to collect, prepare, and disseminate these resources will not be recognized either. It also will be difficult to justify such efforts or new geospatial data collection, preparation, and dissemination efforts without evidence of the impact of reusing geospatial information.

Fortunately, some reuse of geospatial information is properly documented and attribution is provided to credit the sources. Recommendations for properly citing data have been established by members of the scientific community and publishers (Martone 2014). Scientific journals are beginning to require authors to cite data, including geospatial information resources, which have been reused to prepare manuscripts for studies that have been submitted for publication (Mayernik et al. 2015). By identifying the beneficial reuse of geospatial information, as cited in published articles, and counting instances of attributed reuse of specific geospatial products or collections, the impact of reusing such geospatial information resources can be measured. Collecting such attribution metrics provides evidence of the benefits of reusing geospatial information. Furthermore, reporting on the results of assessing the impact of geospatial products also provides evidence on the value of the efforts to collect, prepare, and disseminate these resources. With such evidence, sponsors of geospatial projects, the performing organizations, and their employees can be recognized for sharing geospatial information and justify their efforts to collect, prepare, and disseminate new geospatial information products and services. In addition, such assessments of the reuse of geospatial information can inform the decisions of researchers who are exploring the potential for reusing particular geospatial information products and services.

9.5 An Example of Reuse: The Gridded Population of the World Data Collection

A brief description of the Gridded Population of the World (GPW) series can illustrate the value that can be attained from the reuse of geospatial information. The ongoing development, production, distribution, and reuse of the GPW series also offer a rich history that can demonstrate the value of reusing geospatial information.

The GPW series of data collections have been produced and distributed by the NASA Socioeconomic Data and Applications Center (SEDAC) for more than 20 years. The NASA SEDAC is one of the DAACs of the EOSDIS. SEDAC focuses on human interactions in the environment and serves as a gateway between the social and physical sciences (NASA SEDAC 2017). In collaboration with the Earth science and social science communities, SEDAC develops data products and services, including the GPW series, to meet the evolving needs of its user communities (Downs and Chen 2004). Each collection in the GPW series has represented a significant improvement over the previous release to address the needs for global geospatial information products of georeferenced population data. As a result of subsequent development since the release of GPW version 3 in 2006, SEDAC has created the latest collection in the GPW series, GPW version 4 (Doxsey-Whitfield et al. 2015). After acquiring, curating, developing, and reviewing the data products and services, the GPWv4 series was recently released containing eight datasets along with related geospatial information products and services. The GPWv4 release includes the following datasets: Administrative Unit Center Points with Population Estimates, Population Density Adjusted to Match 2015 Revision UN WPP Country Totals, Population Count, Population Density, Data Quality Indicators, Population Count Adjusted to Match 2015 Revision of UN WPP Country Totals, National Identifier Grid, and Land and Water Area. Summaries of the collection and individual data products appear on their respective landing pages. Metadata provides details about each data product and is accessible from each product landing page and is distributed to catalogs to facilitate discovery for potential reuse. Documentation describing the GPWv4 collection is accessible from the landing pages of the GPWv4 collection and data products to support exploration and reuse (Center for International Earth Science Information 2016a). Map services and a gallery of maps also are accessible from each landing page. An example map, Population Density Grid, 2015: Global (CIESIN 2016b), is reproduced in Figure 9.1.

The GPW collection of data products and services has had a noteworthy impact during the past two decades. There have been well over 1400 citations to GPW data products and services in the published literature since 1996. The citations include references to GPW in peer-reviewed scientific articles, textbooks, published reports, and popular media. In addition, the impact of the GPW collection has been rising. The growth of the impact of GPW datasets, maps, and map services can be observed in the increasing number of citations that appear in the published literature each year. As depicted in Figure 9.2, the number of citations to GPW data products and services has been increasing almost every year from 1996 through 2016. Interestingly, some of the early citations to GPW products and services referenced the Tobler et al. article (1995), which described the initial development of the GPW version 1 product. However, for more than a decade, studies that have used the GPW collection have cited the GPW data products and services. Although in some cases, the references have not specified the particular version of GPW product or service that was used as can be seen in Figure 9.2.

Gridded population of the world, version 4 (GPW v4)

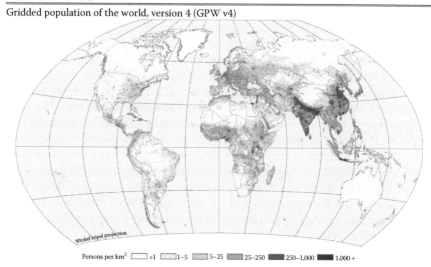

Persons per km² ☐ <1 ☐ 1–5 ☐ 5–25 ■ 25–250 ■ 250–1,000 ■ 1,000 +

FIGURE 9.1
Map, population density grid, 2015: Global. Gridded population of the world, version 4 (GPWv4). Population density consists of estimates of human population density based on counts that are consistent with national censuses and population registers, for the years 2000, 2005, 2010, 2015, and 2020. A proportional allocation gridding algorithm, utilizing approximately 12.5 million national and subnational administrative units, is used to assign population values to 30 arc-second (~1 km) grid cells. The population density grids are derived by dividing the population count grids by the land area grids. The pixel values represent persons per square kilometer. (From Center for International Earth Science Information Network (CIESIN), Columbia University, *Gridded Population of the World, Version 4 (GPWv4): Population Density*, NASA Socioeconomic Data and Applications Center (SEDAC), Palisades, NY, 2016. With Permission.) Map licensed under a Creative Commons 3.0 Attribution License.

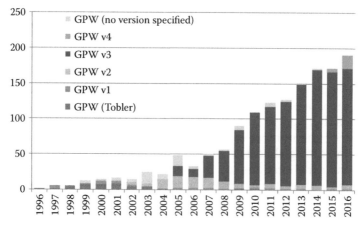

FIGURE 9.2
Number of citations of GPW products and services by year from 1996 through 2016.

9.6 Conclusions and Recommendations

When developing geospatial information products and services for reuse, the creators of the original geospatial information should be involved to ensure that any information that is needed can be utilized to support curation and dissemination for potential reuse. Ideally, the creators of source data and other geospatial information should provide permissions to enable reuse for any purpose, without restrictions. Representatives of potential communities of users also should be engaged to identify their needs for geospatial information products and services. By establishing processes that engage both the creators of source data and potential users, digital repositories, such as scientific data centers and archives, can facilitate reuse by establishing capabilities that address user needs. Representatives of designated user communities should be included in reviews of candidate source data and geospatial information that are being considered for potential reuse. Such representatives also should be included in reviews of geospatial information products and services that have been curated and prepared for dissemination.

Geospatial information products and services should be developed to enable reuse by diverse user communities that represent a range of disciplines and expertise. Disseminated products and services should foster a variety of capabilities to enable the reuse of geospatial information for new purposes, which may be completely different from the purposes for which they were originally created.

The intellectual property rights associated with geospatial information should be clearly described on each product or service in simple language, so that the rights for reuse can be understood. Documentation must be comprehensive and complete to provide sufficient information about geospatial information to foster reuse. Summaries of collections, products, and services should be displayed on landing pages and included in metadata records that are distributed to catalogs that facilitate discovery and exploration for potential reuse of geospatial information. A recommended citation should be displayed on each landing page for geospatial information products and services to encourage citation of the resources in publications that report on the reuse of each product or service.

Geospatial information products and services are increasingly valuable to society. The value of geospatial information is attained when geospatial information products and services are reused. Enabling reuse can increase the value of geospatial information. Recommended practices for enabling the reuse of geospatial information have been described and an example demonstrates how the value of geospatial information products and services can be attained through curation and dissemination.

Acknowledgments

The author appreciates the suggestions for an earlier version of this work that were provided by Robert S. Chen, Director of the Center for International Earth Science Information Network (CIESIN) of Columbia University. The author also appreciates the contributions of Joachim Schumacher, User Services Manager of CIESIN, who provided the annual citation counts and graph for the GPW collection, from which Figure 9.2 was derived. In addition, the author appreciates the comments and suggestions provided by the editors, Françoise Pearlman and Joep Crompvoets, which were most helpful for improving an earlier version of this manuscript. The author also appreciates the support received from the National Aeronautics and Space Administration (NASA) under contract NNG13HQ04C for the Socioeconomic Data and Applications Distributed Active Archive Center (DAAC).

References

Brown, L. A. 1979. *The Story of Maps*. Mineola, NY: Dover Publications.

Campbell, J., and H. Onsrud. 2014. Desirable characteristics of an online data commons for spatially referenced, locally generated data from disparate contributors. *URISA Journal*, 26(1), 35–44.

Center for International Earth Science Information Network (CIESIN), Columbia University. 2016a. *Documentation for the Gridded Population of the World, Version 4 (GPWv4)*. Palisades, NY: NASA Socioeconomic Data and Applications Center (SEDAC). Accessed October 9, 2016. doi:10.7927/H4D50JX4.

Center for International Earth Science Information Network (CIESIN), Columbia University. 2016b. *Gridded Population of the World, Version 4 (GPWv4): Population Density*. Palisades, NY: NASA Socioeconomic Data and Applications Center (SEDAC). doi:10.7927/H4NP22DQ.

Creative Commons. 2013a. Attribution 4.0 international (commons deed). Accessed October 9, 2016. https://creativecommons.org/licenses/by/4.0/.

Creative Commons. 2013b. Attribution 4.0 international (legal code). Accessed February 20, 2017. https://creativecommons.org/licenses/by/4.0/legalcode.

Downs, R.R., and R.S. Chen. 2004. Cooperative design, development, and management of interdisciplinary data to support the global environmental change research community. *Science & Technology Libraries*, 23(4), 5–19. doi:10.1300/J122v23n04_02.

Downs, R.R., and R.S. Chen. 2017. Chapter 12. Curation of scientific data at risk of loss: Data rescue and dissemination. In Johnston, L.R. (Ed.), *Curating Research Data Volume One: Practical Strategies for Your Digital Repository*. Association of College and Research Libraries (ACRL) of the American Library Association (ALA), pp. 263–277. doi:10.7916/D8W09BMQ.

Downs, R.R., R.S. Chen, W.C. Lenhardt, W. Bourne, and D. Millman. 2007. Cooperative management of a long-term archive of heterogeneous scientific data. *Proceedings, Ensuring the Long-Term Preservation and Value Adding to Scientific and Technical Data (PV 2007)*, Oberpfaffenhofen/Munich, Germany, October 9–11. Accessed February 26, 2017. http://www.pv2007.dlr.de/Papers/Downs_CooperativeManagementOfALongTermArchive.pdf.

Doxsey-Whitfield, E., K. MacManus, S.B. Adamo, L. Pistolesi, J. Squires, O. Borkovska, and S.R. Baptista. 2015. Taking advantage of the improved availability of census data: a first look at the gridded population of the world, version 4. *Papers in Applied Geography*, 1(3), 226–234. doi:10.1080/23754931.2015.1014272.

Duerr, R.E., R.R. Downs, C. Tilmes, B. Barkstrom, W.C. Lenhardt, J. Glassy, L.E. Bermudez, and P. Slaughter 2011. On the utility of identification schemes for digital earth science data: An assessment and recommendations. *Earth Science Informatics*, 4(3), 139–160. doi:10.1007/s12145-011-0083-6.

Elwood, S., M.F. Goodchild, and D.Z. Sui. 2012. Researching volunteered geographic information: Spatial data, geographic research, and new social practice. *Annals of the Association of American Geographers*, 102(3), 571–590. doi:10.1080/00045608.2011.595657.

Gervais, M., Y. Bedard, M. Levesque, E. Bernier, and R. Devillers. 2009. Data quality issues and geographic knowledge discovery. In H.J. Miller and J. Han (Eds.), *Geographic Data Mining and Knowledge Discovery*, 2nd ed. Boca Raton, FL: CRC Press, pp. 99–115.

Gewin, V. 2016. Data sharing: An open mind on open data. *Nature*, 529, 117–119. doi:10.1038/nj7584-117a.

Goodchild, M.F., and J.A. Glennon. 2010. Crowdsourcing geographic information for disaster response: A research frontier. *International Journal of Digital Earth*, 3(3), 231–241. doi:10.1080/17538941003759255.

Holden, J.P. 2013. *Increasing Access to the Results of Federally Funded Scientific Research*. Washington, DC: Executive Office of the President Office of Science and Technology Policy. Accessed October 3, 2016. http://www.whitehouse.gov/sites/default/files/microsites/ostp/ostp_public_access_memo_2013.pdf.

Kim, Y., and M. Adler. 2015. Social scientists' data sharing behaviors: Investigating the roles of individual motivations, institutional pressures, and data repositories. *International Journal of Information Management*, 35(4), 408–418. doi:10.1016/j.ijinfomgt.2015.04.007.

Kim, Y., and C.S. Burns. 2016. Norms of data sharing in biological sciences: The roles of metadata, data repository, and journal and funding requirements. *Journal of Information Science*, 42(2), 230–245. doi:10.1177/0165551515592098.

Kratz, J.E., and C. Strasser. 2015. Researcher perspectives on publication and peer review of data. *PLoS ONE*, 10(2), e0117619. doi:10.1371/journal.pone.0117619.

Martone, M. 2014. Joint declaration of data citation principles–FINAL. FORCE11. https://www.force11.org/group/joint-declaration-data-citation-principles-final. Accessed October 21, 2016. https://www.force11.org/datacitation.

Mayernik, M.S., M.K. Ramamurthy, and R.M. Rauber. 2015. Data archiving and citation within AMS journals. *Journal of the Atmospheric Sciences*, 72(4), 1281–1282. doi:10.1175/2015JAS2222.1.

Michener, W.K. 2015. Ecological data sharing. *Ecological Informatics*, 29(1), 33–44. doi:10.1016/j.ecoinf.2015.06.010.

Morris, S. (Ed.). 2013. *Issues in the Appraisal and Selection of Geospatial Data: An NDSA Report*. Washington, DC: National Digital Information Infrastructure and Preservation Program (NDIIPP), Library of Congress. Accessed November 3, 2016. http://www.digitalpreservation.gov/documents/NDSA_AppraisalSelection_report_final102413.pdf.

NASA Socioeconomic Data and Applications Center (SEDAC). Accessed February 24, 2017. http://sedac.ciesin.columbia.edu/.

Onsrud, H.J., J. Campbell, and B. van Loenen. 2010. Towards voluntary interoperable open access licenses for the Global Earth Observation System of Systems (GEOSS). *International Journal of Spatial Data Infrastructure Research*, 5, 194–215.

Piwowar, H.A., and T.J. Vision. 2013. Data reuse and the open data citation advantage. *PeerJ*, 1, e175. doi:10.7717/peerj.175.

RDA-CODATA Legal Interoperability Interest Group. 2016. Legal interoperability of research data: Principles and implementation guidelines. *Zenodo*. Accessed February 25, 2017. https://doi.org/10.5281/zenodo.162241.

Ramapriyan, H.K., R. Pfister, and B. Weinstein. 2010. An overview of the EOS data distribution systems. In *Land Remote Sensing and Global Environmental Change*. New York: Springer, pp. 183–202. doi:10.1007/978-1-4419-6749-7_9.

Sturges, P., M. Bamkin, J.H.S. Anders, B. Hubbard, A. Hussain, and M. Heeley. 2015. Research data sharing: Developing a stakeholder-driven model for journal policies. *Journal of the Association for Information Science and Technology*, 66, 2445–2455. doi:10.1002/asi.23336.

Tenopir, C., E.D. Dalton, S. Allard, M. Frame, I. Pjesivac, B. Birch, D. Pollock, and K. Dorsett. 2015. Changes in data sharing and data reuse practices and perceptions among scientists worldwide. *PLoS ONE*, 10(8), e0134826. doi:10.1371/journal.pone.0134826.

Tobler, W., U. Deichmann, J. Gottsegen, and K. Maloy. 1995. The global demography project (95-6). National Center for Geographic Information and Analysis, Santa Barbara, CA. Accessed October 25, 2016. http://www.ncgia.ucsb.edu/technical-reports/PDF/95-6.pdf.

Trapp, N., U.A. Schneider, I. McCallum, S. Fritz, C. Schill, M.T. Borzacchiello, C. Heumesser, and M. Craglia. 2015. A meta-analysis on the return on investment of geospatial data and systems: A multi-country perspective. *Transactions in GIS*, 19(2), 169–187. doi:10.1111/tgis.12091.

Ubukawa, T., A. de Sherbinin, H. Onsrud, A. Nelson, K. Payne, O. Cottray, and M. Maron. 2014. A review of roads data development methodologies. *Data Science Journal*, 13, 45–66. doi:10.2481/dsj.14-001.

Section III

Measuring Economic and Social Values, Benefits, and Impacts

Methods and Implementation Examples

10

A Review of Socioeconomic Evaluation
Methods and Techniques

Alan Smart, Andrew Coote, Ben Miller, and Richard Bernknopf

CONTENTS

10.6.4   Value of Remote Sensing from LANDSAT............................ 188
10.6.5   Value of Remote Sensing from Space for Decision-Making..... 188
10.7   Summary... 188
10.8   New and Emerging Techniques... 189
References.. 192

10.1 Introduction

This chapter reviews methodologies in economics for valuing the benefits of geospatial data and services. It explores the meaning of socioeconomic value and reviews the methodologies underpinning value assessments. It explains the theoretical background to economic welfare, gross revenue, and value-added analyses. Different methods for economic impact assessment are discussed. These include benefit–cost analysis, and Computable General Equilibrium (CGE) modeling. Examples of the application of these techniques are discussed, and the issues that arise when applying these techniques are canvassed. It notes that the methodology selected should be appropriate to the context in which it is to be applied, including the decision support required by policy-makers and suppliers of geospatial data.

10.1.1 Methodology

The use and application of geospatial information have delivered significant economic and social benefits to society. It is anticipated that these benefits will increase in future as geospatial information is further integrated into emerging data systems and technologies.

Understanding the value of geospatial information has become important to decision-makers in government and private organizations to support the case for investment in and maintenance of spatial data infrastructure (SDI). It is also important for policy development with respect to the capture, curation, and dissemination of geospatial data. Most significantly, it underpins the understanding of the importance of geospatial infrastructure to society in general.

Numerous valuations of geospatial information have been undertaken over past 20 years. These have drawn on varying techniques, including economic welfare theory, gross revenue estimates, value-added analysis, and general equilibrium modeling. Each methodology has advantages and disadvantages.

It is important for policy-makers and decision-makers alike that the general economic principles underlying the different valuation techniques are understood. This chapter canvasses the main economic principles underlying these techniques and discusses their application in practice.

10.1.2 The Meaning of Value

Assessing the value of geospatial information is a complex task. Valuing the contribution made by open geospatial data provided at no cost is even more complex, because there is no market in which it can be bought or sold.[*]

A starting point for estimating its value is to clarify what is meant by the term *value*. Fundamental geospatial data are intermediate goods that enable other activities through value-added services. To understand their value, we need to explore the value that suppliers and users draw from the data.

For a government agency, this could be as narrow as a financial benefit (e.g., realized future savings). For a policy decision-maker, it could be as wide as the expected benefits that would accrue to society as a whole from the use of the data.

A suggested framework for considering different concepts of value is provided in Figure 10.1. Values are divided into use and non-use values. Use values comprise direct use values (such as the value of goods and services),

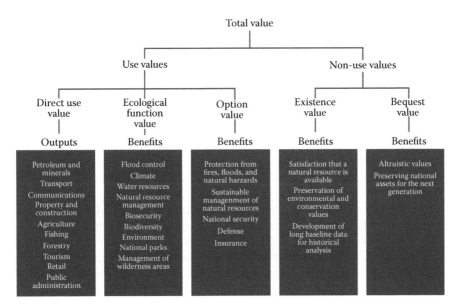

FIGURE 10.1
The nature of socioeconomic value. (Based on a conceptual framework from Young, 1992. With permission.)

---

[*] Open data is a term used to refer to the accessibility of data, indicating that it is available with little or no restrictions of its use. Although not part of its formal definition, it is commonly used to describe data that is free at the point of delivery.

ecological values (such as biodiversity or sustainable rivers and streams), and option values (such as insurance against the costs of natural disasters).

Non-use values can be considered as existence values (valuing the existence of a coral reef but never visiting it) and bequest values (preserving the value of assets for later generations). Although non-use values are conceptual, they are real in the minds of many in society and potentially become policy issues for this reason.

There is also a long history of valuation of intangible benefits derived from geospatial data by using willingness-to-pay methodologies.

There has been limited use of option valuation in assessing the value of geospatial applications; however, there are examples in areas such as astronomy and research and development. There appears to be little in the way substantive studies of non-use values for geospatial information in the literature.

10.2 Current State of the Art

Several techniques have been applied over the past 20 years to estimate the value of geospatial information. A representative sample is discussed in the following sections.

10.2.1 Welfare Analysis

Welfare analysis is a theoretical conceptual economic model that describes the economic value of a good or a service. In an efficient market, the economic welfare of a good or a service to society is measured by consumer and producer surplus. The conceptual base for consumer and producer surplus is the supply and demand or the market model depicted in Figure 10.2.

The market supply curve (which comprises the summation of individual firm supply curves) indicates the costs of extra production, that is, the costs to society of producing an extra unit of a good or service. Firms aim to operate on the upward sloping part of their marginal cost curve, above the minimum average variable cost. The upward slope reflects diminishing returns to factor inputs, which means that it costs more to produce each additional unit of output.

The market demand curve (which comprises the summation of individual demand curves) indicates the maximum amount that consumers are willing to pay for incremental increases in the quantity of the good or service. The demand curve slopes downward to the right because more consumers are willing to purchase the good as the price falls. This concept is generally known as diminishing marginal utility.

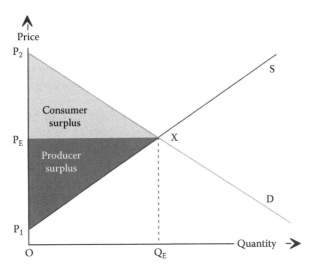

FIGURE 10.2
Standard concepts of producer and consumer surplus. (From Marshall, 1890; Hanley, 1997. With permission.)

The interaction of demand and supply determines the market price (P_E) for a good and the quantity that is produced in any given time period (Q_E).

This market model provides the basis for identifying and estimating the net economic value to consumers and producers, referred to as consumer surplus and producer surplus, respectively.

Consumer surplus is the difference between what consumers would be willing to pay for a good or service (the total benefit to the consumer) and what they have to pay (the cost to consumers). In Figure 10.2, consumer surplus is represented as the area between the demand curve and the price line (P_2XP_E).

Producer surplus is the difference between the revenue received for a good or a service (total benefit to producers) and the costs of the inputs used in the provision of the good or the service[*] (economic cost to producers). In practical terms, it is the net revenue (before tax) that is earned by a producer of goods and services. In Figure 10.2, it is the area between the price line and the supply curve (P_1XP_E). If the diagram is specified in annual terms, the sum of the shaded areas will represent the annual value of the product to society.

10.2.1.1 Application to Fundamental Geospatial Data

The application of the economic welfare theory to empirical analysis of the value of fundamental geospatial data requires many assumptions about

---

[*] Spatial data would usually be classified as a good. However the provision of spatial data might be regarded as a service.

the real nature of the markets under examination. Its application to decision support for investment in geospatial data must account for two important issues. First, the custodian of the data is generally a government organization, where the price for access is set by a policy decision—not by the market. Second, the data can exhibit public good characteristics, which has implications for the value that is generated for society.

Under an ideal open data policy, the price for fundamental spatial data would be set at the marginal cost, which is close to or equal to zero when supplied through web-based applications. This is illustrated in Figure 10.3. The value to consumers of this arrangement is the consumer surplus, shown as the shaded area in the diagram.

Welfare analysis has been used on a number of occasions to estimate both the value of geospatial data services in general and the relative economic benefit of different pricing policies.

Pollock (2008) estimated the value of moving from average cost pricing to marginal cost in a report prepared for HM Treasury; the theoretical framework used by Pollock is illustrated in Figure 10.4.

The average cost curve is the upwardly concave curved line identified in the diagram. Average cost includes an allowance for the original cost of acquiring the data and the annual cost of maintaining and supplying the data. The marginal cost is the horizontal line P_EX. For open data, the short run marginal cost of supply is constant and close to zero for web-based applications. This outcome is a special case that arises for government-supplied goods, where governments have a supply monopoly. The increase

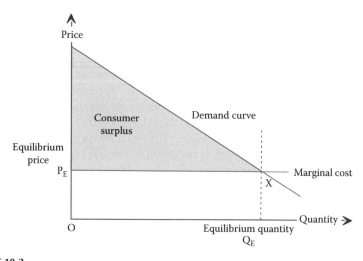

FIGURE 10.3

Demand and supply curves in the case of fundamental geospatial data. (An example of this approach can be found in Pollock, R., *Models of Public Sector Information Provision by Trading Funds*, Cambridge University, Cambridge, 2008. With permission.)

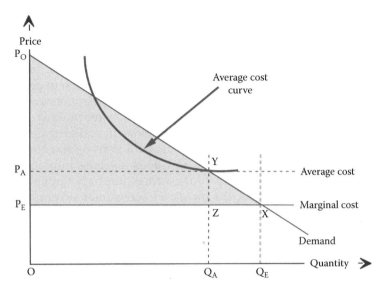

FIGURE 10.4
Demand and supply curves in the case of fundamental geospatial data. (From Pollock, R., *Models of Public Sector Information Provision by Trading Funds*, Cambridge University, Cambridge, 2008. With permission.)

in consumer surplus that results from changing from average cost pricing to marginal cost pricing is the area $P_A YZP_E + YXZ$ and the corresponding loss in producer surplus $(P_A YZP_E)$.* Pollock adjusts the loss of producer surplus by a factor α, which represents the extra costs of raising revenue through taxation (the marginal excess burden).† Hence, the value of changing from average cost pricing to marginal cost pricing is the increase in consumer surplus less the decrease in producer surplus multiplied by α, as shown in the following formula.

$$\begin{pmatrix} \text{Value of changing from average} \\ \text{to marginal cost pricing} \end{pmatrix} = YXZ - \alpha\left(P_A YZP_E\right) \qquad (10.1)$$

If the marginal excess burden of taxation (α) = 0, the economic value of changing from average to marginal cost pricing becomes the area XYZ.

Under open date policies there is often functioning market for foundation geospatial data supplied by government and hence no observable

---

* In this model, drawn on by Pollock, the declining average cost curve and constant marginal cost curve are representative of production of a good with a specific initial fixed cost, with each additional unit of the good produced for the same cost after the initial outlay. This is a characteristic of a good that involves collection and processing of data with the cost of distribution of each unit of output constant.
† The marginal excess burden is equal to 1 minus the marginal cost of public funds. For a discussion, refer to (Dahlby, 2008).

price–quantity trade-off that would enable one to estimate the demand curve. The Pollock study used evidence from the United Kingdom and literature to estimate the price elasticity of demand for geospatial data.[*] Using comparative studies, the report assumed a price elasticity of demand for Ordnance Survey (OS) data of −2. This means a 1% reduction in price would produce a 2% increase in the volume of data demanded by consumers.

The study made other adjustments for innovation, for the cost of government funds, and an adjustment for the time delay in realizing the benefits. The most important of these was an adjustment to recognize the fact that welfare analysis is a static analysis and does not take into account the dynamic effect of innovation. To address this, the study applied a multiplier λ to the results.[†] That is, the increase in consumer surplus was multiplied by λ to capture the downstream effects plus the dynamic effects of innovation. With these assumptions, the Pollock report estimated the economic value to society of moving from average cost pricing to marginal cost pricing to be £168 million, whereas the net cost to government was estimated to be around £12 million—a net benefit to society of £156 million.

A subsequent study for the Australian and New Zealand Land Information Council in 2010 used a similar conceptual model to estimate the economic value of different pricing policies for selected geospatial datasets (ANZLIC, 2010). In this study, a willingness-to-pay approach was used to estimate the price elasticity of demand for selected foundation spatial datasets. Users were interviewed to assess the value that they placed on certain data and their likely demand response to different price points.

The study estimated a price elasticity of −1 and a multiplier of 1. Using these assumptions, this study estimated that the value of moving from average cost to marginal cost pricing was A$3.3 million for Victorian topographic data, A$1.4 million for Western Australian topographic data, A$1 million for Western Australian aerial photography, and A$4.7 million for national topographic data.

Houghton (2011) applied a similar welfare analysis using increases in downloads of geospatial data released by Geoscience Australia following the introduction of free online data, accessed through the internet. Houghton estimated the price elasticity of demand for scheduled datasets to be −1.3. Using this estimate along with download data and estimates of agency and user cost savings, Houghton estimated the total increase in consumer surplus of moving from cost recovery to freely available data to be A$60.2 million over the period from 2001–2002 to 2005–2006.

---

[*] The price elasticity of demand of a good or service is the ratio of the percentage change in quantity demanded for a given percentage change in price.
[†] In the study, λ was set at 3.

10.2.1.2 Issues with Welfare Analysis

Welfare analysis is generally best suited to evaluating a product or service that is uniform in quality and availability. This is not a major drawback for consideration of defined datasets such as addresses or topography. However, it is perhaps less useful for analyzing the socioeconomic value of a package of fundamental datasets across the economy.

The nature of the demand curve is also critical to the examination of consumer and producer surplus. Estimates of elasticity of demand based on two price–quantity observations provide little evidence of the shape of the demand curve between or beyond those observations.

Welfare analysis is also a static analysis. It does not (without the use of multipliers) take into account changes in demand patterns, innovation, competition, changes in data quality, or resource shifts in the economy resulting from changes in the use of the data. To some extent, this can be addressed through the use of multipliers.

Welfare analysis is useful for comparing changes in socioeconomic impacts of different pricing policies, as long as the range of change along the demand curve is not large. It is less helpful when estimating socioeconomic value along the total demand curve because of difficulties in estimating its shape.

10.2.2 Estimates of Gross Revenue

Some studies in the past have used gross revenue to estimate the size of the geospatial sector. For example, in 2013, Oxera Consulting estimated that the global revenues from geoservices ranged between US$150 billion and US$270 billion per year as one indicator of the size of the sector (Oxera, 2013). However, such approaches can be challenging. The treatment of the geospatial sector in the standard industry classifications of national accounts is, in many cases, inadequate for the purpose of estimating gross revenues. The sector is generally accounted for partly in the professional services sector and partly in the information technology (IT) sector in standard industry classifications. Extracting a realistic estimate of the total revenue for the sector from national accounts requires considerable judgment, for which there is generally little data.

Revenue estimates indicate the size of the transactions being generated by an industry sector, but as the Oxera report notes, they do not indicate the full economic contribution of the industry.

10.2.3 Value-Added Approaches

Value-added represents the value of output produced by an organization less the cost of inputs used to produce that output. In practice, it largely reflects the return to capital (profits) accruing to firms plus salaries and wages paid to employees. Gross value added (GVA) across sectors of the economy makes up the bulk of the gross domestic product (GDP).*

The Oxera report also estimated that the GVA by the geoservices sector was $113 billion compared to a GVA of all sectors in the global economy of $70 trillion, suggesting that geoservices account for roughly 0.2% of global GDP (Oxera, 2013). Such comparisons can help place the contribution of each sector in context. Gross value-added approaches are far more rigorous than general descriptions of market size when questions of economic contribution are concerned. However, they are less useful when examining the economic impact of certain investments or of policy change.

10.2.3.1 Value Added Along Supply Chains

Value-added analysis can be undertaken along a supply chain to enrich the analysis of the relative contributions from organizations involved in the supply of and use of geospatial information. This extends the estimate of the value added beyond that immediately associated with the geospatial sector alone. This approach can provide a more realistic estimate of the wider contribution of geospatial systems to the supply chain.

The Oxera report cited above estimated GVA along a value chain for geospatial services. Such an approach was also adopted by Oxera in 1999 to estimate the economic contribution of ordnance survey (OS) in the United Kingdom (Oxera, 1999). The production chain adopted in the study is shown in Figure 10.5.

This study estimated the value added by each sector along the supply chain that could be attributed to the production and use of OS's geospatial products. The result was a total value added, attributable to OS, its suppliers, and distributors, of around £86 million.

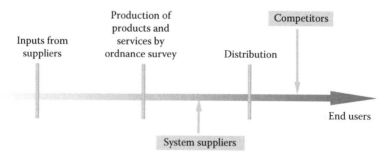

FIGURE 10.5
Production chain assumed by Oxera. (From Oxera, *Economic Contribution of Ordnance Survey in Great Britain*, Oxera Consulting, London, UK, 1999. With permission.)

---

* GDP comprises the sum of gross value added plus taxes less subsidies.

The study also found that between £79 billion and £136 billion of the GVA was dependent to some extent on OS's products and services (between 12% and 20% of GVA for the economy).

The principal sectors included in this estimate were local government, utilities, and the transport sector. The report's authors emphasized that this did not mean that, without geographic information and OS's products and services, the size of the economy would be some £79 billion to £136 billion less. The economy would find other ways to obtain geographic information. Rather, the estimates were used to demonstrate that geographic information, in general, and OS's products, in particular, played a significant role in the economy.

A report published by the Allen Consulting Group estimated that the value added attributable to spatial information and systems in Australia was around $12.5 billion in 2010 (ACG, 2010). This was found to arise mainly in the areas of government administration, property, business services, construction, and mining. This amounted to around 1% of GDP at the time. The same report estimated the GVA in New Zealand to be $1.6 billion or around 1.4% of GDP.

Such approaches provide information about the size of the footprint of the geospatial sector in an economy, but they are dependent on the quality of the underlying input–output tables and estimates of the proportion of each sector's value added that can be attributed to geospatial information systems.

10.2.3.2 Value Chain Analysis

Value chain analysis is another approach to understanding business processes along supply chains. Value chain analysis documents business processes to provide insight into where there are the opportunities for adding value and generating further value and competitiveness.

A hypothetical example of a value chain for property services is shown in Figure 10.6. In this example, provision of property data as part of spatial

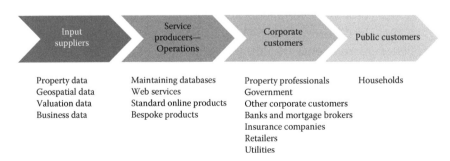

FIGURE 10.6
Hypothetical example of a value chain with reference to property data. (Based on unpublished work by ConsultingWhere. With permission.)

data infrastructure (SDI) delivers savings in time and costs for both government and private-sector service providers.

Value chain analysis can identify wider benefits that occur along the supply chain for other parties. Such analysis can reveal new insights into productivity and employment impacts and identify network effects that are not captured in traditional static analysis. Network effects can lead to further value creation, as outlined in Longhorn and Blackmore (2008).

10.3 Economic Impact Assessment

The discussion in Section 10.2.2 focuses on estimating the size or value of geospatial services as an indicator of their significance to the economy. However, when decisions are being made on investment in geospatial information or policy change, it is necessary to consider economic impact rather than economic significance, that is, what net additional value has been or will be created by an investment or a policy change.

Geospatial information services are enabling technologies that improve the productivity of firms or government services. In many cases, these services foster new applications and create new markets, resulting in extra value in the economy (Bernknopf and Shapiro, 2015). This extra value may come in several forms:

- Cost savings in doing the same things more efficiently
- Delivery of new products or services producing greater value in the use of the resources required to deliver them
- Dynamic savings within and across sectors of the economy, creating new value not previously possible
- Lower costs for governments and regulators in managing environmental, health, and social services
- Better environmental, health, and social outcomes with the resources available

From an economic perspective, the economic impact of geospatial services might be summarized as the ability to deliver more output for a given combination of resources. This concept is illustrated in Figure 10.7, which shows an economy's *production possibility frontier* shifting outward as a result of the use of geospatial data services.

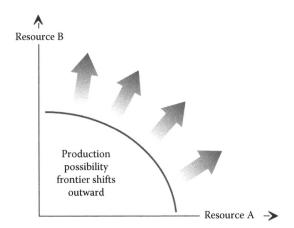

FIGURE 10.7
Geospatial information and the economy's productive capacity. (From ACIL Tasman, 2008. With permission.)

Economic impact assessments attempt to estimate part or all of this total extra value. To do this, the analysis must establish two scenarios:

- A reference case, representing the situation with geospatial data services that are to be assessed
- A counterfactual, representing the situation without these geospatial data services

The difference between the reference and the counterfactual scenarios represents the economic impact of implementing the reference scenario. This is illustrated in Figure 10.8.

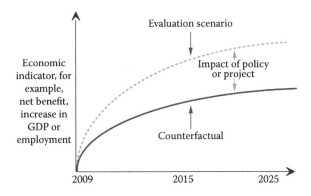

FIGURE 10.8
Economic indicators comparing an evaluation scenario and a reference case.

Defining the counterfactual is extremely important. Houghton (2011) points out that a counterfactual should represent the next best option that would be available in the absence of the reference case.

Common mistakes are to assume that without the reference case, nothing happens. This is rarely the case. Other approaches are likely to exist, but they are usually less efficient than the approach under evaluation, or they deliver benefits at a much later time. To be credible, an economic impact assessment must have a credible counterfactual.

This concept is drawn on an analysis of the value of information of moderate-resolution satellite imagery in a paper published by the U.S. Geological Survey (Forney et al., 2012). The reference case was the use of satellite imagery to support better management of the balance between agricultural production and groundwater quality. The counterfactual was the situation that would have arisen without the use of the imagery.

10.4 Economic Impact Methodologies

In the following segment, three approaches to undertaking economic impact analysis are discussed:

- Benefit–cost analysis*
- Multi-criteria analysis (MCA)
- Computable general equilibrium modeling

Other methodologies discussed are input–output multiplier analysis and real options analysis. These can be relevant to valuing geospatial information but have not been applied extensively in valuations of geospatial information to date.

10.4.1 Benefit–Cost Analysis

Benefit–cost analysis is an empirical form of welfare analysis. It is an investment evaluation technique that can also be used for policy evaluation. Benefits represent the additional value produced as a result of an investment or policy change. This is the additional value that is created under the reference case when compared with the counterfactual. A similar approach is taken to costs.

An investment or policy change is considered to be economically justified when the net benefit (total benefits less total costs) is equal to or greater than zero. Benefit–cost analysis generally involves developing a time series of benefits and

---

* In some countries, this technique is referred to as cost–benefit analysis.

costs and using discounting techniques to bring the cash flows back to a common date. The discount rate reflects the opportunity cost of capital.

Cash flows are discounted back to a reference year (usually the date of evaluation or the commencement of a project) to calculate the present value of the net cash flows.

The present value of a monetary value $A(n)$ accruing in a future year n is discounted according to the following formula:

$$\text{Present value of } A(n) = \frac{A(n)}{(1+r)^{n-1}} \tag{10.2}$$

where r is the discount rate.

Net present value (NPV) benefit can be calculated with the following formula:

$$\text{Net present value} = \sum_{1}^{n} \frac{B(n)}{(1+r)^{n-1}} - \sum_{1}^{n} \frac{C(n)}{(1+r)^{n-1}} \tag{10.3}$$

where:
$B(n)$ is the benefit in year n
$C(n)$ is the cost in year n

The results can be expressed as either an NPV, a benefit–cost ratio (BCR), or an internal rate of return (return on investment [ROI]). The ROI is the discount rate that equates the present value of benefits to the present value of costs. It is a popular metric and is well recognized. However, it also exhibits some technical limitations and needs to be treated with care.[*]

Benefit–cost analysis can require complex calculations and careful treatment of uncertainty. A primer for undertaking benefit–cost analysis has been published by the National Aeronautics and Space Administration (NASA) (NASA, 2013). This sets out the steps and approaches required for application of benefit–cost analysis to Earth observations (EO) from space. However, the approach also has general application to the task of assessing socioeconomic value of geospatial services.

Benefit–cost analysis generally focuses on a subset of benefits rather than attempting to estimate the benefit delivered across all applications, as illustrated in Figure 10.9. This figure assumes that the geospatial data under evaluation are supplied free, as it might apply under a full open data policy. The total value of the data is the area under the demand curve (consumer surplus). In such cases, the results obtained from the selected subset are likely to capture a lower bound estimate of the total benefits that accrue across all users.

---

[*] The ROI can have more than one solution where cash flows switch between positive and negative over time. It also assumes that the borrowing and reinvestment rate is equal to the ROI. This is unrealistic where the ROI departs significantly from the borrowing or reinvestment rate. There are techniques to address these problems, but this is beyond the scope of this chapter.

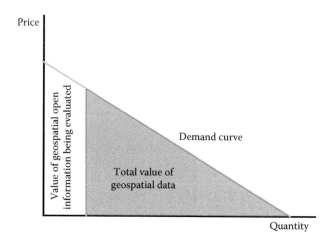

FIGURE 10.9
Value of open geospatial data. (From Raunikar, R.P., Pers. comm., 2011. With permission.)

10.4.2 The Benefit–Cost Analysis Process

The following points outline the steps in a conventional analysis (TIDE, 2015):

1. *Define scope and evaluation criteria*: It is important to define the clearly defined scope based on the customers' expectations and the evaluation criteria, such as the period over which the costs and benefits are to be evaluated (often referred to as the project life cycle), the discount rate to be adopted, and the form of presentation of the results.

2. *Strategic context and implementation options*: These set out the strategic case for investment and shortlisted options. The number should be kept down to no more than three or four options of varying complexity and include a *do nothing* or *do minimum*, scenario, which is important, as the reference situation might have negative impacts, particularly if the project is required to respond to regulatory requirements.

3. *Identify project impacts*: All costs and benefits resulting from the project's implementation should be identified. In doing so, it is important to understand the causal relationship between the measure and its various impacts (positive and negative). Usually, impacts for public investments will include the impacts on the organization itself, other public-sector organizations, citizens, and businesses. For a comprehensive approach, spillover effects, where economic effects in one context lead to impacts in a seemingly unrelated context, may also be relevant. Local guidance maybe required as to whether citizen benefits will be considered valid. In some instances, it maybe the case that only benefits to the organization will be taken into account.

4. *Prioritizing measurement effort*: Each impact is considered, and the most significant ones, either in terms of monetary or socioeconomic consequences, are identified. Criteria for prioritization of the potential impacts need to be agreed (Hubbard, 2015). Clearly, the more the impacts involved, the more the effort required. Furthermore, if the investment can clearly be demonstrated with a small number of positive impacts, then it is often easier to explain to decision-makers and assert credibility.

5. *Quantify monetary valuation of impacts*: Instruments for measurement of market impacts, such as increases in productivity, are often evaluated by directly measuring performance gains by reducing the work required for particularly manually intensive processes or removing duplication in data acquisition. In other cases, measurement may only be possible by an indirect method, such as measuring something that has been shown to have strong correlation with the impact in question. For instance, in the insurance market, there is a well-established correlation between the time required by the customer to get an online quote and the likelihood of acceptance—the shorter the period, the greater the sales conversion rate.

    Different ways to monetize nonmarket effects are covered later in Section 10.4.3. An additional consideration is what is referred to as apportionment. As the geospatial system and its information are usually only a component of an impact, a defensible method (commonly, expert opinion) of assigning a percentage of the impact to it must be determined.

6. *Create financial model*: Many standard models are available. Software packages such as TRN4 are widely used in the transport sector. However, simple spreadsheet packages, such as Microsoft Excel, are usually sufficient. They support the necessary calculation of indicators such as net present value (NPV), BCR, and internal rate of return. Often, the NPV is used to justify the adoption or rejection of a project. The BCR is often used to rank different projects in order of benefits per unit of invested capital, as it allows comparisons across different project types, sizes, and durations.

7. *Sensitivity analysis*: As impact values are associated with predictions of future behavior, they are innately uncertain. The risks of being smaller or larger than predicted must be taken into account. This is achieved by sensitivity analysis, whereby the impacts on the financial model are varied to provide upper and lower bound indicators, based on the so-called pessimistic and optimistic assumptions.

8. *Presentation of results*: It is an unfortunate fact that many analyses fall down at this stage. Rarely will decision-makers have the time or inclination to study the analysis in any detail, so summarizing

results is vital. Explain how the investment will benefit the organization or society at large, make it succinct, remove technical detail, and
focus on the most compelling arguments.

Benefit–cost analysis is a fundamental tool for the development of business
cases. It has many uses beyond simple assessment of benefits and costs.
Sensitivity analysis can provide information to better manage downside risk
and prioritize options.

In some circumstances, it can be appropriate to undertake a cost-effectiveness
analysis rather than a full cost–benefit analysis. This is often used when assessing alternative approaches to meeting legislative and regulatory mandates such
as those applied in environmental or health standards. This does not place a
value on the mandates but determines the most cost-effective solution to meeting them (Macaulay, 2006).

There are two issues that require attention in the application of benefit–
cost analysis. The first arises in highly uncertain environments. Benefit–cost
analysis assumes that all costs are locked in over the full project period. This
does not allow for the possibility of sequential investment decisions over
time that can lower or defer costs with further experience.

The second is the potential for portfolio approaches to investment. In
some circumstances, more optimal outcomes can be found through assessment of a portfolio of investment options rather than evaluating a single
investment option. An example of this is in evaluation of water supply
investments, which involve a network of supply sources. In such circumstances, it can be more appropriate to apply benefit–cost analysis to alternative portfolios of investments rather than evaluating a single investment
option.

10.4.3 Valuation of Tangible Benefits

Tangible benefits are those that can be quantified. They can be described
in terms of monetary or physical values such as productivity, employment,
or even time saved through better use of operating systems. However, for
estimates of economic value, it is necessary to express benefits in monetary
terms. For revenue assessments, it requires an estimate of market price and
quantity sold. Benefits can be both direct and indirect.

10.4.3.1 Direct Benefits

Direct benefits are the value of the benefits with reference to market outcomes from provision of a good or a service. Increases in output or reductions in inputs are quantified and combined with market prices to estimate
monetary value. Market prices may need to be adjusted for subsidies, tax, or
monopoly pricing.

10.4.3.2 Indirect Benefits—Defensive Expenditure or Substitute Cost Approaches

In some cases, benefits can also be quantified in terms of time saved, complaints reduced, clients serviced, or reduction in exposure to natural hazards. Although many such benefits can be difficult to price, they can often be estimated in monetary terms by using substitute costs as a proxy for value.

An example of indirect benefits realized through better use of information is contained in a benefit–cost assessment of geological maps undertaken for the U.S. Geological Survey (Bernknopf, 2004). The report noted that information from mapping data is important for the following:

- Better management of water quality
- Mapping of groundwater
- Managing natural hazards such as landslides or volcanic activity

In each case, the value of the information contained in geological maps was in reducing the probability of environmental or other damage costs through better decision making. The report used two case studies to illustrate this. The first described how better decisions could be made on the location of landfill sites. Information on permeability of soils provided regulators with more accurate information on the potential for contamination of soils around landfill sites. This enabled regulators to be more precise about areas of environmental sensitivity. The value was reduced loss of property values, as a result of more effective location of landfill sites.

The second case study addressed the use of mapping data to better locate the Washington Bypass, a new arterial highway. Mapping data enabled better prediction of land slide potential, which reduced the mitigation costs for slope failures. Benefits were estimated as information for planning highway alignments.

Estimating changes in defensive expenditure is another way of estimating benefits. For example, the benefits of improved flood control can be estimated from the reduction in the expected annual average damage costs from future flood events. Such an approach was used in an assessment of the value of earth observation (EO) from space in Australia (ACIL Allen, 2015). In this study, the reduction in the average annual damage from floods, fires, cyclones, and extreme weather by using remote sensing from EO satellites was estimated to be $213 million in 2015.

In these examples, benefits were calculated as avoided expenditure or reduced damage costs. There are many academic and other studies of the cost of incidents, such as fires, floods, and earthquakes, that can provide useful data for such studies.

10.4.4 Valuation of Nonmarket (Intangible) Benefits

Generally, when consumers want to consume or benefit from a particular good or service, such as a hamburger or a car wash, they purchase the product or service from a supplier. The agreed-upon price for this exchange is a useful piece of information. By the fact that the trade occurred, we know that the seller was willing to provide the good or service at that price and the consumer was willing to pay for the good or service at that price. How much individuals are willing to pay for a particular good or service is a common measure of the value of that good or service.

Nonmarket benefits are any goods or services that individuals can consume or benefit from, without purchasing directly. Examples of such public goods and services include public parks and data that are made freely available. Individuals might also benefit from goods or services purchased by others. For example, when one person purchases a vaccination to avoid an illness, he or she also provides the benefit of not passing that illness to others with whom they interact. When one person spends the time and money to take public transit rather than driving, others benefit from being exposed to less air pollution.

In both cases, because the beneficiaries did not pay for the good or service directly, we do not observe an amount that they are willing to pay for the benefit that they received. Those individuals still have an amount that they would be willing to pay for that good or service, but it is unobserved. The value of the good or service can be thought of as the sum of all the observed and unobserved amounts that beneficiaries are willing to pay. This section focuses on methods for estimating the value of a good or service when the willingness to pay of most or all beneficiaries is unobserved, particularly with respect to cases where the good or service is geospatial information.

Geospatial information may directly increase the utility through the use or Non–use values discussed in Section 10.1.2. We assume that the value of information is determined solely by these observed or unobserved benefits provided to individuals. Rather than assigning any value to the existence of the information in and of itself, the information has value only when an individual does or could benefit from it, either directly or indirectly.

One mechanism through which geospatial information might provide these benefits is known as *Bayesian updating*. In this process, agents have a prior belief about the probability of some outcome, such as drought, and make investments of time and money with respect to their belief. In a Bayesian model, new geospatial information provides value by enabling individuals to update their prior beliefs with a new, more accurate perception of the probability of an outcome such as drought. More accurate beliefs about probabilities enable individual agents to make more efficient decisions about how to invest their time and money. Any resulting gains from this more efficient decision-making can be directly attributed the new geospatial information. The Monty Hall problem offers a classic example of how new information can enable agents to update their beliefs about probabilities, resulting in more efficient decision-making.*

It is also important to not double count the value of information. End users' willingness to pay for the final data product that they receive incorporates both the value of the original raw data and any additional processing or analysis of the data that has occurred before receipt by the end user. An intermediate user, such as an organization that sells processed or analyzed data, also values the geospatial information, because users further down the line are willing to pay for processed or analyzed data. Summing the willingness to pay of both end users and intermediate users would double count some individuals' value of the raw geospatial information.

Federal governments finance the collection of some geospatial information. Taxpayers finance government collection of geospatial information; the information is often made freely available, while in other cases, access to the information may require paying a fee. Other geospatial information is collected by private entities and is sold at market prices. Hence, some geospatial information is directly linked to market prices but other information is not. Many individuals who benefit may not have purchased any services. Although these users have not purchased the information, it has still provided value to them.

It is important to accurately estimate the value of information, in order to determine the efficient level of investment. Rather than collecting revenue from the sale of information to finance the costs of collection, governments collect taxes across all potential users to cover the expenses associated with collection. In order to ensure that resources are being allocated efficiently, it is important to assess the value of publicly distributed information, in order to determine the extent to which public resources should be invested.

10.4.4.1 Methods for Calculating Nonmarket Benefits

There are two general approaches used to estimate the value of nonmarket benefits. The first is *revealed preference,* and the second is *stated preference* methods. These are discussed in detail in the following sub-sections.

---

* The Monty Hall problem is a game based on a former game show that demonstrates the case for re-evaluating decisions as new information comes to hand. Players are asked to choose one of three doors, of which one contains a car while the other two contain a goat. The player chooses one door, but before it is opened, the host then looks behind the other two doors and opens one of them that reveals a goat. The host then asks the player if he wishes to change his decision. It turns out that the player who switches doors has a 2/3 chance of winning the car on average, while the player that does not switch has a 1/3 chance of winning the car. The given probabilities depend on specific assumptions about how the host and contestant choose their doors but the example demonstrate the value of new information on probability assessments.

10.4.4.1.1 Revealed Preference

Revealed preference methods require observable data about behaviors or information related to the value of the nonmarket good or service. Revealed preference methods tend to estimate the value of information by comparing outcomes for a *treatment group*, which has access to the nonmarket benefit, with a *control group*, which does not have access to the nonmarket benefit. Ideally, which individuals have access to the nonmarket benefit is randomized. If the treatment and control groups are not randomly assigned, then it is important to control for nonrandom differences between the two groups. In some cases, researchers extrapolate the value of nonmarket benefits from the value of similar goods or services or use information about the value of similar goods or services as a proxy for the unobserved value of nonmarket goods. Well-known revealed preference methods that are popular for estimating the value of a nonmarket good or service include benefits transfer, travel cost, hedonic pricing, and various control-and-treatment group methods.

10.4.4.1.2 Benefits Transfer

Benefits transfer estimates the value of a nonmarket good or service, based on the estimated value of a similar good or service. For example, one could estimate the value of a new park, based on a study that was done on the value of a similar park in an adjacent area. Benefits transfer has the advantage of being simple and cheap to implement, and for this reason, it is commonly used by government agencies that are required to perform a large number of cost–benefit analyses in a limited budget. The main downside of benefits transfer is that the accuracy of the estimate depends on the existence of a well-done study on a very similar nonmarket good or service.

Benefits transfer studies typically follow one of two main approaches. The first case, known as *average value transfer* or *unit value transfer*, looks at prior studies to establish the average value of a single *unit* of the nonmarket good or service. Examples include regional estimates of the average value per person per day of camping, swimming, and sightseeing resources (Rosenberger and Loomis, 2001).

The second approach, known as *benefit function transfer* or *value function transfer*, involves using a model to estimate the value attributed to different aspects of the nonmarket good or service. In some cases, prior studies (such as hedonic pricing studies described below) may have already estimated the value associated with different aspects of the nonmarket good or service. In other cases, researchers can examine the results from a large number of existing studies to identify which aspects of the nonmarket good or service are associated with higher willingness to pay.

The function transfer method is generally considered to produce more accurate estimates, because it does not assume that the nonmarket good has the same value as the average value in other studies.

As an example of a benefit transfer study, Eade and Moran (1995) used geospatial data on the location of roads, rivers, and ecological factors to estimate the direct and indirect use values associated with land in the Rio Bravo Conservation Area in northwestern Belize. They used the existing studies to estimate the value of each area of land, based on the properties of that 50 m^2 unit of land, and then pieced these units together into a map that showed the geospatial variation in the economic value of the region.

10.4.4.1.3 Travel Cost

Another common method for estimating the value of a nonmarket good or service is the *travel cost method*. The heart of this methodology is the idea that for nonmarket goods or services that require travel or other effort to use, the amount of money that individuals are willing to spend to access that good or service is a good indication of how valuable the good or service is to them. Typically, this approach requires some information about the users of the good or service, such as the location from which they are traveling. Information about alternative goods or services can also be useful.

There are several ways to implement the travel cost method in practice. The simplest option is to gather information about the location from which users have traveled. The researcher then estimates the cost of travel from various locations or distances or preferably collects this information from the users. If the researcher also has information about the number of individuals from each location that uses the nonmarket good or service per year, then it is straightforward to multiply the total travel costs for each location by the number of users from each location to measure the total willingness to pay of society as a whole.

One downside of the travel cost method is that we do not directly observe the user's willingness to pay. Rather, we know that the users are willing to pay at least as much as it costs them to use the nonmarket good or service. The true willingness to pay might be higher. However, it is also difficult to be sure that the cost of travel truly reflects the individual's marginal cost of using the nonmarket good or service. An individual who traveled from far away may have been brought to the area for several reasons and would not have been willing to pay the travel cost for that particular good or service alone. That individual's true willingness to pay would be lower than that estimated using the travel cost method. Because of these problems, it is difficult to be certain that the travel cost method has not under-estimated or over-estimated the true value of the nonmarket good or service.

Some of these problems can be diminished. Bateman et al. (1996) showed how geospatial information system software can be used to improve the accuracy of location and distance measures needed for this methodology. Using trips, including Bryce Canyon National Park, Mendelsohn et al. (1992) showed how the travel cost method can be adapted to account for multiple-destination trips. However, it remains difficult to ensure that the travel cost method produces an accurate estimate of the willingness to pay for a particular good or service, leaving some authors such as Randall (1994) to conclude that the travel cost method is only useful in supporting the results of other techniques.

10.4.4.1.4 Hedonic Pricing

Often, we know the final value of a multi-faceted good or service, but we do not know how much each of the various factors contributes to that value. The most common example is a house; we know the total sale price of the house but do not know how much of that value comes from having 2000 square feet and three bedrooms and how much of that value comes from the home being located near good schools or a park. *Hedonic pricing* looks at the properties and total values of many houses, in order to infer how much each aspect of the house contributes to the total value. This can be used to evaluate the value of a nonmarket good or service, such as a park or a school's quality. The researcher can estimate the amount by which having a nearby park or high-quality school increases the value of surrounding property, and the sum of these contributions to the value of property reflects the total value of the nonmarket good or service to the homeowners.

In order to perform a hedonic pricing study, the research must have detailed data about housing prices and characteristics. The simplest approaches to hedonic pricing often involve a structural model, with assumptions about how factors relate to each other, such as a linear regression model, and then, the research estimates those parameters. However, there are a variety of more flexible nonparametric estimation methodologies, which are less likely to misspecify how parks, school, or other factors relate to the housing price. For example, see McMillen and Redfearn's (2010) approach to estimating access to public transit in Chicago.

10.4.4.1.5 Control-and-Treatment Group Methods

A final, broad category of revealed preference methods for estimating the value of nonmarket good and services involves comparison of the control and treatment groups. At the heart of this method is a comparison of a *treatment group*, which has access to the nonmarket good or service, with a *control group*, which does not have access to the nonmarket good or service. Ideally, these two groups are otherwise similar. If not, the researcher will need to control for other differences between the groups.

One such approach for estimating the value of geospatial information is to look at the trend over time for a particular outcome, such as a stock price or gross revenue, and to look for changes in that outcome after the nonmarket benefit becomes available. This could mean looking for changes in agricultural revenue after new satellite data first becomes available or looking for changes in the stock price of certain companies after the quality of data available to them improves. In order to attribute changes to the availability of the nonmarket good, we require eliminating the impacts of other factors, which may also be changing at the same time. This can involve detrending data, such as to control for seasonal variation. It also requires us to ensure that other unrelated shocks do not occur at the same time. For example, if a company brings on a new CEO who institutes several new changes, including increased use of geospatial information, it is difficult to know how many changes in revenue are due to the increased use of geospatial information and how many of them are due to other changes made by the new company leadership.

For example, Roll (1984) examined how the errors in temperature and rainfall forecasts in Orlando, Florida, affect the price of orange juice futures. In this case, the nonmarket benefit is access to more accurate weather forecasts. Roll found that an unexpectedly low temperature increases the current price of a contract to deliver the orange juice several months later. The price increase reflects an expectation that cold weather will damage crops, decreasing the quantity of oranges available. Figure 10.10 shows how this price change relates to the financial magnitude of the loss caused by the unexpectedly low temperature. Roll (1984) estimated the difference between *old price* and *new price* caused by an unexpected temperature shock of a given size. The shaded area represents the revenue lost due to the poor weather, which is the maximum loss potentially prevented by improved information. Although geospatial information cannot prevent this loss entirely, better information can be used to inform when to invest in techniques to mitigate frost damage.

Sometimes, we want to know how geospatial information influences the behavior of individuals or companies in order to determine how much they value that information. Suppose we want to look at how the amount or quality of geospatial information available to an individual or organization is correlated with some behavior, such as their decisions about what goods to purchase or how to spend their time. Suppose we want to know if giving

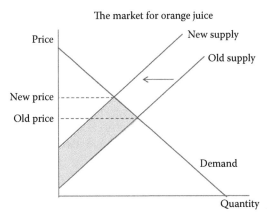

FIGURE 10.10
How a change in price related to the total financial loss.

farmers better-quality weather information causes them to plant different crops. We might worry that the farmers that always have access to better information might always plant different crops than farmers that normally have access to lower-quality weather information, for reasons unrelated to information quality. For example, perhaps farmers with better information also have higher incomes or better equipment, or regions with better weather information have different climates. Household-level fixed effects remove the influence of differences between households. Region-level fixed effects remove the influence of difference between regions but do not control for differences between households within a single region.

Fixed effects can be used to control for constant differences between groups in order to focus on the effect of changes that occur over time. For example, Rosenzweig and Udry (2013) used household-level fixed effects, among other control variables, to show that farmers with access to higher-quality forecasts are more likely to increase their planting-stage investments in response to a rain forecast. The improvements in farmer's financial well-being showed the value of access to the nonmarket service of higher-quality weather forecasts. Miller (2016) used state-, year-, and month-level fixed effects, among other control variables, to show that areas with weather warning systems have fewer fatalities and injuries from tornadoes than would have occurred without the weather warning system. The value of the weather warning system can be expressed in terms of the number of prevented injuries and fatalities.

10.4.4.1.6 Stated Preference

Stated preference methods use surveys or experiments to determine how much individuals would be willing to pay in hypothetical or laboratory settings. These approaches are particularly valuable when the data to support revealed preference methods are unavailable, as is often the case with public goods. Stated preference methods may ask people to rank or choose between several alternative options in order to determine preferences over those options. Alternatively, stated preference methods may ask individuals for the amount they are willing to contribute to a particular public good. Just as revealed preference methods rely on assumptions about the appropriateness of the extrapolation or control group, stated preference methods rely on assumptions about the survey design, ensuring that individual's responses in hypothetical or laboratory settings are reflective of the decisions that those individuals would make in the unobserved interactions with the public good. Well-known stated preference methods that are popular for estimating the value of a nonmarket good or service include contingent valuation and choice experiments.

10.4.4.1.7 Contingent Valuation

This approach to stated preference studies is to essentially survey individuals, asking how much a particular nonmarket good or service is worth to them. This approach is known as *contingent valuation*, and it relies heavily on good survey design and encourages truth telling by limiting incentive, which might cause respondents to strategically report a value other than the true value of the good or service. The survey must carefully describe exactly what good or service in question is being studied, so that all respondents understand what they are evaluating. Moreover, the respondents themselves must be selected carefully. A random sample is ideal, and it is important to consider response rates and what type of individuals will respond. An in-person survey conducted by going door to door during the day might be more likely to miss individuals who are away at work. A survey conducted online might be less likely to get responses from individuals who are less computer savvy.

One concern with contingent valuation is that respondents might over-state or under-state their true willingness to pay. In a meta-analysis of literature comparing stated preferences with other estimates, Carson et al. (1996) found that, on average, the ratio of stated preference to revealed preference estimates for the same nonmarket good or service is 0.89, meaning that stated preference approaches tend to elicit smaller estimates of willingness to pay. However, Murphy et al. (2005) found that, on average, stated preference methods result in 35% larger estimates of willingness to pay, which suggests that most studies are over-stating the true willingness to pay.

10.4.4.1.8 *Choice Experiments*

Another approach to stated preference methodologies is known as a *choice experiment*. In these studies, instead of asking for dollars directly, ask individuals to choose between or rank preferences over different baskets of nonmarket goods or services. There may be costs associated with these options, such as choosing between spending $200 on trail maintenance in Park A versus $250 on trail maintenance in Park B. By presenting each respondent with randomly different prices for the same bundles shown to all respondents, the researcher can estimate how much the average respondent values one option over another. The logic extends to comparing sub-groups of respondents and to considering more than two options.

Boxall et al. (1996) discussed some of the advantages of choice experiments over contingent valuation. In particular, choice experiments offer more flexibility to the researcher after the study is complete and more control over the variety of responses that might get recorded. Careful design of a choice experiment can allow the research to consider several different aspects of the nonmarket *good or service and interactions between those aspects.*

10.4.5 Multi-Criteria Analysis

One of the major criticisms of benefit–cost analysis is that it does not give a complete picture of the socioeconomic benefits of an investment, because it focuses on quantifiable impacts. In an MCA, it is possible to incorporate both quantitative and qualitative criteria. The performance of a policy or project is assessed by scoring, ranking, and weighting the impacts rather than by expressing the impacts in monetary terms.

The impacts' performance values can be based on projections, literature, or experts' opinions, whereas the criteria's importance-weighting values are assigned in consultation with stakeholders. The final scores for each measure can be used to rank the measures, which, in turn, can be used to identify a limited number of options for detailed assessment.

As a very brief summary, the classic steps in a multi-criteria analysis are as follows:

1. Define the project/policies to be assessed: Alternative options are identified and described, including a *status quo* case.
2. Agree to assessment criteria: These may include sub-criteria and maybe aggregated according to the beneficiaries, for example, public sector, businesses, and citizens.
3. Assign weights to the criteria: Importance values are assigned to reflect their relative importance to stakeholders.
4. For each criterion, objectively assess the impact: If quantitative data are available, it is used; the fallback is to create a performance scale and use expert opinion to rank the options.
5. Calculate the impacts: The numerical score for criterion is multiplied by the weight.
6. Rank the options based on the calculated scores.
7. Apply sensitivity analysis: See previous discussion on benefit–cost approach.
8. Presentation.

The criticisms of MCA center on its subjectivity. Selection of the analysis criteria and determination of their weighting are subject to bias. Similarly, individuals or groups involved in the analysis will have their own interests, leading to selection bias. To minimize the risk of both types of bias, various additional techniques can be introduced, such as the Delphi method to normalize expert judgments (Hsu, 2010).

For a more detailed discussion on the MCA methodology, see Dodgson (2009).

10.4.6 Computable General Equilibrium Modeling*

An alternative approach to estimating the benefits of new technologies is to estimate the economic impact on the value that they add to the economy.

---

* The authors are indebted to Guy Jakeman of ACIL Allen Consulting for assistance with discussion on CGE modeling and input–output multiplier analysis.

As discussed in Section 10.2.3 above, value added by a firm is the difference between the value of final goods and services sold and the cost of inputs required to provide those goods and services. It is the core component of GDP and gross national product (GNP). Changes in these measures can also indicate the overall economic impact of the use and application of new technologies.

There will always be winners and losers from shifts in technology and services—some tasks or jobs may, for example, become redundant—but the question is whether, overall, society can produce more and better outputs with the same inputs. This means that the productivity of the economy as a whole is greater.

The CGE models are a class of models that look at the economy as a complete system of interdependent agents (including industries, private households, governments, savers, investors, importers, and exporters). Typically, each agent within the model undertakes some form of optimizing behavior according to standard neo-classical cost-minimizing or welfare-maximizing principles.

The models explicitly recognize that changes, or shocks, impacting any one agent in the models can have repercussions throughout the entire economic system. Accounting for such repercussions is essential for assessing the potential economic impacts of a new project, policy, or other economic shock. Producers in most CGE models act according to production functions, which provide the capability to model economy-wide impacts of technology shifts, such as the introduction of new geospatial information technologies (*what-if* or *with-and-without* scenarios).

A CGE model is therefore a representation of all markets in an economy. The model solves a suite of prices by commodity and factor to clear all markets (balance supply and demand). Most CGE models are based on social accounting matrices drawn from national accounts. The number of sectors potentially available differs based on the model and available data but is typically around 60–90 for international models and 60–150 for single-country models. However, in practice, databases are often aggregated to around 30 or 40 sectors to improve computational run time.

Analyses of changes brought about by the introduction of a new technology are generally made by solving the CGE model for a specifically defined evaluation scenario that includes the new technology and a reference scenario that excludes the new technology. The difference between the two scenarios provides the estimates of the economic impacts of the new technology. The CGE models can produce estimates of impacts on a wide range of economic indicators, including the following:

- GDP and GNP
- Incomes
- Trade

- Consumption
- Investment
- Employment

Thus, it is possible to compare the accumulated impact of geospatial services on the economic aggregates between two scenarios representing different levels of access to geospatial information and services.

To produce an accurate CGE analysis, it is necessary to first estimate the likely improvement in performance of sectors of the economy that are affected by the use of geospatial information. This can be modeled in different ways but generally involves estimating the productivity impacts of specific applications from case studies. Productivity impacts can be improvements in the productivity of capital, labor, or any input in the production function for the sector.

These case studies are then augmented with information from research studies of levels of adoption across the sector in question. Rogers (2003) illustrates the proportion of the population that adopts an innovation as a probability density function consistent with a normal distribution. The phases of innovation adoption are characterized as innovators, early adopters, early majority, late majority and laggards. Using the point in time that half of the population adopts an innovation as the reference, 68% of the population is equally divided between early majority and late majority. Going more than one standard deviation but less than two standard deviations earlier than the midpoint we find the early adopters representing 13.5% of the population. Innovators representing 2.5% of the population adopt the innovation more than two standard deviations earlier than the mean adoption time. Lastly, the laggards representing 16% of the population adopt one standard deviation after the mean adoption time. Figure 10.11 organizes the probability density functions as geospatial adoption waves for the series of innovations associated with geospatial information (ConsultingWhere, 2010).

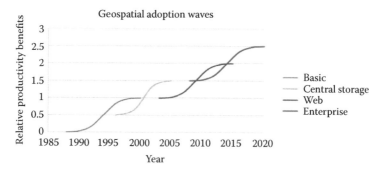

FIGURE 10.11
Adoption curves. (From Rogers, E.M., 2003; ConsultingWhere, 2010. With permission.)

It is necessary to assess what stage the adoption phases are for each application.

In practice, geospatial systems frequently come in waves as different technologies combine to produce new products or services. For example, the four adoption curves in Figure 10.11 show each wave of adoption building on a previous wave to produce incrementally higher levels of productivity.

The first wave represents the introduction of basic geographic information system (GIS) services in an organization. The second wave involves the central storage of these systems, so that more than one person in the organization can use the GIS. The third stage represents the migration to web-based application, and the final stage represents migration to enterprise-based applications. This finding was based on work undertaken in England and Wales for the Local Government Association (ConsultingWhere, 2010).

Estimates of the productivity impacts for an application and the level of adoption are combined to provide productivity shocks for a sector. These data are then entered into the CGE model, which produces a new equilibrium, showing the impacts on macro-economic aggregates such as GDP, consumption, trade, investment, and employment. An example of results for a study undertaken in Australia in 2008 is shown in Table 10.1.

This study estimated that the value of spatial information across the Australian economy ranged from $6.43 billion to $12.57 billion in 2007. The first estimate (scenario 1) was based on observed applications that could be quantified from case studies. The second estimate (scenario 2) was based on examples that had been provided but were estimated on the basis of evidence.

Results can also be reported by sector, as shown in Figure 10.12 from a study on the value of precise Global Navigation Satellite System (GNSS) positioning technologies (ACIL Tasman, 2012). The CGE model can produce increases in output by specific sectors in the economy. For example, Figure 10.12 shows that the sectors that benefited most from precise satellite positioning services were the grains industry, mining industry, and the construction industry. The CGE analyses has been applied in various studies for a variety of regions, including Australia, New Zealand, Great Britain, England, Wales, and Canada, with some analyses applied at the state or provincial level.

In 2011–2012, Ordnance Survey in the UK (OS) commissioned a study to assess the broader economic value of releasing OS data through its new OS OpenData Platform (Ordnance Survey, 2013). The report found that making nine OpenData, sets free at the point of entry would directly improve the level of productivity in the economy and higher overall levels of output. A CGE model was used to estimate the impact on macro-economic aggregates

TABLE 10.1

Economic Impacts of Spatial Information—Two Scenarios for Australia (2008)

	Scenario 1				Scenario 2			
	Productivity Only		Productivity Plus Resources		Productivity Only		Productivity Plus Resources	
	Percentage	A$ billion	Percentage	A$ billion	Percentage	A$ billion	Percentage	A$ billion
GDP	0.51	5.31	0.61	6.43	0.99	10.31	1.20	12.57
Household consumption	0.50	2.89	0.61	3.57	0.93	5.39	1.16	6.78
Investment	0.51	1.43	0.61	1.73	0.98	2.78	1.20	3.39
Capital stock	0.56	–	0.72	–	1.05	–	1.38	–
Exports	0.45	0.98	0.58	1.26	0.80	1.73	1.07	2.30
Imports	0.39	0.89	0.52	1.18	0.72	1.64	1.98	2.23
Wages	0.50	–	0.60	–	0.92	–	1.12	–

Source: ACIL Tasman, 2008. This modeling used the Tasman Global CGE model maintained by ACIL Allen.

Note: The resources scenarios took into account the increase in minerals and petroleum resources arising from geoscience and geospatial systems.

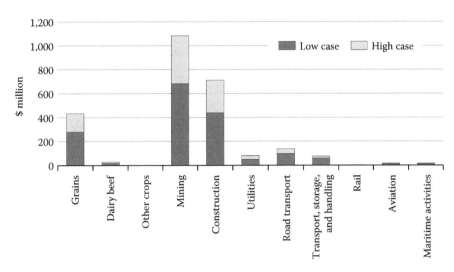

FIGURE 10.12
Example of output change by sector from the CGE modeling. (From ACIL Allen, 2013.)

of this approach. Specific estimates have been reported in Rogawski et al. (2016) and include that following:

- Ordnance Survey OpenData will directly deliver a net £13 million– £28.5 million increase in GDP by 2016. The main components of this increase are net productivity gains (£8.1 million–£18.2 million) and additional tax revenues (£4.4 million–£8.3 million).

- The GDP increase is net of £3.7 million per year, applied as a negative to UK exports to account for OS OpenData being integrated into products of companies paying taxes abroad. Despite this loss of export income, overall, the value of exports to the economy increases by £6.1 million–£10.3 million as other sectors of the economy expand.

- The increased exports will enable UK residents to purchase more foreign goods, increasing real imports by £3.6 million–£7.1 million.

- Real national disposable income (real GNP) will increase by £10.2 million–£24.1 million by 2016, indicating an increase in the economic welfare for British society as a whole.

- Real consumption will increase by £8.1 million–£20.3 million, and real investment will increase by £2.3 million–£5.1 million as result of OS OpenData.

The strength of the CGE analysis is its ability to account for productivity improvements across more than one sector of the economy and for resulting resource shifts between sectors. However, the result are only as robust as the quality of the case studies. This method requires extensive and robust case studies and estimates of levels of adoption to build credible results.

10.4.7 Input–Output Multiplier Analysis

The direct impacts on the economy of a particular industry are represented by the value added by that industry, as discussed in Section 10.2.3. The intermediate inputs used by an industry (e.g., software to support distribution of geospatial data) can be sourced either from within a national economy or from foreign economies. If purchased from within the national economy, the portion of value added embodied in the intermediate input is indirectly associated with the activity of that industry. In a national context, input–output tables and associated *input–output multipliers* can be used to estimate the indirect economic contributions of intermediate goods.

An input–output model is a quantitative economic technique that uses a representation of the interdependencies between different branches of a national economy or different regional economies to estimate the contribution of a particular sector or activity to the economy and is often used to draw conclusions about the potential impact of changes in economic activity. The modeling recognizes the impact of inter-industry transactions involved in the production of a specific good or service (Sargento, 2009).

An input–output approach estimates how many goods and services from other sectors are needed (*inputs*) to produce each dollar of *output* for a specific sector being examined. These tables can be used to develop multipliers that represent the total amount of a good or service that must be produced to meet a final demand of that good or service, assuming that the production technology and all prices remain constant. These multipliers can then be used to estimate regional or national impacts of an increase in output from a specific sector.

Input–output tables may be generated as part of the national accounts. In some circumstances, economic and research agencies produce national and even regional input–output models. The U.S. Bureau for Economic Analysis, for example, produces a regional input–output modeling system that includes regional multipliers that can be used to assess wider regional impacts of a change in one sector (Ambargis, 2011).

The input–output multipliers allow rigorous and credible analysis of the economic footprint or *contribution* of a particular facility, industry, or event for the region of interest. Because they contain a strong understanding of the size and inter-linkages of various industries within an economy, input–output multipliers have often been used for analyzing the potential economic impact of various types of economic change—they are used for *what-if* analysis? However, caution needs to be exercised in their application for this purpose, because they rely on a range of limiting assumptions, not least of which is an inability to change relative prices in response to demand and a theoretically unlimited supply or loss of factors to meet the modeled change in supply.

The assumptions underlying input–output models typically mean that they predict bigger economic effects compared with other modeling techniques. Misuse of input–output multipliers for the purpose of impact analysis has led to skepticism of their general use in favor of other tools such as the CGE modeling discussed previously. Notwithstanding this, they are still suitable for understanding the economic linkages between a given facility or industry to gain an appreciation of the wider interactions of the industry beyond its direct contribution.

Input–output multiplier *analysis* has not been widely used for estimating the economic impacts of geospatial systems.

10.4.8 Real Options

An important issue for assessing the value of investment in geospatial information is the fact that it will almost certainly yield access to growing volumes of information, where only some of its potential applications and value can be currently identified. Similar to many R&D investments, investments in geospatial information can be expected to lead to growing time series of data whose future value is necessarily highly speculative. These time series will embed options for future applications, including in ways not currently envisaged.* It would seem reasonable and appropriate to recognize these options as *cream on top* of the immediately planned applications. Failure to recognize this could be expected to imply systematic bias toward undervaluing the investments—and could well result in underinvestment.

This approach—recognizing *option value* on top of planned use value—is worth considering if there is to be an appropriate balance between planning whether to invest and the level and form of investment in these systems. It may not be necessary to attempt any precision in quantifying the option value, but recognition of the existence of this extra value, and some characterization of whether and how it could prove important over time, could comprise a sound approach to planning and justifying forward investment strategy.

A further characteristic of the emerging use of geospatial information systems is the impact of exogenous technological change on its use. These external changes are difficult to predict, but history has shown that they can

---

* The potential value of data that might not be immediately recognized is discussed in Houghton (2011).

find future valuable uses as other technologies emerge. One example of the use of geospatial systems has been the development of autonomous mining vehicles and robotic mining that have emerged through the convergence of GIS, sensors, vehicle control systems, and precise global navigation satellite system (GNSS).

Real options thinking can help in deciding whether there is value in preserving an option for the use of data at a later date. It can also help with decisions of how long to preserve the option and when preserving that option should be curtailed.*

There is no evidence in the literature of real options being applied to valuations of geospatial information. However, it has been used in Australia for valuing investments in R&D, including for the Square Kilometer Array telescope in Western Australia and for research into geoscience and marine science in Australia. This research has not been published.

10.5 Comparison of Methodologies

The methodologies outlined previously describe a range of approaches to estimating economic value. Choice of appropriate approach depends on the context and needs of decision-makers.

Welfare analysis, using estimates of consumer and producer surplus, is generally appropriate for valuation of specific geospatial data products or for estimating the economic value of different pricing policies for access to data. However, it is a static analysis and does not, without some adjustment, take into account dynamic effects and resource transfers.

Estimates of revenues or turnover are useful indicators of the size of an industry or sector. However, they are not a true indicator of economic contribution, as they ignore the cost side of the equation.

Value-added analysis incorporates the cost of inputs and, in turn, provides an indication of the contribution of an activity to GDP. Value added is the main component of GDP and is therefore consistent with the national accounts.

Value chain analysis can be useful to illustrate the linkages between geospatial data supply chain and related and supporting industries. Such linkages are important determinants of international competitiveness.

Benefit–cost analysis, a form of welfare analysis, is suitable for assessing economic impacts of investment decisions or policy change and is a well understood technique in government and industry. Cash flow analysis and discounting techniques are important aspects of this approach, as they

---

* The authors are indebted to David Campbell of ACIL Allen Consulting for inputs and commands on real options.

convert cash flows accruing over time to an equivalent present value, allowing comparison between projects with different time frames and cash flows.

Benefit–cost analysis is a partial analysis that does not take changes elsewhere in the economy into account. Multipliers have also been used in the past to allow for these changes. However, as mentioned previously, multipliers can be difficult to estimate.

Cost-effectiveness analysis, a subset of benefit–cost analysis, can be appropriate when evaluating the best approach to meeting legislative or regulatory mandates. This does not place a value on the regulatory mandates (which may be hard to quantify) but assesses the least cost solution of meeting them.

CGE modeling addresses these problems by taking resource shifts into account. It is a more rigorous method of estimating the economy-wide benefits when significant resource shifts are likely. The CGE modeling is heavily dependent on data requiring extensive surveys and case studies.

Input–output multiplier analysis can be useful for a rigorous and credible analysis of the economic footprint or *contribution* of a particular facility, industry, or event. However, this approach assumes that prices remain fixed, import shares remain fixed, and there are no constraints on resources of labor or capital.

Real options approaches can be useful in assessing potential value of investments in times of high uncertainty. However, they have not been used extensively in analyzing the value of geospatial information to date.

These observations are summarized in Table 10.2.

10.6 Case Studies

This section outlines a number of case studies to illustrate the application of some of the techniques described previously. The case studies vary in complexity and focus and illustrate different approaches to economic impact assessment in different contexts. Further, more detailed examples are provided in Chapters 12 through 17.

10.6.1 Benefit–Cost Analysis of the National Map

An early example of a benefit–cost analysis can be found in a 2004 cost–benefit analysis of the National Map undertaken by the U.S. Geological Survey (Halsing, 2004). This report compares the state of the world with the National Map (reference case) and without it (the counterfactual). It allows for the fact that uses of spatial data are likely to increase over time, in part as a function of the National Map, and accounts for variation in the ability of customers to use its data. It also assesses three

TABLE 10.2

Comparison of Methodologies

Methodology	Application	Comments
Welfare analysis—estimates of consumer and producer surplus	Evaluation of a clearly defined service such as mapping or address data	Requires estimates of willingness to pay if there is no market price for the service
Revenue or turnover estimates	Provides an indication of the size and significance in the economy	Not a precise indicator of economic contribution
Value-added approaches	Provides an indication of the contribution of geospatial information activities to GDP	Can be applied along supply chains and shows important relationships between components of the supply chain
Benefit–cost analysis	Evaluation of a clearly identified investment or policy change	A well-understood technique among policy makers. Cash flow analysis and discounting techniques provide a rigorous means of comparing projects. This is a partial analysis that does not take into account resource shifts elsewhere in the economy
Cost-effectiveness analysis	Useful when assessing legislative and regulatory requirements where benefits are intangible	Does not provide an indication of the value arising from regulatory requirements but indicates the least-cost option of meet them
Computable General Equilibrium Modeling	Useful when regional or economy-wide impacts are required	Takes into account resource constraints and resource shifts in the economy. Results are dependent on data from surveys or case studies
Input-output analysis	Evaluation of the wider economic contribution from a specific activity	Analysis is based on input-output tables from official national accounts. It is a static analysis that does not take into account resource constraints in the economy
Real options	Assessment of potential value in situations of high uncertainty, with the potential for sequential decision making as more information becomes available	Can be useful in dealing with high uncertainty. Has had little application in the assessment of the value of geospatial information and systems

scenarios: two with different levels of implementation of the National Map, and the counterfactual.

Benefits were estimated as the NPV of a user's ability to improve a decision's effectiveness/efficiency with the use of spatial information or to use spatial information in a way that would not otherwise be feasible.

Examples of applications include the following:

- Creating an emergency evacuation plan
- Designating critical habitat for an endangered species
- Conducting property tax assessments
- Researching land cover change and deforestation
- Verifying insurance claims

An average improvement in net benefit per application was calculated and a model built to estimate the distribution of total values and a mean total value. The total value could be updated as information on the use of an application of the National Map evolved. The ability to do post-implementation analysis is an important aspect of this work.

The study estimated that the National Map could have an NPV of $2.05 billion in 2001. Sensitivity testing was done to assess the robustness of the findings. The testing showed that net benefits remained positive in all tested scenarios and identified those scenarios that did the most damage to the NPV. Such analysis is critical for policy formulation under uncertainty.

10.6.2 Value of Spatial Information in Catalonia

Another example of a study of the value of geospatial information is provided in a report on the "socio-economic impact of the Spatial Data Infrastructure of Catalonia," edited by Craglia (2008), for the Joint Research Centre of the European Commission. This study examined the benefits and costs of enhanced SDI by surveying 20 local government authorities participating in the Catalan SDI, 3 local authorities not participating in the SDI, 15 end-user organizations, and 3 large institutional users of geographic information.

Net benefits were based on time saved in the organizations surveyed. Time was valued at €30 per hour, being equivalent to the cost of senior technical staff. The report estimated the net economic benefits in 2006 to be €2,371,000 compared with the costs of €1,231,000 incurred over the 2005–2006 period. This represents a payback period of just more than 6 months in 2006. The study did not include a longer-term cash flow analysis that would most likely have delivered a higher NPV. Important political and social benefits were also identified, including reducing the digital divide and improving communication between local authorities.

10.6.3 Value of a Spatial Data Infrastructure in Lombardy

A further report of the value of an SDI to the regional authority of Lombardy in Italy was undertaken by Campagnelli and Craglia (2012), also on behalf of the Joint Research Centre of the European Commission. This study surveyed organizations outside public administration, finding cost savings of 11%–12% in the exercise of technical practice and time savings of 17%–19% arising from open access to the regional SDI data and services. These savings were estimated to amount to €3 million per annum.

Both this and the previous studies were ex-poste analyses, but their strength lies in the thoroughness of the surveys of the target organizations, which provides verifiable and quantifiable examples of applications.

10.6.4 Value of Remote Sensing from LANDSAT

An example of a complex benefit–cost analysis can be found in Forney et al. (2012). This paper outlines a benefit–cost analysis of the use of remote-sensing information from LANDSAT in the application of agricultural production and in maintaining groundwater quality. The value of information from satellite remote sensing was achieved through better-informed decisions, based on remote sensing from LANDSAT, leading to higher net crop production without sacrificing water quality in aquifers. Data from LANDSAT was compared with ground-based methods.

The study found that the expected additional NPV accruing from increased production with remote sensing from LANDSAT was $38.1 billion.

10.6.5 Value of Remote Sensing from Space for Decision-Making

EO and other remotely data are collected in many forms and at many scales. However, the data require translation to deliver information for systematic use in decisions. Economists regard the EO and other geospatial information as an intermediate good that provides a link between economic sectors in an economy. When EO can be linked to other types of models and data to create new geospatial information, decisions involve spatiotemporal change. The two case studies provided in Chapter 6, Section 6.5, demonstrate how observations of land use and land cover can be coupled with other models and data to create information for decisions.

10.7 Summary

Although each of the methodologies outlined in this chapter have applications to the field of geospatial information, the appropriate methodology to apply in each case should be guided by the context in which the results will be used and the requirements of decision-makers. The two most commonly used methodologies for assessing economic impact are benefit–cost analysis and the CGE modeling. Benefit–cost analysis (and its variants) will continue to be essential for project-scale investments, because it is widely understood and so offers a mechanism for comparison of heterogeneous investment choices.

For national-scale assessment of economic impact, CGE analysis allows for the interaction between sectors and export–import effects to be modeled more effectively and provides results in the form of GDP and other macro-economic indicators, including employment. However, it relies on the historical performance of the world economy, which means that it is more suitable for the assessments of the current situation and the medium-term future.

10.8 New and Emerging Techniques

As we ask ourselves about the path forward, it is useful to quickly summarize what we have learned to date about the value of spatial data and its application to geospatial information. By following the spatial data and geospatial information value chain from the initial data collection through satellites and in situ sensors, through transformation into information and knowledge, we learned that to be of value, the resulting knowledge must be used by and considered useful to its intended recipients, be they citizens, scientists, or decision-makers. Earlier in this chapter, a range of economics concepts, methods, and models about valuing geospatial information were discussed. To that purpose, the application of spatial data was considered, together with human processes to provide an integrated framework for addressing considerations of the earth environment and associated societal challenges. The future based on those concepts and methods presented and described can be viewed as a broadening of this integration into modeling ensembles to help inform decisions that are made under uncertainty. These models would be employed in use cases and eventually in decisions, as geospatial data archives expand and deepen. As these cases evolve, the socioeconomic value of the geospatial data and information also becomes clearer and greater. A brief description of the types of decision-oriented uses for geospatial data is presented as follows:

Ensemble modeling is the process of integrating two or more related though separate analytical models and then synthesizing the results into a single indicator of a range of values. Modeling ensembles utilize probability distributions and take advantage of Bayesian methods to improve the accuracy of predictive analysis (Reiners et al., 2002; Liu et al., 2008; Zaitchik et al., 2008).

The EO can be combined with other earth science data in a data ensemble, both horizontally and vertically, for use in resource allocation decisions. Horizontal integration occurs when there is a merger between at least two types of geospatial data at the same scale to solve a decision problem.

Vertical integration would include the application of the spatiotemporal data in a spatial infrastructure that is tiered via a hierarchy of remotely sensed data types and resolutions to provide optimal decision-relevant input.

One application of a modeling ensemble is to evaluate the improvement in allocating public resources for disaster assistance in a Bayesian decision model (Bernknopf et al., unpublished manuscript). The analysis was to estimate the economic value of merging the Gravity Research and Climate Experiment (GRACE)-derived total water storage (TWS) indicators with the U.S. Drought Monitor (USDM). The evaluation involved the application of the USDM to assess the societal benefits of the indicators in administering a legislative mandate (Agricultural Act of 2014 (H.R. 2642; Pub. L. 113–79). The GRACE observations are assimilated into a Land Surface Model* to improve the accuracy of TWS estimation with models of hydrological processes that are used to enhance the satellite observations (Zaitchik et al., 2008). The result is three drought indicators that have been used for drought monitoring in the USDM and, consequently, in disaster relief decisions (a summary is contained in Chapter 14). In the modeling ensemble, USDM drought severity categories are assigned, with and without the GRACE drought indicators. The results were then compared in an econometric model to estimate the societal benefits of reducing misclassification errors and incorrect assignments of financial assistance that result during severe droughts with the EO. This example incorporates both horizontal (hydrologic data and process models and economic data and process models) and vertical (EO and other precipitation-related indices) integration.

A second example of horizontal and vertical integrations is in applications of nonmarket valuation. The economic valuation process could take advantage of visualization tools that can provide support for cost-effective elicitation of consumer preferences for ecosystem goods and services. This type of input would be helpful in expanding the use of statistical meta-analysis, which combines revealed and stated preference methods for nonmarket valuation. Current revealed preference models such as travel cost or asset valuations could be improved by maintaining the heterogeneity of spatial characteristics. In addition changes to ecosystem services and their socioeconomic consequences could be improved with spatiotemporal visualisations of local community impacts in stated preference models (Brookshire et al., 2016). Merging of econometric models with geospatial data would be relevant to issues that focus on regional preferences and the use of benefit transfers.

Ensembles could be used to combine theory-based models (deductive approaches) and empirically based models (inductive approaches) in a spatial infrastructure. Coupled human and natural systems that use agent-based

---

* A Land Surface Model is a unidimensional computational model that describes ecological processes joined in many ecosystem models, hydrological processes found in hydrological models, and flow common in surface models, using atmospheric models. The model examines interactions of the biogeophysics and biogeochemistry of the land and atmosphere, and effect on the surface of the land (https://en.wikipedia.org/wiki/Land_Surface_Model_(LSM_version_1.0).

models have the capability to combine deductive and inductive methods (An, 2012). An agent-based model is a bottom-up modeling approach to explore emergent collective phenomena that result from interactions of individual agents with heterogeneous characteristics who make independent and cooperative decisions. This example could take advantage of convolving EO with crowdsourced data, which has been used to support community decision-making. The horizontal integration of the technologies allows for both the distributed collection of in situ geospatial data and EO to be used in original analyses and case studies.

Vertical integration is different from horizontal integration, which occurs when EO and other remote-sensing data are combined at different scales to be merged in an information structure.* Vertical integration of geospatial information has the capability for use in monitoring of land and sea changes that affect the quality and quantity of investments in mitigating losses to public infrastructure. Geospatial information is currently used for monitoring land use, and land cover changes could be used to forecast hazardous events. An example of such an approach has been used to identify the impacts of land subsidence in China, Taiwan, and the United States (Farr et al., 2016; Hwang et al., 2016). This example highlights the need for merging a variety of EO and other remotely sensed data to be integrated with hydrogeological models that can represent spatial variations in the subsurface (Hwang et al., 2016).

In the subsidence application in California, Hwang et al. (2016) applied altimeter data from TOPEX/POSEIDON, JASON-1 and JASON-2 (JPL/NASA) and ENVISAT (European Space Agency), groundwater support and Landsat (U.S. Geological Survey, Earth Resources Observation and Science Center), and SPOT satellite images (Center for Space and Remote Sensing Research, National Central University). In addition, GRACE data that detected mass losses attributable to severe groundwater storage declines related to land surface changes were provided, for example, the San Joaquin Valley in California. It is suggested that the approach described would produce cost-effective measurement of land deformation that could minimize the effect of regulating groundwater extraction on public infrastructure. Evaluation of choosing which information to include in decisions would apply econometric analyses and techniques. The statistical model could be used to estimate the optimization cost of information in an investment decision (Conlisk, 1988, 1996).

To support these types of integrated models, analytical decision support tools will need to be developed, with a combination of theory-based decision frameworks that utilize subjective probability and rapid response ontologies based on machine learning and data-mining algorithms. Ensemble models that are integrated horizontally and vertically can and, most likely, will provide enrichment of the use of geospatial information beyond data collection and delivery.

---

* An information structure is a stochastic mapping from objective states of the world to a set of subjective signals (Caplin and Dean, 2015).

References

ACIL Allen (2015). *The Value of Earth Observation from Space*. Canberra, Australia: Geoscience Australia.

ACIL Tasman (2008). *The Value of Spatial Information in Australia*. Melbourne, Australia: Cooperative Research Centre for Spatial Information.

ACIL Tasman (2013). *The Value of Precise Positioning in Australia*. Canberra, Australia: Department of Industry, Science, Research and Tertiary Education.

AGC (2010). *Size of the Spatial Industry*. Canberra, Australia: Australian and New Zealand Land Information Council Allen Consulting Group.

Ambargis, B. (2011). *Input Output Models for Impact Analysis*. Washington, DC: Bureau of Economic Analysis of the Department of Commerce.

An, L. (2012). Modeling human decisions in coupled human and natural systems: Review of agent-based models. *Ecological Modelling*, 229, 25–36.

ANZLIC (2010). *Economic Inplications of Spatial Data Pricing Policies*. Melbourne, Australia: Australian and New Zealand Land Information Council.

Bateman, I.J., G.D. Garrod, J.S. Brainard, and A.A. Lovett (1996). Measurement issues in the travel cost method: A geographical information systems approach. *Journal of Agricultural Economics*, 47(1–4), 191–205.

Bernknopf, R. (2004). *Societal Value of Geological Maps*. Washington, DC: U.S. Geological Survey.

Bernknopf, R., D. Brookshire, Y. Kuwayama, M. Macauley, M. Rodell, A. Thompson, P. Vail, and B. Zaitchik (unpublished manuscript). The value of remotely sensed information: The case of GRACE-enhanced drought severity index. *Weather, Climate and Society*.

Bernknopf, R., D. Brookshire, M. Macauley, G. Jakeman, Y. Kuwayama, H. Miller, and L. Richardson (2016). Societal benefits: Methods and examples for estimating the value of remote sensing information. In Morain, S., M. Renslow, and A. Budge, Eds., *Manual of Remote Sensing*, 4th ed. Bethesda, MD: American Society for Photogrammetry and Remote Sensing.

Bernknopf, R., and C. Shapiro (2015). Economic assessment of the use value of geospatial information. *International Journal of Geographic Information*, 4, 1142–1165. doi:10.3390/ijgi4031142.

Boxall, P.C., W.L. Adamowicz, J. Swait, M. Williams, and J. Louviere (1996). A comparison of stated preference methods for environmental valuation. *Ecological Economics*, 18(3), 243–253.

Brookshire, D., R. Bernknopf, D.R. Adhikari, C. Babis, C. Broadbent, and V. Tidwell (2016). Net Resource Assessment (NetRA): A collaborative effort between the USGS Science and Decisions Center, the Science Impact Laboratory for Policy and Economics (University of New Mexico) and Sandia National Laboratories Final Report.

Campagnelli, M., and M. Craglia (2012). The socio economic impact of spatial data infrastructure of Lombardy. *Environment and Planning B: Planning and Design*, 39, 1069–1083.

Caplin, A., and M. Dean (2015). Revealed preference, rational inattention, and costly information acquisition. *American Economic Review*, 105, 2183–2203. doi:10.1257/aer.20140117.

Carson, R.T., N.E. Flores, K.M. Martin, and J.L. Wright (1996). Contingent valuation and revealed preference methodologies: Comparing the estimates for quasi-public goods. *Land Economics*, 72, 80–99.

Conlisk, J. (1988). Optimization cost. *Journal of Economic Behavior and Organization*, 9, 213–228.

Conlisk, J. (1996). Why bounded rationality? *Journal of Economic Literature*, 34, 669–700.

ConsultingWhere (2010). *The Value of Geospatial Information in England and Wales*. London, UK: Local Government Association.

Craglia, M. (Ed.) (2008). The socio-economic impact of the spatial data infrastructure of Catalonia. JRC, European Commission, Luxembourg.

Dahlby, B. (2008). *The Marginal Cost of Public Funds – Theory and Application*. Cambridge, MA: MIT Press.

Dodgson, J.S. (2009). *Multi-criteria Analysis: A Manual*. London, UK: Department of Communities and Local Government.

Eade, J., and D. Moran (1995). Spatial economic valuation: Benefits transfer using geographical information systems. *Journal of Environmental Management*, 48(2), 97–110.

Farr, G., C. Jones, and L. Zhen (2016). Progress report: Sunsidence in California, March 2015–September 2016. http://www.water.ca.gov/waterconditions/docs/2017/JPLsubsidencereportfinalforpublicdec2016.pdf.

Forney, W.M., R. Rauniker, R. Bernknopf, and S. Mishra (2012). An economic value of remote sensing information—Application agricultural production and maintaining groundwater quality. U.S. Geological Survey Boston, Reston, VA.

Halsing, T. (2004). *Benefit Cost Analysis of the National Map*. Washington, DC: U.S. Department of the Interior.

Hanley, N., F. Shogren, and B. White (1996). *Environmental Economics—In Theory and Practice*. London, UK: MacMillan.

Houghton, J. (2011). *Costs and Benefits of Data Provision*. Melbourne, Australia: Centre for Strategic Economic Studies Victoria University.

Hsu, C.-C. (2010). The Ohio State University and Brian A. Sandford, Oklahoma State University.

Hubbard, D. (2015). *How to Measure Anything: Finding the Value of Intangibles in Business*. Wiley.

Hwang, C., Y. Yang, R. Kao, J. Han, C.K. Sum, D. Galloway, M. Sneed, W-C. Hung, Y-S. Cheng, and F. Li (2016). Time-varying land subsidence detected by radar altimetry: California, Taiwan and north China. *Scientific Reports*, 6, 28160. doi:10.1038/srep28160.

Longhorn, R.A., and M. Blakemore (2008). *Geographic Information, Value, Pricing, Production, and Consumption*. Boca Raton, FL: CRC Press, pp. 45–46.

Liu, J., S. Liu, T. Loveland, and L.L. Tieszen (2008). Integrating remotely sensed land cover observations and a biogeochemical model for estimating forest ecosystem carbon dynamics. *Ecological Modeling*, 219, 361–372.

Macauley, M. (2006). The value of information: Measuring the contribution of space-derived earth science data to national resource management. *Space Policy*, 22, 274–282.

Marshall, A. (1890). *Principles of Economics*. London, UK: Macmillan and Co.

McMillen, D.P., and C.L. Redfearn (2010). Estimation and hypothesis testing for nonparametric hedonic house price functions. *Journal of Regional Science*, 50(3), 712–733.

Mendelsohn, R., J. Hof, G. Peterson, and R. Johnson (1992). Measuring recreation values with multiple destination trips. *American Journal of Agricultural Economics*, 74(4), 926–933.

Miller, B.M. (2016). *The Not-so Marginal Value of Weather Warning Systems.* San Diego, CA: University of California.

Murphy, J.J., P.G. Allen, T.H. Stevens, and D. Weatherhead (2005). A meta-analysis of hypothetical bias in stated preference valuation. *Environmental and Resource Economics*, 30(3), 313–325.

NASA (2013). *Measuring the Socioeconomic Impacts of Earth Observation—A Primer.* Washington, DC: National Aeronautics and Space Administration.

Ordnance Survey (2013). *Assessing the Value of Open Data to the Economy of Great Britan.* London, UK: Ordnance Survey.

Oxera (1999). *Economic Contribution of Ordnance Survey in Great Britain.* London, UK: Oxera Consulting.

Oxera (2013). *What is the Economic Impact of Geo Services.* London, UK: Oxera Consulting.

Pollock, R. (2008). *Models of Public Sector Information Provision by Trading Funds.* Cambridge, UK: Cambridge University Press.

Randall, A. (1994). A difficulty with the travel cost method. *Land Economics*, 70, 88–96.

Reiners, W., S. Liu, K. Gerow, M. Keller, and D. Schimel (2002). Historical and future land use effects on N_2O and NO emissions using an ensemble modeling approach: Costa Rica's Caribbean Lowlands as an example. *Global Biogeochemical Cycles*, 16, 1068, doi:10.1029/2001GB001437.

Rogawski, C., S. Verhulst, and A. Young (2016). *Great Britain's Ordnance Survey, in the Global Impact of Open Data.* Sebastopol, CA: O'Reilly Media.

Rogers, E.M. (2003). *Diffusion of Innovations.* New York: New York Free Press.

Roll, R. (1984). Orange juice and weather. *The American Economic Review*, 74(5), 861–880.

Rosenberger, R., and J. Loomis (2001). Benefit transfer of outdoor recreation use values: A technical document supporting the Forest Service Strategic Plan (2000 revision). Gen. Tech. Rep. RMRS-GTR-72. Fort Collins, CO: U.S. Department of Agriculture, Forest Service, Rocky Mountain Research Station, 59 p.

Rosenzweig, M., and C.R. Udry (2013). Forecasting profitability (No. w19334). Cambridge, MA: National Bureau of Economic Research.

Sargento, A.L.M. (2009). Introducing input-output analysis at the regional level: Basic notions and specific issues. Urbana, IL: University of Illinois.

TIDE (Transport Innovation Deployment for Europe) (2015). Methodologies for cost-benefit impact analysis in urban transport innovations. Huging, H., K. Gelnsor, and O. Lah, Wuppertal University for European Union.

Young, M.D. (1992). *Sustainable Investment and Resource Use: Equity, Environmental Integrity and Economic Efficiency.* Paris, France: Parthenon Press, Carnforth and UNESCO.

Zaitchik, B., M. Rodell, and R. Reichle (2008). Assimilation of GRACE terrestrial water storage data into a land surface model: Results for the Mississippi River Basin. *Journal of Hydrometeorology*, 9, 535–548. doi:10.1175/2007JHM951.1.

11

Qualitative and Other Social Science-Oriented Methods to Assess the Importance of Geospatial Information

Miriam Murambadoro and Julia Mambo

CONTENTS

11.1 Introduction and Approaches Used
in Noneconomic Social Sciences

Geospatial information can be used to enhance societal understanding of how different components of the social ecological system* relate to each other. Geospatial information was previously represented only on maps, which contained series of geographic information overlaid and presented on a flat piece of paper. Advances in digital technology and Geographical Information System have revolutionized the way in which geographic information is collected, stored, analyzed, and presented. In recent years, there has been increasing demand for geospatial information and tools such as the Geographical Information System to manipulate and display geospatial information. Geospatial research has the ability to bring the best available knowledge to address some of the current and future problems such as climate change, environmental management, and meeting energy demands.

A lot of scientific information has emerged out of the research done by several institutions at local, regional, and international levels to increase the understanding of the world's pressing problems. As government and other funding institutions continue to receive requests for funding for Earth observation (EO) and other research, there is a need to assess the benefits of such studies that compete with other public service demands. This is leading to a growing interest in social value, value-added, and outcomes measures as alternatives to narrow interpretations of monetary value for money. Assessing the value of geospatial information is not an easy task, given that defining the term *value* is also difficult and is based on different belief systems. People's values are related to the meaning that they assign to objects, conditions, traditions, and so on.

Qualitative research and specifically participatory action research are the key approaches that can be used to improve understanding of the different value that people or institutions place on geospatial information as well as the associated societal benefits. Value systems comprising different value priorities would essentially inform the way in which people perceive their environment and any new information/knowledge that is presented to them. These values also govern the nature and content of the issues that decision-makers focus on, and these decisions are made from within particular frames of reference based on particular value orientations (De Wit 2016, in prep.).

When geospatial information is applied in decision-making, it has the potential to support many activities in the society; therefore, understanding the value and benefit of geospatial data is essential, and the benefits can be better explained by assessing how these data are used in the decision-making process. When scientific data such as geospatial data are analyzed, then it has value and can be used to inform decisions. If this data is not used,

---

* Social ecological systems are complex, integrated systems in which humans are part of nature (Berkes and Folke 1998; Resilience Alliance 2017).

then its value to society is also limited. Macauley (2006) argues that there are certain factors that give information value, and these include the cost for decisions-makers to use the information.

First, if the cost of producing the information is higher than the expected gains and there are additional costs for processing and interpreting the data to make it usable for decision-making, then it has low value.

Second, the value of information depends on whether the information is in a format that the decision-makers can use, is accurate, and has been validated to make decision-makers indifferent toward other sources of information. For example, there are other substitute sources of information such as the aerial photography, which can be used to monitor certain land uses instead of satellites, which need further processing and interpretation to be used (Macauley 2006, p. 275). In developing countries in Sub-Saharan Africa, decision-makers often lack the technical skills required to interpret geospatial data and apply it in tools used for planning and decision-making. Constraints of using the geospatial information need to be few for the information to have value.

Third, the value of information to decision-makers and users depends on what is at stake, that is, what would be affected by making the wrong decision or value of the goods and services at risk. For example, a hotel on the beach front will be willing to pay for information on the sea level rise and extreme weather events, depending on the value of the hotel and what it stands to lose.

Lastly, the value of information is influenced by how uncertain decision-makers are regarding a decision on a particular issue. For example, the value of information on hydrological observations depends on how uncertain decisions-makers are on water quality and availability to meet the societal water needs.

At times, users of information in the public and private sectors are able to identify the value of information and also know their knowledge gaps but are not always able to find the information to fill in these gaps. In most instances, the information is also presented in a format that has no value to them, and as such, it is not used. The following section looks at qualitative participatory action research approaches that can be used to assess the value and societal benefits of geospatial information.

11.1.1 Participatory Approaches

Participatory approaches include communities, institutions, or individuals to actively participate in understanding a phenomenon or addressing a particular issue. Data that are collected by using such nonscientific methods is subjective, and participants decide what knowledge and/or experiences they can share and what to withhold. The roots of participatory action research come from the work of different scholars, including Kurt Lewin (1944), who introduced the term *action research* as a tool to study social systems while imparting changes that are driven by the users of the knowledge created (Gillis and Jackson 2002). Another scholar who contributed to participatory action research is Paulo Freire, who argued that personal and social changes can be achieved through a process of critical reflection by different actors to take action and change the repressive elements in that social space (Maguire 1987; McIntyre 2002; Selener 1997).

Selener (1997) identified the following seven attributes of participatory research:

- The phenomenon under study or research problem needs to be defined, analyzed, and solved by, and/or in partnership with, the community.
- It strives to transform societies and individuals involved in the study.
- The participants actively contribute to the entire research process.
- It provides a more reliable and accurate analysis of social reality.
- It is an inclusive process that engages minority and marginalized groups.
- It allows participants to identify and appreciate the resources they have that can be used to promote self-reliant development.
- The researcher can become part of the participants as they engage in dialogue to better understand the area under research (through interviews and facilitated group discussions), and this process promotes learning for all those who are involved.

The various methods used in participatory action research are qualitative in nature and include methods and techniques of observing, documenting, analyzing, and interpreting characteristics, patterns, traits, and values embedded in humans under study (Gillis and Jackson 2002; Leininger 1985). Qualitative approaches are generally not best suited for large statistical descriptions of the population. Instead of predicting and controlling the subjects under study, qualitative research strives to describe and understand the subject in its entirety or components of the subject (McDonald 2012; Streubert and Carpenter 1995). Therefore, participatory approaches can be used to understand the different values that individuals and institutions place on geospatial information, which can be used, for example, to improve the management and protection of terrestrial, coastal, and marine resources as well as to enhance understanding of the water cycle and management of water resources.

11.1.2 Benefits of Participatory Approaches

Maguire (1987) illustrated that there are three types of change that occur in participatory research (Box 11.1). These are as follows:

1. Development of critical consciousness of the researcher and the participants

2. Improvement in the lives of research participants

3. Alteration of societal structures and relationships (McDonald 2012, p. 38)

Participatory methods allow people/participants to identify positive and negative aspects of their socio-ecological system. Appreciative Inquiry (Box 11.1) is one of the tools that can be used for collective inquiry into the best of what is there so as to imagine what could be and ensures collective design of the future (Kessler 2013) below

BOX 11.1   ENGAGING STAKEHOLDERS IN SELF-DETERMINED CHANGE

Appreciative Inquiry (AI) model is a participatory tool that can be used to engage participants to identify the shared problem and design the solution. The AI model can be used in group discussions such as workshops with stakeholders to understand a problem and allow participants to contribute to the designing of the solutions based on what is good and has worked. The example given below shows how the AI model can be used in a workshop with participants to enhance climate change adaptation in a particular area:

Definition: Stakeholders define climate change risk and impacts in a particular area.

Discovery: What is good and has worked to enhance climate change response. (*This would include interventions by the internal and external stakeholders.*)

Dream: What structures/mechanisms and resources might be needed for your community to become more resilient to the impacts of climate change and global change? (*This can include human resources that have indigenous knowledge.*)

Design: Based on what has worked in the past and present for a mechanism/structure to assist with climate change response to be useful, what should it be able to do? (*This will include discussions where people share their experiences on what has worked and how to redesign it to address the current problem.*)

Destiny: How to make it happen?

The AI model provides an opportunity for researchers to co-produce and integrate scientific information with tacit knowledge held by participants/users of information, and this enhances the uptake of information in planning and decision-making.

11.1.3 Social Oriented Methods Used in the Social Sciences

Various participatory methods in qualitative research can be used to assess the societal benefits of geospatial information and the value of this information. These include focus group discussions, participant observation, and interviews. These will be discussed in detail in the next section. In participatory research, the role of the researcher is that of facilitating a process within a community where participants are engaged in discussions to change their situation. A mix of these methods can be used in one study so as to enhance the validity of the research.

The researcher or research team (that can be a team of researchers from different disciplines) needs to agree on which methods would be appropriate for the study (Audouin 2011). The specific methods selected need to be appropriate to the resources available, be appropriate to the nature of the knowledge to be gained, enable an understanding of practice and context, and be transparent while ensuring that the assumptions made in the study are explicit (Audouin 2011; Flyvbjerg 2001).

11.1.4 Focus Group Discussions

Focus group discussions are a common method used to gather information. This method has its origins in sociology, but it has evolved to be used in other fields such as marketing and in other new fields that require gathering information from a group of stakeholders (Box 11.2). The information gathered is usually related to questions about the participants' experiences, ideas, or an event and involves the use of each person's reality. By gathering people

---

BOX 11.2   USING FOCUS GROUP DISCUSSIONS TO UNDERSTAND CHANGES IN ECOSYSTEM COMPOSITION AND DIVERSITY IN A PARTICULAR AREA

Focus group discussions can be used to understand changes in ecosystem composition and diversity. The researcher/facilitator should give a presentation and provide participants with some background information on the research as well as on defining the terminologies that will be used in the discussion (e.g., biodiversity and ecosystem). This allows all participants to have a shared understanding of the terminology and avoid any misunderstanding or ambiguity associated with the terms. The facilitator should present the specific questions for discussion and allow participants to share their experiences as they respond to the questions (e.g., have you noted any changes in ecosystem diversity in the past 40 years). The information collected here could be presented as narratives, accompanied by geospatial maps that show areas where change has occurred.

together, focus group discussions assist to create an environment that encourages spontaneous expression of each participant while interacting with other participants (Freitas et al. 1998). Focus group discussions may include a series of meetings for groups of people with similar research interest. Group discussions maybe referred to as workshops, in some contexts. Focus group discussions are facilitated by a moderator/facilitator whose job is to stimulate discussions with comments and/or questions on the subject being discussed (probing). Box 11.3 below provides a summary on how to organise a focus group discussion. The outcome of focus groups discussions can be captured as transcripts of the discussions and may include the reflections, comments, and/or observations from the moderator (Freitas et al. 1998).

Focus group discussions are best applied in situations where there is a need to:

- Generate new ideas for investigation, or take action in new fields
- Develop hypotheses premised on the perceptions of the participants
- Evaluate various research situations or study populations
- Develop drafts of interviews and questionnaires
- Supply interpretations of the participants' results from initial studies
- Generate additional information for a study on a wide scale

BOX 11.3   ORGANIZING FOCUS GROUP DISCUSSIONS

When organizing a focus group discussion, there are four key steps that the researcher needs to take when using this method:

1. Clarify what the group needs to accomplish and the best method to be used, including the type of questions that are appropriate, for example, using opened-ended questions, which would encourage the discussion.

2. Find a good facilitator who is able to facilitate the discussions, is knowledgeable of the research topic, relates with the participants, and, most importantly, is able to work well with the researcher to meet the research objectives. The facilitator needs to show respect for all participants; be modest and humble about his or her abilities, knowledge, and insight; and ensure that communication is effective, that is, the messages and information are clearly conveyed and received through active listening. A facilitator should integrate information and contributions from different knowledge types (e.g., scientific, experiential, and traditional) into the process and analysis.

(Continued)

> **BOX 11.3 (Continued)   ORGANIZING**
> **FOCUS GROUP DISCUSSIONS**
>
> 3. Use an appropriate method to capture discussion points, so as to ensure that all the points raised or discussed are well represented. With consent from participants, a recorder can be used, or a person can be assigned with taking discussion notes.
> 4. Find the right mix of participants and representative sample, as this is crucial for the outcome of the research.
>
> Other logistical considerations for focus group discussion include ensuring that the venue, time, and other logistics set up the right environment for the discussions.

Focus group discussions are an ideal tool for the collection of information, given that it combines two techniques of gathering information, interviewing individuals and group observations by the researcher, enabling the collection of substantial data or information under a short space of time. However, this method has its advantages and disadvantages, which are highlighted in Table 11.1.

TABLE 11.1

Advantages and Disadvantages of the *Focus Group*

Advantages	Disadvantages
• It is comparatively easier to carry out	• It is not based on a natural atmosphere
• It enables to explore topics and to generate theories	• The researcher has less control over the data that is generated
• It produces opportunity to gather data from the group interaction, on the researcher's topic of interest	• It is not possible to know if the interaction under study is the result of a group or individual behaviour
• It has high "face validity" (data)	• The data is more difficult to analyze
• It is cheaper to implement compared with other methods	• The comments need to be interpreted within the context of the social environment created for the group
• It gives speed in the supply of the results (in terms of evidence of the meeting of the group)	• It takes effort and experience to assemble and interview the groups
• It allows the researcher to increase the size of the sample of the qualitative studies	• The discussion should be conducive for dialogue

Source: Based on Krueger, R.A., *Focus Groups: The Practical Guide goes Applied Research*, SAGE Publications, Thousand Oaks, 1994; Morgan, D.L., *Focus Groups the Qualitative Research*, SAGE Publications, Beverly Hills, CA, 1988.

11.1.4.1 Specific Research Tools That Can Be Used within a Focus Group Discussion

There are several research tools that can be used within a group of participants to assess the societal benefits or values, and these are not limited to
those highlighted in Box 11.4.

BOX 11.4 PARTICIPATORY TOOLS THAT CAN BE USED WITHIN A GROUP TO UNDERSTAND THEIR VALUES OR SOCIETAL BENEFITS INCLUDE THE FOLLOWING

- *Value Orientation Method (VOM)*: This tool can be used to identify
differences in core values across cultures and will assist in understanding the differences in how each culture views the world or
certain aspects of it. This kind of understanding is essential to
avoid conflict between different cultural groups of people and
is useful in cases where services need to be provided (Gallagher
2001). The VOM is an important tool that will assist to understand and identify the core values of cultures and the differences.

- *Participatory mapping*: This is an innovative and interactive
method that is accessible and allows groups or individuals
to visually cross-examine qualitative research questions in a
discussion setting (Emmel 2008). Participatory mapping can
be used to map changes in the environment over time, especially for issues such as deforestation and desertification, using
local knowledge. The maps enable participants to describe and
represent their knowledge/values on a map through drawing,
which is followed by a discussion.

- *Anecdote circles*: These are narrative techniques that can be used
in a focus group, and they are facilitated to elicit stories rather
than judgment and opinion. To assess the value of geospatial
information, participants are given a specific question, and each
one is given a chance to tell a narrative/story to reveal how this
is used in his or her organization/community and what people
value and how they benefit from using this information.

- *Dotmocracy*: It is an established facilitation method for collecting and prioritizing ideas among a large number of people. It is
an equal opportunity and participatory group decision-making
process. To assess the value of geospatial information, participants can be asked to write down the different sources of
information that they use for planning and decision-making.
Participants are then given an opportunity to apply dots under

(Continued)

**BOX 11.4 (Continued)   PARTICIPATORY TOOLS THAT CAN BE
USED WITHIN A GROUP TO UNDERSTAND THEIR VALUES
OR SOCIETAL BENEFITS INCLUDE THE FOLLOWING**

each information source to show which ones they value. This
can be followed by a discussion to probe why people made
those choices.

- *Resource map*: This can be used to provide useful information
  on participant's perceptions of what is useful to them and the
  resources that they use for planning and decision-making.

- *Appreciative enquiry model*: The tool allows the participants to
  reflect and search for the best in people, their organizations,
  and the relevant world around them.

- *Prioritization*: There are a number of methods that can be used
  for prioritization. These include the preference ranking, pair-
  wise ranking, and Delphi technique. These methods can be
  used in different ways to elicit participants to determine for
  themselves how the criteria should be weighted. This procedure provides a formal mechanism to introduce individual
  preferences and values into the decision-making process.

11.1.5 Participant Observation

Participant observation is a tool mainly used in qualitative research, social
sciences, and other disciplines to collect data about people, processes, and
cultures (Kawulich 2005). According to DeWalt and DeWalt (2002), observation utilizes the five senses to collect information, and this can include
active looking, improving memory, informal interviewing, writing detailed
field notes, and patience. Researchers use observation to learn about people under study and their activities in their own environment, which can
include learning what participants do on a daily basis and observing their
routine and other activities. For the method to be successful, the researcher
needs to build trust with the participants and develop a good relationship
with them. The researcher should be able to blend in with the participants,
and he or she needs to remain objective (Box 11.5). Participant observation
can be done during normal conversations, interviews, focus group discussions, and other data collection methods (Kawulich 2005).

Similar to other research methods, participant observation has advantages and disadvantages, and some of these are discussed in Table 11.2.
Participant observation may be used to complement interviews as illustrated in Box 11.6.

The validity of this method can be further strengthened by using additional information from document analysis, surveys, questionnaires,

BOX 11.5 PARTICIPANT OBSERVATION

Participant observation is a method that can be used to gather information by the following methods:

- Assessing the nonverbal expression of feelings
- Determining who interacts with whom
- Grasping how participants communicate with each other
- Checking how much time is spent on various activities and what seems valuable to them (Schmuck 1997).

TABLE 11.2

Advantages and Disadvantages of Participant Observation

Advantages	Disadvantages
• Provides access to the *backstage culture*, whereby the researcher has access to much more material than researchers who are making observations from outside a situation. Research participants are more willing to share their lives, feelings, values, and goals more freely with someone inside their circle • It provides opportunities for viewing or participating in unscheduled events • It facilitates the development of new research questions or hypotheses based on observation	• The technique is based solely on the researcher for collection and analysis and can be constrained by the researcher's bias or misinterpretation • Technique makes use of the people's expressions and behavior, some of which are difficult to translate • Sometimes, the researcher may not be interested in what happens out of the public eye and that one must rely on the use of key informants • Different researchers gain different understanding of what they observe, based on the key informant(s) used in the study • Problems related to representation of events and the subsequent interpretations, especially if the observer has become too personally involved, affect the outcome of the study

Source: Kawulich, B.B., *Forum: Qual. Res.*, 6, Art. 43, 2005.

interviews, and other qualitative methods used to develop and test theories (De Walt and De Walt 2002). Like in any other research, the setting of the environment, the selection of the participants, and the questions to be asked contribute to the success of the observations.

11.1.6 Interviews

Interviews are a commonly used method for gathering information, and the number of people that can be interviewed ranges from one to a small group (Box 11.7). Interviews can also be conducted over the telephone or by using Internet-based communication services that allows for voice and video calling (e.g., Skype and Facebook). The interviewer is regarded as

BOX 11.6   USING PARTICIPANT OBSERVATION TO ADVANCE KNOWLEDGE ON MANAGING ENERGY RESOURCES

Participant observation can be used to complement interviews for example in a study aimed at matching the supply and demand of energy in a community. The researcher can spend time in the community observing how energy is used and the different sources of energy currently used. There is a need to get permission from the relevant authority to spend time with the community and visit their homes. A researcher is an observer and should stay objective, so that he or she can capture and interpret data without bias. Through participant observation, researchers are able to gain insights into uses and actions that participants may not be able to share or unwilling to share as well as observe some of the issues that are raised during the interviews or discussions.

BOX 11.7   INTERVIEW STRUCTURES THAT CAN BE USED TO IDENTIFY THE VALUE OF GEOSPATIAL INFORMATION

There are different formats that can be used to conduct an interview, and this can be influenced by the type of questions asked and the objectives of the study. Interviews can be structured, semistructured, and unstructured.

- *Structured interviews* use predetermined questions, and there is little or no opportunity for follow-up questions to probe the responses.
- *Semistructured interviews* contain numerous key questions that assist in defining the areas or subject to be investigated, while giving the interviewer a platform to probe and the interviewees a chance to provide more details to their responses (Gill et al. 2008).
- *Unstructured interviews* can be used in an investigative situation, where no prior information exists about the subject (Gill et al. 2008). It requires little organization and resources and do not contain any defined theories or ideals. The interviewer/researcher is guided by the responses of the participants, which can be time consuming and problematic to keep the focus on the study objective.

TABLE 11.3

Advantages and Disadvantaged of Using Interviews

Advantages	Disadvantages
• Excellent for in-depth testing of research hypotheses and insights • Semistructured can elicit rich qualitative information • Semistructured and unstructured interviews allow the researcher to explore and probe responses	• Expensive and time consuming. • Respondents may be reluctant to share personal information and values with a stranger • Possibility of interviewer bias. especially in semistructured and unstructured interviews • It rarely yields useful quantitative data • Not ideal for large groups of people

---

**BOX 11.8   ASSESSING THE VALUE OF EARLY
WARNING INFORMATION TO FARMERS**

Interviews with participants (e.g., farmers) can be used, for example, to illustrate the value of early warning information available to support communities with extreme weather events such as drought. A semi-structured interview can be used to allow participants to share their experiences with regard to early warning information and the value of this information in informing their farming decisions (what to plant and when to plant).

---

part of the research instrument and needs to be well trained to ask the questions in an appropriate manner and to respond to or clarify issues for the participant should the need arise (Valenzuela and Shrivastava 2008).

Interviews also have advantages and disadvantages. These are highlighted in Table 11.3. Box 11.8 is showing how this method can be applied.

11.1.7 Conclusion

Participatory methods illustrated previously allow for dialogue with different stakeholders, allowing them to share their different values, perceptions, dreams, strengths, and challenges experienced by individuals and groups. In using these methods, there is no single, objective reality, but rather, there are multiple realities based on subjective experiences, values, and circumstances, where none is more important than the other (Wuest 1995, p. 30; as cited in McDonald 2012). It is important to also note that the prevalent value orientations of any individual and the culture, to which a person belongs, can, in many cases, act as barriers to intercultural or intersystem communication. Given that people have different values and perceptions on what is valuable to them, what passes as common sense or valuable in one culture or system might appear deviant in another.

Users of geospatial information often use their knowledge in different ways (Bielak et al. 2008), and as such, there is a need to understand the different user groups. Dunlop (2009) argues that decision-makers often fail to use or appreciate the value of scientific information (e.g., geospatial information), not because they are ignorant to epistemic knowledge but because, at times, they struggle with interpreting this information, so that they can apply it. These decision-makers also possess their own knowledge, which may complement the knowledge from the *earth observation science*. Therefore, scientists also need to understand the policy-making and implementation process in order for them to package and disseminate knowledge, so it can influence policy and implementation at the community level. Sutherland et al. (2012) suggest that a better understanding of the science–policy–practice interface can help improve the way in which scientific evidence and advice is developed, interpreted, and communicated, so that is makes a significant societal impact.

Research participants can include community members, representatives of institutions, policy-makers, and decision-makers, who all have different values and operate in different contexts of power positions in their society. Power relations in the decision-making process influence the decisions that are made on whether to use certain information to inform planning and decision-making. This process is influenced by the values of the stakeholders, and this, in turn, shapes the state of the social ecological system.

11.2 Qualitative Considerations

Participatory approaches are powerful qualitative techniques that provide insight into the nature of human relations and the complexity around them. Questions often arise on the quality of qualitative research and the degree of confidence that the qualitative end product is legitimately useful (Roller and Research Design Review 2017). Unlike quantitative research, which provides a quality framework that illustrates the errors, so that both the researcher and information user can fully appreciate research outcomes and understand what is presented, we are still lacking in qualitative research (Roller and Research Design Review 2017). The quality of qualitative research is highly dependent on the ability of the researchers to genuinely commit to honing their skill as a research instrument. These skills are essential in planning the interviews or group discussion, in designing the questions for discussion, in dealing with group dynamics, and in ensuring participation of all participants. Other factors that influence the quality of qualitative research include the environment in which the research will be conducted, presence of tools such as recorders, and participant dynamics in terms of their social/cultural/political, age, and gender differences.

However, it is important to note that researchers in the social science have not yet reached a consensus on the definition of scientific research in the social sciences, and there are no guidelines on how to compare one qualitative research with another (Porta and Keating 2008; Roller and Research Design Review 2017). One school of thought has argued that the social sciences are still in the pre-paradigmatic stage, where multiple paradigms are being put forward by different schools of thought (Kuhn 1970). Here, it is argued that the social sciences still need to look for uniting principles and standards. A second school of thought posits that social science is in the post-paradigmatic stage, whereby it has identified a set of scientific assumptions that are linked to the post-modern world views. The third group believes that the social sciences are nonparadigmatic, as the discipline will never have one dominant approach or set of explicit standards. In this paradigm, it is argued that the social world needs to be understood in various ways, each of which may be valid for specific purposes. Further to this, the third group posit that the social science field is multi-paradigmatic, with different paradigms either struggling against each other or ignoring each other.

In the following subsections, we discuss some of the qualitative considerations, which include guidelines on validity, reliability, and ethical issues when conducting qualitative research.

11.2.1 Validity and Reliability

The two concepts of validity and reliability emerged through the positivist paradigm, where experimental methods and quantitative measures are used to test hypothetical generalizations and analyze the causal relationship between variables (Golafshani 2003). Validity in quantitative research defines whether the research accurately measures what it was intended to measure or how truthful the research results are (Golafshani 2003, p. 599). Reliability is *the extent to which results are consistent over time and an accurate representation of the total population under study.* Furthermore, when the results of a study can be replicated using the same method, then the research instrument is also considered to be reliable (Joppe 2000; as cited in Golafshani 2003, p. 598). Reliability and validity in quantitative research focus on whether the result is replicable, whether the means of measurement is precise, and if it is measuring what it is expected to measure.

In qualitative research, numerous definitions of reliability and validity are given by many qualitative researchers from different paradigms. Lincoln and Guba (1985) argue that there can be no validity without reliability, and as such, a demonstration of the former is adequate to institute the latter. These two terms are often not used separately in qualitative research. There is consensus that the researcher or research team is the research instrument; hence, some qualitative researchers have argued that the concept of reliability, as used in quantitative research, needs to be redefined to meet the realities of qualitative research (Strauss and Corbin 1990). Others have even argued further that reliability is misleading and has no place in qualitative research (Stenbacka 2001). More acceptable terms such as quality, rigor, and trustworthiness have been adopted as concepts of validity of qualitative research (Davies and Dodd 2002; Lincoln and Guba 1985; Seale 1999; Stenbacka 2001).

11.2.2 Testing Validity and Reliability in Qualitative Research

Validity and reliability of qualitative research can be improved through triangulation, which is a multi-strategy approach, whereby a number of methods are used in a complementary way. Triangulation compensates for the weaknesses of each method and increases the possibility that data gathered can be compared (Cresswell 1994). Gorman and Clayton (1997, p. 96) argue that a researchers should choose the most appropriate data collection methods to ensure a match between the research problem and research question formulation in their particular paradigm (e.g., constructivism vs. realism paradigm). However, it is important that each paradigm has its own assumptions in terms of theoretical frameworks; hence, there is a need to define triangulation from a qualitative research's perspective in each paradigm (Barbour 1998; Golafshani 2003).

11.2.3 Ethical Considerations

In conducting participatory research, the researcher engages with human subjects to get their views, values, and perceptions on a particular phenomenon under research. However, the researcher also has to treat the research participants ethically and respectfully by adhering to some ethical principles. Emanuel et al. (2010) identified seven ethical principles that protect research participants, that is, social value, science validity, fair subject selection, independent review, informed consent, respect for selected participants, and favorable risk–benefit ratio. These are further discussed as follows:

1. Participation of human subjects in research should add *social value*; that is, research should be aimed at improving people's well-being. Researcher should show the impact and relevance of the research on the local community and/or on the community from where participants are drawn.

2. To ensure that people do not waste time and effort, the proposed research should use human participants if this can add *scientific validity*. Research needs to provide results that are useful and increase knowledge on the subject under investigation.

3. *Selection of participants needs to be fair* to those who can participate in the study and to those who will potentially benefit from the research.

4. There should be a favorable *risk–benefit ratio*; that is, any risks of participation should be balanced by benefits to the participants and the valuable new knowledge that will be generated.

5. To ensure that the research is consistent with the ethical principles, the research proposal should be independently reviewed by a group of individuals who are not connected to the research before engaging with the participants. This gives participants trust in the research that they would be treated fairly.

6. The researcher should provide research participants with information regarding the proposed study before they can agree to participate in the study. Participants should not be forced to participate in a study and should be allowed to withdraw at any time. Participants who agree to participate should give their *informed consent*. There are four components of informed consent: First, the participant needs to be *competent*. Second, they should get full *disclosure* about the study, its goals, benefits, and risks. Third, participants should have an *understanding* of what the research is about (use language that the participant understands). Fourth, they should decide to participate *voluntarily*, without undue inducement and coercion.

7. The researcher should show respect to the participants by:

   a. Continuously checking the well-being of participants throughout the research process and removing the participants from the study when it becomes too risky/harmful.

   b. Taking responsibility to keep the participants' personal information (e.g., identity, household income, and medical records) confidential.

   c. Allowing participants to withdraw from the study at any time.

   d. Informing participants of any changes to the study or any new information that they should know, including new risks revealed when the study was underway.

   e. The researcher should share research findings with the participants to show that they are partners in the study. Furthermore, before publishing any work, the participant contribution should be discussed with all those who participated in the research.

   f. The development of the research should continuously be visible and open to suggestions from others throughout the research process (Audouin and Murambadoro 2011; Emanuel et al. 2010).

Qualitative research allows for improved understanding of any social phenomenon. Results here show that there is no one truth but several truths about the social world, depending on the research methods used and assumptions made. This then often makes it difficult to replicate or compare

212
GEOValue

qualitative studies (Porta and Keating 2008, p. 20). Therefore, researchers need to take note of the different paradigms and approaches used within the social sciences, the differences between them, and the extent to which they can be combined (Porta and Keating 2008).

11.3 New and Emerging Techniques in Noneconomic Social Science-Oriented Methods

Section 11.1 provides an overview of some of the social science-oriented methods that have been used for decades to engage with participants to understand the value of geospatial information. These methods include interviews, focus group discussions, and observation. Technology advances, including the onset and improvements in the ease in access to the internet as well as the onset of social media platforms, have revolutionized stakeholder participation and valuation of societal impact in research. These new and emerging techniques are contributing to knowledge gathering for a variety of disciplines, including social science (Bright et al. 2014), and they have changed the scale and context for stakeholder interaction and participation (Voinov et al. 2016).

Social media may be defined *as internet-based tools that allow people to create, share, or exchange information, career interests, ideas, and pictures/videos in virtual communities and networks* (Obar and Wildman 2015). Social media encompasses varied platforms, which include discussion forums, chat rooms, search engines, and microblogging outlets, for example, Global Earth Observation System of Systems and Southern African Science Service Centre for Climate Change and Adaptive Land Management. There is an increasing interest in social media in research for academic, business, and public sectors, as seen by the rise in people/institutions using social media platforms such as Twitter, Facebook, and YouTube. These tools have proved to be very effective in reaching many people in a limited time, and this could be attributed to the open-source approach for the distribution of data (Bright et al. 2014).

New and emerging techniques for stakeholder participation and valuation of societal impact of geospatial information through social media and web applications have transformed research, as it allows researchers to make new connections and reach new audiences. Social media platforms can be used to inform the social science research process and are useful and efficient tools for evaluating societal impact, as participants can provide feedback on key research areas and products. Advances in technology have also made it easier to integrate information in interactive formats through visualization and games to augment participatory experiences (Voinov et al. 2016). There is also socially generated data, which are found on search engines (e.g., Google), which even though not considered social media, can provide important perceptions into the behavior and opinions of the greater public. As the use and role of social media continue to grow, so do more insights on human behavior, which is of great interest to the social sciences; hence, a myriad of private organizations are tapping into these platforms for research purposes (Bright et al. 2014).

One of the key uses of social media emerging in social science research is as an alternative to surveys and opinion polls, especially where there is a need to know what the public is thinking on some specific issues and to anticipate what action the public would take. Therefore, social media can be used as a way of studying people's sentiments about specific issues (e.g., energy efficiency), which was previously done using traditional survey methods. Social media has assisted in answering key questions, in elaborating on key issues by using online surveys for gathering information on public opinion, and in improving the traditional survey methods (Bright et al. 2014).

Like all forms of research methods, using social media also has advantages and disadvantages. Its potential advantages include technical and nontechnical issues (Table 11.4).

Like all qualitative research methods, online-based participant interactions have similar problems to those encountered in face-to-face interviews, group discussion, and focus group discussions. The issues include sampling, ethical problems, validity, reliability, and legitimacy (Sullivan 2012). There is a continuous discussion on whether people present their true selves on virtual text-based environments or whether they present their fake personality. This debate is ongoing and has an impact on the validity and reliability of this information, as some argue that the Internet is a place where people can fabricate whom they feel they should be, which may not be a true reflection of their true self (Sullivan 2012). This presents a problem for social science research, which usually studies peoples' interactions within their environments and depends on people being honest about the matters being researched (Sullivan 2012).

It is important to note that the uses of social media in the social sciences are growing and not all the potential uses (and challenges) have been uncovered yet, especially as technology continues to advance at an unprecedented pace, as illustrated in Figure 11.1, which shows the growth in use of social media, from 2012 to 2014.

Figure 11.1 highlights how powerful social media is as a research tool, given the number of people and information that are present on the different shared platforms in only sixty seconds. New and emerging technologies such as social media and web applications are increasing the population groups that can be reached, which would have otherwise not been possible by using traditional research methods such as face-to-face interviews. Social media, for example, can be an effective tool to reach the existing and potential users of geospatial data and for the evaluation of societal impact as well as for getting key messages to a wider audience (Bright et al. 2014). Therefore, institutions that produce geospatial data can also take advantage of these emerging technologies and use them to assess the value of their information as well identify information gaps/needs.

TABLE 11.4

Advantages and Disadvantages of New and Emerging Technology

Advantages	Disadvantages
• Social media generates huge amounts of data such as statistics, linked studies on topics of interest, social sentiments, background information, news, reports, and trends on all sorts of social topics • The ease of use and, in some cases, the generic nature of the sites have had a wider impact on the public and have created a network effect, which has promoted an increased uptake of the use of the sites, with some people using sites such as Facebook for business, personal, and other social networks • Emerging technology has the potential to gather data and information for qualitative research, especially the ability to interview geographically dispersed people in one sitting, in real time (Sullivan 2012) • Audio and video conferencing can be used for individual interviews, for small groups, and for focus group discussions • Advances in these types of technologies have made communication in the real time possible, using computers, laptops, and smart phones • The anonymity presented by online discussions forums and chatrooms as well as the absence of shared social networks at times allow people to be themselves, and they can express themselves with no fear or reprieve	• The downside to having huge amounts of data from social media is that all this information needs to be filtered to see what is useful for the researcher and how to analyze it, and there are tips that can be used by researchers to filter the information they need, but this can be time consuming (www.investintech.com) • The investment needed in running and maintaining the infrastructure of social media as well as other internet platforms can become costly for the researcher or the organization, especially when it comes to storing the data • Despite the reduction in the costs for data storage, the costs are still substantial for smaller organizations • Substantial computer skills are essential in order to be able to access the data online as well as analyzing the data gathered. There is currently a shortage in these skills, and some organizations are forced to contract out some of these tasks, which is very costly (Bright et al. 2014) • Online tool for audio and video conferencing may face technical issues from time to time • At times, it is difficult to assess the extent of the impact of social media • Compared with face-to-face interactions, where the researcher is able to observe the participants and how they behave, this is difficult to observe on video conferencing or other social media platforms. Presentation of participants on video conferencing is considered an unnatural performance, and the environment is argued to be surreal (Sullivan 2012) • Although emerging technology can reach a wider audience, it is important to also note that it is only the groups that use that particular application that will be reached, and these might not be representative of the sample needed • Methodological issues such as correcting the bias of the sampling problems can be an issue, especially the automatic analysis of emotions expressed in messages or response on social platforms (Bright et al. 2014)

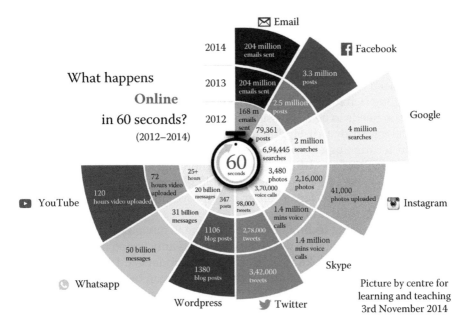

FIGURE 11.1
The changes in the use of social media in 60 seconds between 2012 and 2014. (From Kimmorley, S., INFOGRAPHIC: Everything that happens online in 60 seconds, Retrieved April 18, 2016, from http://www.businessinsider.com/infographic-what-happens-online-in-60-seconds-2015-5, 2015; Courtesy of Centre for Learning and Teaching 2014, Wanchai, Hong Kong. With permission.)

11.3.1 Conclusion

Qualitative research used in the social sciences can play a key role in evaluating the societal benefits of geospatial information. It can be used on its own or to complement quantitative research methods such as surveys. The researcher and facilitator in the participatory action research need to be exposed to and learn from the positive and negative reactions that they receive from the participants and remain critical of this information. People's values inform the decisions that people make, and these values are embedded in their experiential knowledge, which can be shared through qualitative approaches with the help of narratives and focus group discussions. Research participants who can be community members, policy-makers, and decision makers have different values and operate in different contexts of power in their society. Power relations in the decision-making process influence the decisions that are made on whether to use certain information to inform planning and decision-making, and this process is influenced by the values of the decision-makers, and this in turn shapes the state of the social ecological system.

Like quantitative research, qualitative research methods also have challenges with validity and reliability, as they have to deal with people's values and opinions on spatial and nonspatial information. There are also uncertainties with geospatial data that will add onto the challenges of conducting qualitative research, for example, uncertainty on rainfall projections in climate change research, which can affect how policy-makers make decisions on managing water demands. When using participatory methods, there is a need to ensure that all participants values are captured; hence, the use of jargon often found in the different disciplines that produce geospatial information needs to be addressed. Jargon makes it difficult for people who do not belong to that discipline to engage in the discussions. as the concepts and terminologies would constrain their participation. It should also be noted that there are different ways of knowing, and hence, a transdisciplinary approach that incorporates all world views/values through effective participation is argued for, in order to improve understanding on societal benefits of geospatial information.

New and emerging technology will continue to advance the way in which qualitative and quantitative research is done. It is anticipated that innovations in technology will help knowledge producers understand their user needs and provide a platform for direct feedback, which can be used to ensure that geospatial information is packaged to meet different user group needs in terms of accessibility and usability.

References

Audouin, M., 2011. *Transdisciplinary Research for Sustainability: User Guide.* Council for Scientific and Industrial Research (CSIR), Stellenbosch, South Africa.

Audouin, M. and Murambadoro, M., 2011. *Ethics Protocol: Research on Human Subjects.* CSIR Report, Pretoria, South Africa.

Barbour, R.S., 1998. Mixing qualitative methods: Quality assurance or qualitative quagmire? *Qualitative Health Research*, 8(3), 352–361.

Berkes, F. and Folke, C. (Eds.), 1998. *Linking Social and Ecological Systems: Management Practices and Social Mechanisms for Building Resilience.* New York: Cambridge University Press.

Bielak, A.T., Campell, A., Pope, S., Schaefer, K. and Shaxson, L., 2008. From science communication to knowledge brokering: the shift from "science push" to "policy pull." In D. Cheng et al. (Eds.) *Communicating Science in Social Contexts.* New York: Springer Science+Business Media B.V.

Bright, J., Margetts, H., Hale, S. and Yasseri, T., 2014. The use of social media for research and analysis: A feasibility study. Government Social Science Research. Department for Work and Pensions.

Cresswell, J., 1994. *Research Design: Qualitative and Quantitative Approaches.* Thousand Oaks, CA: SAGE.

Davies, D. and Dodd, J., 2002. Qualitative research and the question of rigor. *Qualitative Health Research*, 12(2), 279–289.

De Walt, K.M. and DeWalt, B.R., 2002. *Participant Observation: A Guide for Fieldworkers.* Walnut Creek, CA: AltaMira Press.

De Wit, B., 2016. (Working Paper). Value based decision making in planning and design of large capital projects. A reference guide for project managers. CSIR, Pretoria, South Africa.

Dunlop, C.A., 2009. Policy transfer as learning: Capturing variation in what decision makers learn from epistemic communities. *Policy Studies*, 30(3), 1–44.

Emmel, N., 2008. Participatory mapping: An innovative sociological method. Other. Real Life Methods. (Unpublished) National Centre for Research Methods. Retrieved from http://eprints.ncrm.ac.uk/540/.

Emanuel E.J., Abdoler, E. and Stunkel, L. 2010. Research ethics: How to treat people who participate in research. National Institutes of Health Clinical Center Department of Bioethics and Foundation for the NIH. Bethesda, MD.

Flyvbjerg, B., 2001. *Making Social Science Matter: Why Social Inquiry Fails and How It Can Succeed Again*. Cambridge, UK: Cambridge University Press.

Freitas, H., Oliveira, M., Jenkins, M. and Popjoy, O., 1998. The focus group, a qualitative research method. ISRC, Merrick School of Business, University of Baltimore (MD, EUA), WP ISRC No. 010298 22 p. Retrieved June 20, 2016, from http://www.ufrgs.br/gianti/files/artigos/1998/1998_079_ISRC.pdf.

Gallagher, T., 2001. The value orientations method: A tool to help understand cultural differences. *Journal of Extension*, 39(6), 165–167.

Gill, P., Stewart, K., Treasure E. and Chadwick, B., 2008. Methods of data collection in qualitative research: Interviews and focus groups. *British Dental Journal*, 204, 291–295.

Gillis, A. and Jackson, W., 2002. Research Methods for Nurses: Methods and Interpretation. Philadelphia, PA: F.A. Davis Company.

Golafshani, N., 2003. Understanding reliability and validity in qualitative research. *The Qualitative Report*, 8(4), 597–607.

Gorman, G.E. and Clayton, P., 1997. Qualitative research for the Information Professional. London Library Association, London, UK.

Joppe, M., 2000. The research process. Retrieved April 23, 2017, from https://www.uoguelph.ca/hftm/research-process.

Kawulich, B.B., 2005. Participant observation as a data collection method. *Forum: Qualitative Research*, 6(2), Art. 43, 1–28.

Kimmorley, S., 2015. INFOGRAPHIC: Everything that happens online in 60 seconds. Retrieved April 18, 2017, from http://www.businessinsider.com/infographic-what-happens-online-in-60-seconds-2015-5.

Krueger, R.A., 1994. *Focus Groups: The Practical Guide Goes Applied Research* (2nd ed.). Thousand Oaks, CA: SAGE Publications.

Kuhn, T., 1970. *The Structure of Scientific Revolutions* (2nd ed.). Chicago, IL: University of Chicago Press.

Leininger, M.M., 1985. *Qualitative Research Methods in Nursing* (2nd ed.). Orlando, FL: Grune and Stratton.

Lincoln, Y.S. and Guba, E.G., 1985. *Naturalistic Inquiry*. Beverly Hills, CA: SAGE.

Lewin, K., 1944. The dynamics of group action. *Educational Leadership*, 1, 196–200.

Macauley, M.K., 2006. The value of information: Measuring the contribution of space-derived earth science data to resource management. *Space Policy*, 22, 274–282.

Maguire, P., 1987. *Doing Participatory Action Research: A Feminist Approach*. Amherst, MA: University of Massachusetts Press.

McDonald, C., 2012. Understanding participatory action research: A qualitative research methodology option. *Canadian Journal of Action Research*, 13(2), 34–50.

McIntyre, A., 2002. Women researching their Lives: Exploring violence and identity in Belfast, the North of Ireland. *Qualitative Research*, 2, 387–409.

Morgan, D.L., 1988. *Focus Groups the Qualitative Research*. Beverly Hills, CA: SAGE Publications.

Obar, J.A. and Wildman, S., 2015. Social media definition and the governance challenge: An introduction to the special issue. *Telecommunications Policy*, 39(9), 745–750. doi:10.1016/j.telpol.2015.07.014.

Porta, D.D. and Keating, M., 2008. *Approaches and Methodologies in the Social Sciences: A Pluralist Perspective*. Cambridge, UK: Cambridge University Press.

Resilience Alliance, 2017. Social ecological systems. Retrieved March 1, 2017, from http://www.resalliance.org/concepts-social-ecological-systems.

Roller, M.R. and Research Design Review, 2017. 13 factors impacting the quality of qualitative research. Retrieved April 18, 2017, from https://researchdesignreview.com/2011/02/28/13-factors-impacting-the-quality-of-qualitative-research/.

Schmuck, R., 1997. *Practical Action Research for Change*. Arlington Heights, IL: IRI/Skylight Training and Publishing.

Seale, C., 1999. Quality in qualitative research. *Qualitative Inquiry*, 5(4), 465–478.

Selener, D., 1997. *Participatory Action Research and Social Change*. New York: Cornell University.

Stenbacka, C., 2001. Qualitative research requires quality concepts of its own. *Management Decision*, 39(7), 551–555.

Strauss, A. and Corbin, J., 1990. *Basics of Qualitative Research: Grounded Theory Procedures and Techniques*. Newbury Park, CA: SAGE Publications.

Streubert, H.J. and Carpenter, D.R., 1995. Qualitative Research in Nursing: Advancing the Humanistic Imperative. Philadelphia, PA: J. B. Lippincott Company.

Sullivan, J.R., 2012. Skype: An appropriate method of data collection for qualitative interviews. *The Hill Top Reviewer*, 6(1). Retrieved July 7, 2016, from http://scholarworks.wmich.edu/cgi/viewcontent.cgi?article=1074&context=hilltop review.

Sutherland, W.J., Belligan, L., Bellingham, J.R. et al. 2012. A collaboratively derived science-policy research agenda. *PLoS ONE*, 7(3), e31824.

Valenzuela, D. and Shrivastava, P., 2008. Interview as a method for qualitative research (presentation). Retrieved July 8, 2016, from http://www.public.asu.edu/~kroel/www500/Interview%20Fri.pdf.

Voinov, A., Kolagani, N., Mc Call, M.K., Glynn, P.D., Kragt, M.E., Ostermann, F.O., Pierce, S.A. and Ramu, P., 2016. Modelling with stakeholders—Next Generation. *Environmental Modelling and Software*, 77, 196–220.

Wuest, J., 1995. Breaking the barriers to nursing research. *The Canadian Nurse*, 91(4), 29–33.

12

Qualitative and Other Social Science-Oriented Methods to Assess the Importance of Geospatial Information—Qualitative Considerations and Methods

Serene Ho

CONTENTS

12.1 Introduction

> The paradigm of geospatial information is changing; no longer is it used just for mapping and visualisation, but also for integrating with other data sources, data analytics, modelling and policy-making. Once the geospatial data is created, it can be used over and over again to support a wide range of different applications and services.

The quote above comes from a document published by the United Nations Global Geospatial Information Management initiative titled "Future trends in geospatial information management: the five to ten year vision" (UN-GGIM 2015). It suggests that the current and future value, impact, and benefit that can be derived from geospatial information is likely to lie in how it is implemented. Indeed, there is a longstanding acknowledgment of the value of spatial data in decision-making: such is the value proposition of national spatial data infrastructures, where the integration of heterogeneous spatial datasets across source, scale, and technology boundaries offers improved basis for decision-making (Wallace et al. 2006; Crompvoets et al. 2010). The website of the Australia and New Zealand Cooperative Research Centre on Spatial Information (CRC-SI) provides some common examples in which various types of geospatial information—most of them are public-sector information—are utilized in supporting everyday activities and decision-making (Table 12.1).

In a climate where economic rationalism continues to dominate policy-making, the ability to quantify value, benefits, and impacts of geospatial information, particularly in monetary terms, will continue to have a sure and certain role. However, this section marks a departure, albeit one that is intended to be complementary: as its starting point, it takes the position that value cannot always be measured—nor should it be. For practitioners and users of geospatial information, an understanding of the multiple and layered ways in which geospatial information creates, or can potentially create, value is perhaps of greater relevance. This suggests a space for qualitative approaches to come to the fore and this section aims to illustrate how some of the key philosophies and associated methods can help producers of spatial data construct a richer understanding and description of the value, benefits, and impacts of geospatial data.

TABLE 12.1

Types of Spatial Data and Potential Value in Daily Use

Spatial Data Type	Examples of Value, Benefit, and Impact
Bathymetry	Provides marine navigational information
Big data	Natural resource management modeling, property planning, and privacy
Cadastre	Administers and manages property
Datum	Provides reference points for location in global positioning systems
Geocoded addressing	Applications across emergency response, social services, insurance, telecommunications, navigation, and so on
Geodesy	Understanding changes in the properties of the Earth (e.g., shape and gravity)
Landsat	Global perspective on natural changes and human impact
LiDar	High-resolution maps
Positioning	Provides location and navigation accuracy
Synthetic Aperture Radar	Environmental and resource monitoring
Thematic map	Information atlases

Source: Derived from information at www.crcsi.com.au.

12.2 Framing Value

Notable philosophers, including Aristotle and Marx, have conditioned society to perceive value as something that can be objectively measured: price (or money) is one of the common ways used today to standardize the assessment of value. This is evident in widely used concepts such as *value chain*, developed by Michael Porter in the mid-1980s, where he defined value as the amount that buyers are willing to pay for what a firm provides and *value chain* as a combination of nine generic value-added activities operating within a firm—activities that work together to provide value to customers (Porter 1985).

Korsgaard (1986) found that the prevailing opinion that there are two main types of value, *objective* (where value is implicit in the object of study itself, and human interest does not change this) and *subjective* (where value of the object of study is vested in human interests in the object of study, which determines its utility), is difficult to apply categorically. To overcome this, she proposed a third theory of value, *rational* value, which can only be derived if there is enough practical reason (sensible or otherwise) to realize adoption or implementation of the object of study, that is, an outcome where supply meets demand.

Similarly, Feller et al. (2006) argued that value is firstly subjectively perceived, contingent on contextual conditions; then, this is transformed into a need that may be met through some good or service being provided. However, this transaction also creates value—both to the recipient and to the provider. Consequently, they argue for differentiation between technical (similar to Korsgaard's objective value), organizational, and personal forms of value (more aligned with rational and subjective values, respectively).

In terms of geospatial information, there is evidence of the sometimes murky relationship between supply and demand. For example, in the area of public sector information and public administration, demand for geospatial information or systems might be present, but there might be no coherent supply. For instance, there might be the desire for sustainable methods of recording land rights information, or vice versa, as in the case of the proliferation of three-dimensional (3D) spatial technologies, but at the same time, there might also be difficulty in identifying an effective source of demand. We also see situations where different conceptions of value may be collected together, for example, in the case of establishing land administration systems in developing countries, where demand is constituted by a broad range of users across society and their multiple perspectives of expected *value* may not be sufficiently captured through aggregated means of analysis. Trends in crowdsourced geospatial information have also provided new challenges in value conception and determination, particularly with respect to social value (Feick and Roche 2013).

12.3 The Potential of Qualitative Research

Qualitative research can play a role in addressing some of these gaps in valuation. The term *qualitative* essentially refers to the types of data collected during inquiry, which in turn is an outcome of different systems of methods and philosophies. As such, a broad definition of qualitative research is "any kind of research that produces findings not arrived at by means of statistical procedures or other means of quantification" (Strauss and Corbin 1990, p. 17).

Therefore, by definition, qualitative research is value-oriented, since it seeks to understand how meaning and behavior are socially constructed (Denzin and Lincoln 2005). This can play a key role not only in eliciting and describing broad human interests that motivate demand conditions but also as the tipping point when broad demand gains saliency and transforms into the *practical reason* (as identified by Korsgaard) that results in adoption and implementation (1986). Recognition of this function is leading to the mainstreaming of qualitative dimensions for constructing value. We see this most recently in the European Commission's Value Assessment Tool developed for evaluating and communicating the value of information systems or information technology (IT) projects to government (DIGIT 2016). Within this tool, the Commission emphasizes three aspects of qualitative value: societal, administrative, and user. The details of these aspects are provided in Table 12.2.

Qualitative methods are often used to pursue aims associated with *how* or *why* types of questions, seeking to build a deeper understanding of some

TABLE 12.2

Qualitative Aspects of Value in Information Systems or IT Projects

External Value	Internal Value
Societal value: • Reducing regulatory and administrative burden • Enhancing interactions between government and society and collaboration between governments • Enhancing monitoring and compliance	Political value: • Associated with strategic objectives and overall image
	Administrative value: Efficiency and effectiveness of work
User value • Improving user-friendliness • Improving quality of information and service delivery • Increasing the level of support for users	User value: Comply with information technology governance principles and contribute to, or follow, best practices

Source: Directorate-General for Informatics (DIGIT), VAST guidelines, Brussels, Belgium, 2016.

phenomena, particularly with relevance to the context within which it is embedded and how people ascribe meaning to it (Maxwell 2013). This produces knowledge outcomes that are inherently different to quantitative (i.e., numerical) descriptions of populations—and hence is increasingly argued as being complementary rather than being opposed to quantitative methods—in analyzing complex causal relationships (Garbarino and Holland 2009). Therefore, in line with the theme of this chapter, this section provides a brief overview of qualitative methods and considerations for geospatial researchers and practitioners. The intention here is to support a broader discussion around diversity in geospatial research approaches.

12.4 Quantitative Research in Geospatial Science: Applications and Limitations

Conceptualizing any research design is ultimately a consequence of the personal views and beliefs of the researcher (which influence the philosophy driving knowledge creation) and the research question being pursued. Research, development, and applications related to geospatial information tend to stem from, and continue to be associated with, science and engineering disciplines (e.g., geomatics, software engineering, and information systems), where logical positivism dominates. In turn, this creates a conceptual focus on quantitative empiricism, whose longstanding and widespread use and acceptance in scientific practice have created a profoundly embedded legitimacy (Hirschheim 1985; Alavi and Carlson 1992).

Hence, quantitative methods are commonplace in the spatial domain, often used not only to support or improve statistical descriptions but also to provide a structured way of thinking spatially about a phenomenon. This could rely on demographic attributes, occurrence trends and patterns, spatial variations between locations or groups of people, and so on. However, as with other statistically based analysis, quantitative data can be skewed. A common example is the correlation between the sample size and reliability that the data represents. The oft-cited p-value index reflects how likely the results from a sample have been affected by pure coincidence. Typically, statistically significant studies quote p-values of less than 0.5—the higher the p-value, the more likely the outcomes are related to chance. However, the sample size, being a variable in the calculation, affects the p-value, which means that the p-value can be manipulated by changing the sample size. Of course, there are other statistical tests that can be introduced to improve the reliability and meaning of the data, but this simple example shows how quantitative data can be skewed if it is not used correctly. In addition, quantitative research suffers from other limitations, including divorcing the research output from meaning and purpose, removing contextual variables, lacking applicability at the individual level, and missed opportunities for discovery (Guba and Lincoln 2005).

However, of more interest to us is how traditional methods of inquiry might shift in line with changes to how spatial information and technologies are used and applied. For example, in our times, we are witnessing a rapid move toward individualization. Location, as the basis of building context about individuals for personalized services, is becoming prevalent as a business model—look no further than Google's array of personal and home-based technologies. In this environment, what types of research questions might be of significance? A key proposition for valorizing the use of qualitative methods in the spatial domain is the need for a deeper understanding of the *how* and *why*, that is, a need for explanation or interpretation. Extending our previous example about Google's push toward individually customized location-based services, potential research questions might be the following: How could spatial information and products improve the individual's experience? Why would consumers use individualized spatially enabled products? How might these products confront current norms and values (e.g., privacy expectations)? To answer these questions, we not only need descriptive (quantitative) data about what data or products are being used, but we are also likely to need to understand a range of personally variable dimensions about users that are socially constructed and push some users to perceive these services as valuable, while others perceive it as an imposition on privacy.

These types of data lie outside the scope of quantitative methods. In addition, there is a growing body of evidence around the limitations of such research in addressing the challenges of developing new technologies— that a focus only on external phenomena (e.g., technologies and processes) does not provide sustainable solutions to the challenges of technological innovation in contemporary contexts (Davis and Songer 2002). It also ignores the fact that relationships between people and technologies are constantly being negotiated, even more so today with the rapid advances in technology and information and communication technologies (ICT) and with corollary shifts in response to environmental pressures (Myers 1994). For the application and use of geospatial information, which tends to be implemented within a broader information system, insights into the limitations of quantitatively based research—essentially resulting in simplistic descriptions of cause and effect—are numerous in the broader field of information systems research (e.g., see Orlikowski and Baroudi 1991).

What does this mean for the spatial sciences? It certainly does not mean an abandonment of the epistemological foundations of the discipline. Instead, it is a call for pluralism—in the tradition of a long list of researchers (e.g., Benbasat et al. 1987; Kaplan and Duchon 1988)—that more realistically represents the complexities of issues in the applied sciences. More importantly, given the proliferation of spatially enabled consumer products, qualitative data are essential to design products that are actually fit for purpose. We outline some methodological considerations in the following subsections.

12.5 Qualitative Considerations

For spatial scientists who wish to adopt or incorporate qualitative methods and data into their research, perhaps one of the most significant points of difference to grasp is the philosophical underpinning of qualitative research. Since qualitative research is value-laden, it is incumbent on the researcher to play a mediating role in interpreting the data that he or she has collected—an *insider's view* (or an emic position) versus an *outsider's view* (an etic position) common in the quantitative research (Strauss and Corbin 1990; Guba and Lincoln 1994). Therefore, qualitative researchers must express their ontological and epistemological positions and subsequently their methodological approach relative to their research for it is this transparency that creates validity (Klein and Myers 1999). Ontology generally refers to the nature of the world and this is often dichotomized as objective (social realities that exist and have meaning, independent of actors) and subjective (social realities that exist and are given meaning as a result of actors involved in them) (Bryman 2012). This in turn relates to how knowledge about the world is derived and the relationship between the researcher and the topic, which then directs the researcher's methodology or tools and processes used to produce knowledge about the social reality being investigated (Guba and Lincoln 2005).

Taken together, these three components—ontology, epistemology, and methodology—constitute the research paradigm. For example, an interpretive paradigm is often found in qualitative research, characterized by a pluralistic view of the world, reflecting a fundamental belief that reality, as it is experienced, is a complex phenomenon since it is made up of numerous meanings and interpretations from people's lived experiences (Guba and Lincoln 1994; Schwandt 1994; Hurworth 2005). This indicates an ontology, where reality is socially constructed, thereby reflecting a subjective epistemology and necessitating the use of methods such as interviews, field observations, case studies, ethnography, and so on to address research questions (Creswell 2013). It is also through this process of elucidating world views and ways of knowing that researchers can better understand when to use qualitative or quantitative approaches, or indeed, a combination of both.

Before identifying appropriate research methods, qualitative researchers must also identify a research approach that not only fits with conceptual choices in terms of ontologies and epistemologies but is also appropriate for addressing the research problem. Approaches that are common to qualitative research include ethnography, case studies, and action research. Once this has been selected, the following considerations regarding the selection of research methods will likely apply.

12.5.1 Sampling

Qualitative methods pursue inherently different research questions to those addressed by quantitative methods, which affect participant sampling. The priority here is not necessarily on generalizability (i.e., requiring a statistically significant sample) but on answering the research question (Marshall 1996). There is a tendency for low sample numbers in qualitative studies; as such, the value of such studies lies in offering useful and transferable knowledge (Guba and Lincoln 2005). Miles and Huberman (1994, pp. 27–34) provide a more in-depth discussion on sampling considerations.

Marshall (1996) identified three broad types of sampling that tend to be used in qualitative research: convenience sampling (choosing the most accessible subjects), purposeful sampling (selecting participants [based on a set of criteria] who will be able to answer the research questions most effectively), and theoretical sampling (iterative selection of participants to further examine emergent learning from collected data). Of these, the author contends that purposeful sampling tends to be the most common approach, since it enables a range of practical, intellectual, and theoretical factors to be combined to identify a sample that is most capable of enabling the researcher to address the research question. Indeed, there is a longstanding recognition that researchers will attempt to select the most productive sample, particularly in an early stage (Glaser 1978). In fact, it has been argued that all sampling in qualitative research is purposeful because it is intentional (Patton 1990; Coyne 1997), and different research motivations create distinctions in sampling intention, which result in different sampling strategies (for a more detailed explanation, see Patton 2002).

12.5.2 Data Collection

Common methods for collecting qualitative data are interviews, participant observations, and focus groups. The eventual choice as to which format is most appropriate is a consequence of what type of information is being sought. For example, participant observation could be used to collect information as to how technological prototypes are used and the efficacy of organizational workflows; interviews collect detailed, in-depth information about participants; and focus groups can provide a way to draw slightly broader information, such as information common to a group rather than to an individual.

The types of data collected can span many forms: narrative based, textual, images, or documented through observations, to name some common ones. Methodological considerations here are the development of some form of structure to guide the data collection process so that the focus of inquiry is adequately prepared for, but in a way that allows the researcher to remain open to new ideas being revealed and to be able to explore new directions during the data collection. For example, interviews may be guided by a schedule of topics based on literature review rather than a set list of questions. In addition, the researcher needs to be flexible and be able to adapt to different discourse styles among participants and be aware of interpersonal differences or situational difficulties that are creating bias in the data collection.

Indeed, DeLyser and Sui (2013) were able to show that in the geographical and spatial science disciplines, interviews have persisted as a research method simply because they continue to stand out as a particularly revelatory method. Through a range of examples, the authors demonstrated the value of this qualitative method in uncovering the complexities of human beliefs, motivations, values, and perceptions that influence their behavior. It is likely that these are the very qualities that enhance our understanding of how spatial data can be used and embedded to support decision-making. In land administration, the use of case studies as a qualitative method in doctoral research appears to be common (Çağdaş and Stubkjær 2009).

12.5.3 Data Analysis

Qualitative data can be any data that is not numerical, for example, text, documents, stories, and pictures, and because this does not conform to any type of standardization, it can be difficult to understand how to start analyzing these data types (particularly if more than one format of content is present). Similar to quantitative methods, methodological considerations here first pertain to pattern identification: are there similarities across what participants are saying? The construction of themes from the data is a common practice, where themes can either be theory-driven, a priori, or data-driven (Boyatzis 1998). Themes are constructed through an inductive coding process, where phrases, words, concepts, and ideas are coded and categorized, a process that is iteratively repeated to reduce overlap and redundancy (for more information about coding as a technique, see Boyatzis 1998; Saldaña 2016). To support researchers in their analysis, some frameworks have been developed to provide structure and rigor to the process, such as Spiggle's (1994, pp. 493–496) set of *data manipulation operations*, developed to support the coding and analysis of qualitative data. Maxwell (2009, pp. 236–240) provides a valuable discussion on qualitative data analysis considerations.

Some methodological considerations are the tension between predetermined categories, as identified through literature review, and accommodation of emergent categories through the process of reading and rereading the material. Another consideration is the wholly subjective ability of the researcher to undertake this task, with little variation across different datasets during coding, and also the experience in recognizing different types of themes, including those that are missing. Indeed, a researcher's theoretical sensitivity is also a key consideration (Glaser 1978). This can be improved using multiple coders and comparing results or even by using something as simple as a reflexive journal (Wallendorf and Belk 1989). For example, Ho et al. (2015) and Ho and Rajabifard (2016) used coding techniques to analyze the outcomes of in-depth, unstructured, multi-sectoral stakeholder interviews. This was used to construct a macrolevel narrative of key areas of thematic significance around the (respective) areas of institutional challenges and responses to support the introduction of 3D technologies in land administration.

12.5.4 Data Validation

There is a tendency by practitioners and researchers, especially those who identify with positivist epistemologies, to assume that qualitative methods are less reliable and less valid because they produce data that are not objective or generalizable (Guba and Lincoln 2005). To address this longstanding mistrust, concepts such as *reliability* and *validity*, which are typical in quantitative research, have been redefined as trustworthiness, rigor, and quality for the qualitative context (Golafshani 2003). Consequently, improving these aspects is contingent on improving perceptions of credibility (truthfulness) of the phenomenon. To aid this, triangulation (Creswell and Miller 2000) or mixed-methods research (Creswell 2009) has emerged as a corroborating strategy to reduce or eliminate bias in the analysis.

12.5.5 Time and Costs

For any researcher, time and costs are major factors; it is no different in the adoption of qualitative methods. Time and costs impact methodological aspects in some key ways. Some examples are provided as follows:

- *Sampling*: When sampling, credibility might be traded for convenience and ease of access.
- *Data collection*: Transcription of interviews requires time (if done by the researcher) or cost (if transcription is outsourced).
- *Data analysis*: Coding, as an iterative process, can take weeks, if not months, to be done properly. There are now a range of computer-aided qualitative data analysis software (CAQDAS) options (e.g., NVivo software) to help researchers manage and maintain the integrity of the coding process.
- *Data validation*: Multiple strategies to support data validation exist. The appropriate one may be a trade-off between efficient use of time and feasibility.

12.6 Conclusion

Many geospatial researchers who are untrained in social sciences may consider the conceptual and methodological considerations outlined above to be foreign. Nonetheless, there are clear advantages to giving serious thought to using qualitative methods. For example, the practice of urban planning is one where concepts such as neighborhood character are a function of both physical and social attributes (Dovey et al. 2009), which in turn require the use of both quantitative and qualitative methods for improving planners' understanding. On the other hand, disciplines such as land administration seem to favor the inclusion of qualitative methods (Çağdaş and Stubkjær 2009), but they are rarely used as the basis for theory building.

More generally, the longstanding use of qualitative methods in understanding consumer behaviors is directly relevant for the mass of wearable and personal location-enabled devices. For public administrators everywhere, sharing spatial data through the development of spatial data infrastructures has been both a technical and an organizational challenge, yet technical interoperability is often compromised by the lack of social interoperability, the facilitation of which is often the greater challenge (Nedovic-Budic and Pinto 2001). These examples suggest that in its ability to provide in-depth insight, qualitative methods can offer a critical approach to advancing the use and application of spatial data and technologies.

References

Alavi, M. and P. Carlson (1992). A review of MIS research and disciplinary development. *Journal of Management Information Systems*, 8(4), 45–62.

Benbasat, I., D.K. Goldstein and M. Mead (1987). The case research strategy in studies of information systems. *MIS Quarterly*, 11(3), 369–386.

Boyatzis, R.E. (1998). *Transforming Qualitative Information: Thematic Analysis and Code Development*. Thousand Oaks, CA: SAGE Publications.

Bryman, A. (2012). *Social Research Methods* (4th ed). London, UK: Oxford University Press.

Çağdaş, V. and E. Stubkjær (2009). Doctoral research on cadastral development. *Land Use Policy*, 26(4), 869–889. doi.org/10.1016/j.landusepol.2008.10.012.

Coyne, I.T. (1997). Sampling in qualitative research. Purposeful and theoretical sampling; merging or clear boundaries? *Journal of Advanced Nursing*, 26(26), 623–630.

Creswell, J.W. (2009). *Research Design: Qualitative, Quantitative and Mixed Method Approaches*. Thousand Oaks, CA: SAGE Publications.

Creswell, J.W. (2013). *Qualitative Inquiry and Research Design*. Thousand Oaks, CA: SAGE Publications.

Creswell, J.W. and D.L. Miller (2000). Determining validity in qualitative inquiry. *Theory into Practice*, 39(3), 124–130.

Crompvoets, J., E. De Man and C. Macharis (2010). Value of spatial data: Networked performance beyond economic rhetoric. *International Journal of Spatial Data Infrastructures Research*, 5, 96–119.

Davis, K.A. and D.A. Songer (2002). Technological change in the AEC industry: Asocial architecture factor model of individuals' resistance. *Proceedings of the 2002 IEEE International Engineering Management Conference, "Managing Technology for the New Economy"* (pp. 286–291). Cambridge, UK. August 1820.

DeLyser, D. and D. Sui (2013). Crossing the qualitative-quantitative divide II: Inventive approaches to big data, mobile methods, and rhythm analysis. *Progress in Human Geography*, 37(2), 293–305.

Denzin, N. and Y. Lincoln (2005). Introduction. In Denzin, N. and Y. Lincoln (Eds.), *The SAGE Handbook of Qualitative Research* (pp. 1–32). Thousand Oaks, CA: SAGE Publications.

Directorate-General for Informatics (DIGIT) (2016). VAST guidelines. Brussels, Belgium.

Dovey, K., I. Woodcock and S. Wood (2009). A test of character: Regulating place-identity in inner-city Melbourne. *Urban Studies*, 46(12), 2595–2615.

Feick, R. and S. Roche (2013). Understanding the value of VGI. In Sui, D., S. Elwood, and M. Goodchild (Eds.), *Crowdsourcing Geographic Knowledge* (pp. 15–29). Dordrecht, the Netherlands: Springer.

Feller, A., D. Shunk and T. Callarman (2006). Value chains versus supply chains. *BPTrends*, March 2006, 1–7.

Garbarino, S. and J. Holland (2009). Quantitative and qualitative methods in impact evaluation and measuring results. Issues paper commissioned by the UK Department for International Development.

Glaser, B. (1978). *Theoretical Sensitivity: Advances in the Methodology of Grounded Theory.* Mill Valley, CA: Sociology Press.

Golafshani, N. (2003). Understanding reliability and validity in qualitative research. *The Qualitative Report*, 8(4), 597–606. Retrieved November 9, 2016 from http://nsuworks.nova.edu/tqr/vol8/iss4/6.

Guba, E.G. and Y.S. Lincoln (1994). Competing paradigms in qualitative research. In Guba, E. G. and Y. S. Lincoln (Eds.), *The SAGE Handbook Of Qualitative Research*, Thousand Oaks, CA: SAGE Publications.

Guba, E.G. and Y.S. Lincoln (2005). Paradigmatic controversies, contradictions, and emerging confluences. In Denzin, N. K. (Ed.), *The SAGE Handbook of Qualitative Research* (pp. 191–215). Thousand Oaks, CA: SAGE Publications.

Hirschheim, R. (1985). Information systems epistemology: An historical perspective. In Mumford, E., R. Hirschheim, G. Fitzgerald, and T. Wood-Harper (Eds.), *Research Methods in Information Systems* (IFIP 8.2 Proceedings, pp. 13–36). Amsterdam, the Netherlands: North-Holland.

Ho, S. and A. Rajabifard (2016). Towards 3D-enabled urban land administration: Strategic lessons from the BIM initiative in Singapore. *Land Use Policy*, 57, 1–10.

Ho, S., A. Rajabifard and M. Kalantari (2015). "Invisible" constraints on 3D innovation in land administration: A case study on the city of Melbourne. *Land Use Policy*, 42, 412–425.

Hurworth, R. (2005). Interpretivism. In Mathison, S. (Ed.), *Encyclopedia of Evaluation* (pp. 210–211). Thousand Oaks, CA: SAGE Publications.

Kaplan, B. and D. Duchon (1988). Combining qualitative and quantitative methods in information systems research: A case study. *MIS Quarterly*, 12(4), 571–586.

Klein, H.K. and M.D. Myers (1999). A set of principles for conducting and evaluating interpretive field studies in information systems. *MIS Quarterly*, 23(1), 67–93.

Korsgaard, C.M. (1986). Aristotle and Kant on the source of value. *Ethics*, 96(3), 486–505.

Marshall, M.N. (1996). Sampling for qualitative research. *Family Practice*, 13(6), 522–525.

Maxwell, J. (2009). Designing a qualitative study. In L. Bickman and D. J. Rog (Eds.), *The SAGE Handbook of Applied Social Research Methods* (pp. 214–253). Thousand Oaks, CA: SAGE Publications.

Maxwell, J.A. (2013). *Qualitative Research Design: An Interactive Approach*. Thousand Oaks, CA: SAGE Publications.

Miles, M.B. and A.M. Huberman (1994). *Qualitative Data Analysis*. Thousand Oaks, CA: Sage Publications.

Myers, M.D. (1994). A disaster for everyone to see: An interpretive analysis of a failed is project. *Accounting, Management and Information Technologies*, 4(4), 185–201.

Nedovic-Budic, Z. and J.K. Pinto (2001). Organizational (soft) GIS interoperability: Lessons from the US. *International Journal of Applied Earth Observation and Geoinformation*, 3(3), 290–298.

Orlikowski, W.J. and Baroudi, J.J. (1991). Studying information technology in organizations: Research approaches and assumptions. *Information Systems Research*, 2(1), 1–28.

Patton, M.Q. (1990). *Qualitative Evaluation and Research Methods*. Thousand Oaks, CA: Sage Publications.

Patton, M.Q. (2002). *Qualitative Research and Evaluation Methods*. Thousand Oaks, CA: Sage Publications.

Porter, M. (1985). *Competitive Advantage, Creating and Sustaining Superior Performance*. New York: The Free Press.

Saldaña, J. (2016). *The Coding Manual for Qualitative Researchers*. Los Angeles, CA: SAGE Publications.

Schwandt, T. (1994). Constructivist, interpretivist approaches to human inquiry. In Denzin, N. and Y. Lincoln (Eds.), *The SAGE Handbook of Qualitative Research* (pp. 118–137). Thousand Oaks, CA: Sage Publications.

Spiggle, S. (1994). Analysis and interpretation of qualitative data in consumer research. *Journal of Consumer Research*, 21(3), 491–503.

Strauss, A.L. and J.M. Corbin (1990). *Basics of Qualitative Research*. Thousand Oaks, CA: Sage Publications.

UN-GGIM (2015). Future trends in geospatial information management: The five to ten year vision (2nd ed.). New York: United Nations.

Wallace, J., I.P. Williamson, A. Rajabifard and R. Bennett (2006). Spatial information opportunities for government. *Spatial Science*, 51(1), 79–99.

Wallendorf, M. and R.W. Belk (1989). Assessing trustworthiness in naturalistic consumer research. In E.C. Hirschman (Ed.), *SV—Interpretive Consumer Research* (pp. 69–84). Provo, UT: Association for Consumer Research.

13

Remote-Sensing Supports Economic Inference in Developing Countries

Jacob Hochard and Evan Plous Kresch

CONTENTS

Geospatial data have value when it is analyzed and the findings are used to support policy decisions. This value for researchers is increasing at pace with technological development, as remote-sensing technologies enable large-scale data collection in areas where on-the-ground data collection is otherwise infeasible. In developing countries, where secondary data sources are often unavailable, incomplete, or measured poorly, remotely sensed and georeferenced data are being leveraged to inform policy questions affecting society's most vulnerable populations.

We begin with a brief introduction to the research applications of remotely sensed data products, with a focus on data accessibility and spatiotemporal coverage. We then examine recent trends in the use of remotely sensed data to support economic inference in developing countries. We discuss two distinct ways in which geospatial data availability has supported empirical research in the context of a developing country. First, geospatial data have improved or enabled the measurement of human outcomes (e.g., economic activity, poverty, and deforestation) and exposure variables (e.g., temperature, precipitation, and storm events) that affects human outcomes. Second, geographic relationships inherent to geospatial data have been exploited to support causal inference, when human outcomes and exposure are jointly determined. We summarize a series of recent and policy-relevant contributions, spanning the general interest economics and scientific literatures, which exhibit the value of geospatial data in supporting policy-relevant research in data-poor environments.

13.1 Introduction

Satellite-derived datasets are perhaps the most common remotely sensed data being used to support economic inference in developing countries. Passive satellite sensors are designed to *receive and measure* energy from the Earth's surface, such as thermal imaging or nighttime lights, to determine the physical characteristics of the Earth, while active sensors reflect and measure radiation, such as those used in topographic mapping (Dunbar 2015).

Remote-sensing data are augmented by Light Detection and Ranging (LiDAR) systems, which are active sensors mounted to aircraft that measure distance using laser light. In addition to manned aircraft, LiDAR systems as well as multispectral, hyperspectral, and radar sensors have been used with unmanned aerial vehicles, or drones. This drone sensing has been used to support community monitoring of natural occurrences, such as tropical forests (Paneque-Gálvez et al. 2014), forest fires, or canopy height (Tang and Shao 2015). Other applications of drone-based remote sensing include coastal wetland mapping, flood surveillance, and oil-spill tracking (Klemas 2015).

Nonaerial mobile data collection devices, such as cell phones and global positioning systems, have also been used to conduct population mapping (Deville et al. 2014) and optimize the use of public transportation. They may soon exploit real-time human health data to support a mobile health (mHealth) and clinical care industry globally (Steinhubl et al. 2013; Steinhubl et al. 2015).

Collection of remotely sensed data are generally more cost-effective (Mumby et al. 1999) and precise (Seelan et al. 2003) than the data collected by large-scale surveys. Satellite-derived data, which have countrywide, regional, or global coverage, are often comparable across geopolitical boundaries. This allows the researcher to avoid the quality control problems that arise when synthesizing national surveys or census data from multiple sources. As such, remotely sensed data have been used to conduct global analyses of land degradation (Bai et al. 2008; Barbier and Hochard 2016), coastal human settlements (Small and Nicholls 2003), terrestrial drought (Mu et al. 2013), incidence of poverty (Elvidge et al. 2009), and electric power consumption from economic activity (Elvidge et al. 1997).

The temporal resolution of remotely sensed products also exceeds what is feasible when using survey methods. As an example, the globally comprehensive Normalized Difference Vegetation Index, which is derived from the Moderate-Resolution Imaging Spectroradiometer (MODIS) instrument and is used to monitor droughts and agricultural productivity, has a spatial resolution (250 meters) that exceeds even the most nuanced and detailed survey attempts (Carroll et al. 2004). The MODIS land products,* such as

---

* Available from Oak Ridge National Laboratory's Distributed Active Archive Center for Biogeochemical Dynamics (ORNL DAAC) at https://daac.ornl.gov/.

surface temperature, land cover, and vegetation indices, are available for 8-day, 16-day, and annual frequencies at resolutions ranging from 250 to 1000 meters. Similarly, the National Ocean and Atmospheric Administration Defense Meteorological Satellite Program's global nighttime lights products are available as annual 30 arc second grids from 1992 to 2013, which have been used to estimate gross domestic product (GDP) (Ghosh et al. 2010), electrification rates (Elvidge et al. 2011), and global poverty rates (Elvidge et al. 2009).

The broad spatial and temporal coverage of remotely sensed data products, combined with widespread public accessibility through clearinghouses, such as the Food and Agriculture Organization's (FAO) GeoNetwork and Global Agro-Ecological Zones (GAEZ) Portal, has contributed to a greater reliance on remotely sensed data to examine economic questions in the context of a developing country.

13.2 Measurement of Economic Outcomes

Income levels across countries are one of the most important indicators of economic well-being globally, as they trend closely with life satisfaction (Deaton 2008). Traditionally, income data were widely and publicly available through national accounts. Unfortunately, and for a variety of reasons, income data in least developed countries are often unavailable, incomplete, or poorly measured. The inability to observe changes in households' incomes consistently over time makes it difficult to gauge whether livelihoods are improving. As such, the precise targeting of policy interventions becomes challenging for those communities that may have the greatest need for public support.

Henderson et al. (2012) examine how growth in relatively high-resolution nighttime lights coverage relates to the measurement of national income accounts for countries with low- and high-quality data. To improve the measurement of GDP growth, an optimal weighting that incorporates information from national accounts data alongside satellite-observed changes in luminosity is estimated and applied. The method, albeit of little empirical value in countries with high-quality national accounts data, is particularly well-suited for growth analysis in the developing countries that are known to have data quality issues. The authors found that growth in nighttime lights trends approximately 1:1 with income growth among low- and middle-income countries. For developing and developed countries alike, the authors exhibited the application's ability to predict highly local—that is, specific cities, rural versus urban, and coastal versus noncoastal—growth rates, which are generally unavailable in national surveys or census data.

Blumenstock et al. (2015) also applied remote-sensing technology to gain insight into the indicators essential for examining human progress—poverty and wealth. Here, the wide availability of mobile phones in Rwanda were used to collect information that predicts wealth (and has the capacity to predict poverty). The authors focused specifically on an individual's *digital footprint*, which includes phone calls (e.g., timing, intensity, and directionality), text messages, mobility, and contact networks. These indicators were used to predict a variety of individual-specific wealth variables (e.g., asset ownership and housing characteristics) collected from follow-up phone interviews. The authors underscored that the ability to track an individual offers socioeconomic insights at a higher resolution than when relying on the district-level data publicly available in Rwanda. Among others, noted applications for this method include the precise targeting of aid and information to vulnerable populations that may be overlooked when relying only on district-level data.

Jean et al. (2016) predicted on-the-ground asset wealth and consumption expenditure by combining satellite imagery with machine learning algorithms. The authors, relying on only publicly available data, overcame the Blumenstock et al. (2015) challenge of acquiring proprietary datasets. The approach used a convolutional neural network (CNN) that has been trained to identify edges and corners present in satellite imagery. In doing so, the CNN identifies edges and corners within daytime imagery that are then used to explain nighttime luminosity. Locations with sharp edges or sides that also predict nighttime luminosity are then identified as particularly economically meaningful activity. Identifying such locations enabled a strong prediction of household consumption expenditure (37%–55% of variation) and asset wealth (55%–75% of variation). Importantly, the others showed that, without substantial loss of predictive power, the model can be trained in a different country than it is used to predict economic outcomes.

Remotely sensed data have also been used to examine the effects of geopolitical districting on environmental outcomes. Burgess et al. (2012) employed MODIS data to construct an annual panel dataset of tropical deforestation from 2001 to 2008, with a 250-meter × 250-meter resolution across Indonesia. Under the premise that illegal deforestation is driven by incomplete enforcement at lower-level administrative jurisdictions, the authors examined how deforestation rates are impacted by geopolitical redistricting throughout Indonesia. The argument that local officials are charged with the enforcement of logging rules suggests that illegal logging operations may be sanctioned at a local level. By examining deforestation activity before and after a district splits (189 districts in 2000 and 312 in 2008), the authors estimated that the subdividing of a district increases the overall deforestation by 8.2%. The finding, which argues that local community management of natural resources can be consistent with overexploitation, stands in stark contrast with a larger literature that suggests that community management of natural resources elicits greater conservation by ensuring that local managers are also local stakeholders.

13.3 Measurement of Exposure Variables

In addition to standard economic outcome variables used in the development economics literature, remotely sensed data are increasingly being used in the environmental economics literature as an improved method for measuring the effects of exposure to natural phenomena from economic activities, such as pollution. Using these estimates, researchers can better understand the effects to both humans and the environment of these externalities. We describe three high-impact and recent articles that rely on remotely sensed data to measure exposure variables that influence economic outcomes.

To better understand the impact of climate variables on economic growth, Dell et al. (2012) used 1900–2006 globally and monthly gridded precipitation and temperature data with a 0.5 × 0.5-degree resolution. The authors improved on past work that correlates temperature and economic activity between countries by predicting a battery of within-country economic outcomes, using year-to-year changes in temperatures and precipitation. Such a rich panel dataset was also used to examine whether temperature and precipitation impact growth rates of income, in addition to income levels. Several findings are presented. First, the authors estimated that a 1°C increase in temperature reduces economic growth in poor countries by approximately 1.3% points. Importantly, these estimates assume that no adaptation to temperature increases occurs and the growth effects are therefore persistent. Evidence also suggests that the growth impact is rooted in reductions to agricultural and industrial output as well as to a decrease in political stability. Such an impact on economic growth may explain income gaps between rich countries, which tend to have cooler temperatures, and poor countries, which are generally located in warmer regions.

Beyond temperature and precipitation effects, climate predictions of increasingly severe storm events are also of public policy concern in developing countries (Knutson et al. 2010; Lin et al. 2012). Hsiang and Jina (2014) examined the paths of 6700 cyclones from 1950 to 2008 and monitored exposed countries' economic growth paths preceding and following exposure to storm events. Here, cyclone paths were determined using a combination of *ground, ship, aerial,* and *satellite-based observations*. The researchers found that cyclone impacts reduce economic growth and that those reductions are permanent. Two decades after the exposure, per capita incomes were 7.4% lower than that predicted in the absence of a storm event, which equates to nearly 4 years of lost economic development. The authors combined their estimates with scientific projections for climate change. This analysis suggests that by the persistent and adverse nature of cyclone exposure on economic growth, the social cost of anthropogenic climate change is $9.7 trillion greater than the conventional estimates.

Remote-sensing technology has also been used to examine exposure to air and water pollutants in the context of a developing country. Greenstone and Hanna (2014) collected data from 572 air quality-monitoring stations located in 140 Indian cities, with unbalanced readings ranging from 1987 to 2007. Similarly, across 424 Indian cities, the authors collected water quality data from 489 monitors, with readings ranging from 1986 to 2005. The authors found improvements in air quality, which can be linked to statistically insignificant but economically meaningful reductions in infant mortality across India. However, regulations on water pollution did not appear to yield improved water quality. Focusing on a country with notoriously poor institutions, the researchers suggested that air quality improvements are linked to a strong domestic prioritization for clean air initiatives, whereas water quality issues, among the Indian citizenry, do not share the same support.

The use of remotely sensed data has improved the measurement of economic outcomes, which shed light on the state of livelihoods in areas that escape the reach of national surveys or census initiatives. Similarly, remote-sensing technology has been used to examine how public policies curb or mediate the impact of environmental quality or climate trends that may affect livelihoods. Together, improved measurement of human outcomes and human exposure improve our capacity to conduct empirical research in data-poor areas. Remotely sensed data have also been useful in a third capacity. Geographic relationships inherent to geospatial data have been exploited to support causal inference when human outcomes and exposure are jointly determined.

13.4 Supporting Identification Strategies

In addition to the measurement of final outcomes, such as exposure and economic development, remotely sensed data have also been useful as a tool for identification in causal inference. For example, geographic relationships inherent to geospatial data have been exploited to support causal inference when human outcomes and exposure are jointly determined.

The construction of dams in India is the primary form of public investment in irrigation. Technologically, the public benefits of dams accrue downstream, where mediated water flow serves agriculture. Assessing the impact of 4000 such investments across India is challenging empirically, because many factors, such as local agricultural productivity, industrial composition, population density, and poverty trends, determine the location where dams are placed. For this reason, the assessment of downstream effects of these large public investments requires additional information on dam placement that is unrelated to downstream human outcomes. Here, Duflo and Pande (2007) recognize that dam construction is best suited on relatively flat or moderately steep segments of a river. The authors exploit variation in river gradients, derived from a digital elevation model, to identify areas suited topographically well for dam construction. Throughout India, findings suggest that agricultural production increases and poverty decreases downstream of major dams, while poverty increases in those districts that receive a dam. By focusing on dam investments driven by topographic elements, rather than socioeconomic trends, the authors are able to improve the credence of their results for informing public policy.

Moreover, in India, coastal population vulnerability to storm events is a major public policy concern. India's Bay of Bengal coast, which was struck by a major cyclone in 1999, enjoys expansive mangrove forests that shelter coastal populations by buffering winds and reducing storm surge height. Das and Vincent (2009) investigated whether these storm protection services translated into the protection of human life. Empirically, attributing lower death rates to wider seaward mangroves is made difficult by the presence of unobserved or difficult-to-measure physical coastal features. Specifically, coastal areas sheltered naturally from storms also serve as suitable habitat for mangrove growth. Disentangling storm protection services arising from coastline topography and natural infrastructure, such as mangrove forests, requires information on each location's propensity to harbor mangrove growth. The authors incorporated mangrove width controls from 1944, derived from historical aerial photographs, to ensure that vegetation, not the physical properties of the coastline, was responsible for the protection of life. Findings show that 0.0148 lives are saved per hectare of mangroves (~1 life per 68 hectares), which, based off of the coast of near-mangrove agricultural land, translates to approximately 11.7 million rupees, or $270,000 1999 USD, per saved life.

Ebenstein (2012) attempted to measure the human health benefits from improved water quality in China. Ingestion of polluted water is likely to cause adverse health outcomes, such as digestive cancers, but parsing this effect out from other unobserved factors, such as access to health care and local industrial agglomeration, is empirically challenging. In this case, the author exploited variation in water quality, based on precipitation rates and geographic locations, which are plausibly exogenous to health outcomes. Precipitation rates around a disease occurrence influence the potency of ingested pollutants by diluting surface water. Similarly, water downstream of a tributary is likely to be more polluted than that tributary's headwaters. The author estimated that a single grade reduction in water quality increased digestive cancer rates by 9.7%, which sheds light on public policy options. Specifically, an additional $500 million investment in wastewater treatment saved approximately 17,000 lives per year.

In Indonesia, sea piracy is a public policy concern that worsens in the absence of alternative income-generating opportunities. Axbard (2016) employed satellite data on local fishing conditions to examine the extent to which fishing opportunity, or lack thereof, fuels piracy activities. Several factors that influence fishing conditions but are plausibly exogenous to piracy activities, such as abundance, migration patterns, distribution, and growth of fish, were exploited to better understand determinants of piracy attacks in the Indonesian waters. Compelling evidence suggests that favorable fishing conditions reduce such attacks by 40%, which is robust to a series of robustness checks. The authors join a growing literature by highlighting the potential for climate effects to increase criminal activity.

13.5 Conclusion and Points of Caution

The multifaceted uses of remotely sensed and georeferenced data have supported policy relevant research in the context of a developing country. Although these data have tremendous potential to support empirical work, several points of caution are worth noting. Many remote-sensing *end products* stem from a thorough processing of raw satellite-derived products. For example, the popular Oak Ridge LandScan product offers high-resolution and globally rasterized annual population data that are becoming increasingly popular for use in empirical economics research (Wendland et al. 2010; Nunn and Puga 2012; Damania et al. 2017). Although the product's resolution is standardized across the global mapping, the underlying unit of census variation varies between and within countries. When using such a product *in tandem* with other environmental or socioeconomic sources, careful attention must be paid to the construction of each end-product source and how potential measurement errors may correlate across variables or propagate through empirical models.

Data availability voids in some of the world's poorest, remote, and vulnerable locations have created a call for increased reliance on remotely sensed sources. However, sample selection becomes of increasing importance when individuals privately acquire the sensors that are collecting information. In many cases, such as the reliance on mobile phone data to predict within-district asset ownership in Rwanda (Blumenstock et al. 2015), empirical insights garnered are richer than those when relying on available survey sources. However, and as Blumenstock et al. (2015) noted, results are not necessarily generalizable to broader populations, that is, nonmobile phone users. Similar sampling challenges arise when using data from environmental monitoring devices, which, even when broad coverage is available, are often sited based on factors other than coverage of a representative populace.

References

Axbard, S. (2006). Income opportunities and sea piracy in Indonesia: Evidence from satellite data. *American Economic Journal: Applied Economics* 8(2): 154–194.

Bai, Z.G., D.L. Dent, L. Olsson, and M.E. Schaepman. (2008). Global assessment of land degradation and improvement. 1. Identification by remote sensing. Wageningen, the Netherlands: International Soil Reference and Information Centre (ISRIC).

Barbier, E.B. and J.P. Hochard. (2016). Does land degradation increase poverty in developing countries? *PLoS One* 11(5): e0152973.

Blumenstock, J., G. Cadamuro, and R. On. (2015). Predicting poverty and wealth from mobile phone metadata. *Science* 350(6264): 1073–1076.

Burgess, R., M. Hansen, B.A. Olken, P. Potapov, and S. Sieber. (2012) The political economy of deforestation in the tropics. *The Quarterly Journal of Economics* 127(4): 1707–1754.

Carroll, M.L., C.M. DiMiceli, R.A. Sohlberg, and J.R.G. Townshend. (2004). 250m MODIS normalized difference vegetation index, 250ndvi28920033435, Collection 4, University of Maryland, College Park, MD.

Damania, R., C. Berg, J. Russ, A.F. Barra, J. Nash, and R. Ali. (2017). Agricultural technology choice and transport. *American Journal of Agricultural Economics* 99(1): 265–284.

Das, S. and J.R. Vincent. (2009). Mangroves protected villages and reduced death toll during Indian super cyclone. *Proceedings of the National Academy of Sciences* 106(18): 7357–7360.

Deaton, A. (2008). Income, health, and well-being around the world: Evidence from the Gallup World Poll. *The Journal of Economic Perspectives* 22(2): 53–72.

Dell, M., B.F. Jones, and B.A. Olken. (2012). Temperature shocks and economic growth: Evidence from the last half century. *American Economic Journal: Macroeconomics* 4(3): 66–95.

Deville, P., C. Linard, S. Martin, M. Gilbert, F.R. Stevens, A.E. Gaughan, V.D. Blondel, and A.J. Tatem. (2014). Dynamic population mapping using mobile phone data. *Proceedings of the National Academy of Sciences* 111(45): 15888–15893.

Duflo, E. and R. Pande. (2007). Dams. *The Quarterly Journal of Economics* 122(2): 601–646.

Dunbar, B. (2015). What are passive and active sensors? *NASA.* NASA. May 5, 2017. https://www.nasa.gov/directorates/heo/scan/communications/outreach/funfacts/txt_passive_active.html.

Ebenstein, A. (2012). The consequences of industrialization: Evidence from water pollution and digestive cancers in China. *Review of Economics and Statistics* 94(1): 186–201.

Elvidge, C.D., K.E. Baugh, E.A. Kihn, H.W. Kroehl, E.R. Davis, and C.W. Davis. (1997). Relation between satellite observed visible-near infrared emissions, population, economic activity and electric power consumption. *International Journal of Remote Sensing* 18(6): 1373–1379.

Elvidge, C.D., K.E. Baugh, P.C. Sutton, B. Bhaduri, B.T. Tuttle, T. Ghosh, D. Ziskin, and E.H. Erwin. (2011) Who's in the dark—Satellite based estimates of electrification rates. *Urban Remote Sensing: Monitoring, Synthesis and Modeling in the Urban Environment* 211–224.

Elvidge, C.D., P.C. Sutton, T. Ghosh, B.T. Tuttle, K.E. Baugh, B. Bhaduri, and E. Bright. (2009). A global poverty map derived from satellite data. *Computers & Geosciences* 35(8): 1652–1660.

Ghosh, T., R.L. Powell, C.D. Elvidge, K.E. Baugh, P.C. Sutton, and S. Anderson. (2010). Shedding light on the global distribution of economic activity. *The Open Geography Journal* 3(1): 147–160.

Greenstone, M. and R. Hanna. (2014). Environmental regulations, air and water pollution, and infant mortality in India. *The American Economic Review* 104(10): 3038–3072.

Henderson, J.V., A. Storeygard, and D.N. Weil. (2012). Measuring economic growth from outer space. *The American Economic Review* 102(2): 994–1028.

Hsiang, S.M. and A.S. Jina. (2014). The causal effect of environmental catastrophe on long-run economic growth: Evidence from 6,700 cyclones. No. w20352. National Bureau of Economic Research.

Jean, N., M. Burke, M. Xie, W.M. Davis, D.B. Lobell, and S. Ermon. (2016). Combining satellite imagery and machine learning to predict poverty. *Science* 353(6301): 790–794.

Klemas, V.V. (2015). Coastal and environmental remote sensing from unmanned aerial vehicles: An overview. *Journal of Coastal Research* 31(5): 1260–1267.

Knutson, T.R., J.L. McBride, J. Chan, K. Emanuel, G. Holland, C. Landsea, I. Held, J.P. Kossin, A.K. Srivastava, and M. Sugi. (2010). Tropical cyclones and climate change. *Nature Geoscience* 3(3): 157–163.

Lin, N., K. Emanuel, M. Oppenheimer, and E. Vanmarcke. (2012). Physically based assessment of hurricane surge threat under climate change. *Nature Climate Change* 2(6): 462–467.

Mu, Q., M. Zhao, J.S. Kimball, N.G. McDowell, and S.W. Running. (2013). A remotely sensed global terrestrial drought severity index. *Bulletin of the American Meteorological Society* 94(1): 83–98.

Mumby, P.J., E.P. Green, A.J. Edwards, and C.D. Clark. (1999). The cost-effectiveness of remote sensing for tropical coastal resources assessment and management. *Journal of Environmental Management* 55(3): 157–166.

Nunn, N. and D. Puga. (2012). Ruggedness: The blessing of bad geography in Africa. *Review of Economics and Statistics* 94(1): 20–36.

Paneque-Gálvez, J., M.K. McCall, B.M. Napoletano, S.A. Wich, and L. Pin Koh. (2014). Small drones for community-based forest monitoring: An assessment of their feasibility and potential in tropical areas. *Forests* 5(6): 1481–1507.

Seelan, S.K., S. Laguette, G.M. Casady, and G.A. Seielstad. (2003). Remote sensing applications for precision agriculture: A learning community approach. *Remote Sensing of Environment* 88(1): 157–169.

Small, C. and R.J. Nicholls (2003). A global analysis of human settlement in coastal zones. *Journal of Coastal Research* 19(3): 584–599.

Steinhubl, S.R., E.D. Muse, and E.J. Topol. (2013). Can mobile health technologies transform health care? *JAMA* 310(22): 2395–2396.

Steinhubl, S.R., E.D. Muse, and E.J. Topol. (2015). The emerging field of mobile health. *Science Translational Medicine* 7(283) 283rv3–283rv3.

Tang, L. and G. Shao. (2015). Drone remote sensing for forestry research and practices. *Journal of Forestry Research* 26(4): 791–797.

Wendland, K.J., M. Honzák, R. Portela, B. Vitale, S. Rubinoff, and J. Randrianarisoa. (2010). Targeting and implementing payments for ecosystem services: Opportunities for bundling biodiversity conservation with carbon and water services in Madagascar. *Ecological Economics* 69(11): 2093–2107.

14

Agricultural Case Studies for Measuring the Value of Information of Earth Observation and Other Geospatial Information for Decisions

Richard Bernknopf

CONTENTS

14.1 Introduction

In this example, geospatial information is applied by using Earth observation (EO) data to the agricultural sector. Two case studies that involve the quantitative estimation of the value of information (VOI) in specific adaptation and mitigation decisions are summarized. The first case study focuses on adapting land use to sustain drinking water quality and to avoid an increase in the contamination of groundwater by agrochemicals. The second case study concentrates on mitigating drought disasters by determining farmer eligibility for financial assistance by the U.S. Department of Agriculture (USDA). In both of these instances, the EO data are transformed into information and is convolved with other science-based indicators as well as with socioeconomic data to assist individuals and communities. The two case studies demonstrate the relevance and benefit of EO as a frequent, objective, and timely monitoring system of terrestrial conditions for cost-effective resource management.

14.2 Expected Societal Benefits of a Decision with and without Earth Observation

The EO data and other remotely data are collected in many forms and at many scales. However, the data require translation to deliver information for systematic use in decisions. Economists regard EO and other geospatial information as an intermediate good that provides a link between economic sectors in an economy (Bernknopf et al. 2016, in press). The two case studies demonstrate how observations of land use and land cover can be coupled with other models and data to create information for decisions. When EO can be linked to other types of models and data to create new geospatial information, as in the two case studies summarized here, the decisions involve spatiotemporal change.

In the case studies, the economic value for geospatial information depends on what is at stake in a decision and how uncertain a decision-maker is. The VOI is defined as the gains that result from making better decisions that are based on additional information in the presence of uncertainty. The economic value is measured as the expected change to an economy that could occur with EO compared with a baseline without EO. Economic benefits accrue to society if the spatial data are more precise to help inform government operations to function efficiently (i.e., at less social cost).

The EO data are digital, which makes it possible to build a comprehensive archive of global land cover and land use. The data require translation to information for systematic use in decisions. The economic value is derived by delivering strategically relevant information to decision-makers. The case studies are focused on specific policy-relevant decisions that take advantage of the spatiotemporal attributes of EO.

Below are two retrospective analyses. In the first example, the Landsat* archive is used to evaluate the societal benefits to adapt agricultural land management to reduce nonpoint-source groundwater contamination. For the second example, the Gravity Recovery and Climate Experiment (GRACE) provide data to assess the economic loss due to the misspecification of eligibility for drought disaster assistance and insurance that is evaluated in a specific drought policy.

14.3 Measuring the Incremental Value of Information of Earth Observation

The microeconomics approach to estimating VOI (described in Chapter 10) depends on the mean and spread of uncertainty surrounding a decision (Macauley 2006). It is defined as the economic gain that results from making better decisions, with less uncertainty from additional information, and depends on (a) the reduction in uncertainty for the decision-maker that is provided by the additional information and (b) what is at stake as an outcome of the decision (Bernknopf et al. 2016, in press).

Studies about the valuation of Global Earth Observation System (GEOS) information have focused primarily on potential benefits of GEOS information (Williamson et al. 2002; Macauley 2006; Macauley and Diner 2007). Potential societal benefits from Moderate-Resolution Land Imagery (MRLI) include cost savings in natural resource allocation, environmental regulation, and reduced damage to public goods (Kalluri et al. 2003; Isik et al. 2005; Macauley and Diner 2007). A few examples have attempted to quantify the benefits of information from GEOS in monetary values (Bouma et al. 2009; Macauley et al. 2010; Macauley and Laxminarayan 2010). For example, Macauley et al. (2010) developed an expenditure-based VOI estimation model to derive a value for Landsat data from the economic value of accurately estimating forest carbon offsets. The case studies described in this chapter are focused on specific applications of the value in use for two types of EO—MRLI (Landsat) and gravity field measurements (GRACE). Analysis involves a comparison of the economic benefits that a decision-maker is

---

* The Landsat Program is a series of Earth-Observing satellite missions jointly managed by NASA and the U.S. Geological Survey (http://landsat.usgs.gov/what_is_landsat.php; accessed November 17, 2016).

able to obtain in a scenario with the EO data relative to a baseline without the EO data. The comparison has three parts:

1. Development of a quantitative representation of a decision without EO and a representation of how this information is improved when EO is available.
2. Application of a decision-making model to predict the choices that a decision-maker could make based on the baseline and the improved information.
3. Estimation of the VOI, as in any benefit–cost analysis (see Chapter 10), is the difference between the economic benefits that result from a decision made with EO and the ones that result from a decision made without it.

In the examples, it is assumed that the EO is available as open-access data from the public sector. Furthermore, it is assumed that the decision-maker usually has some information that provides support for subjective probabilities of the outcome of a decision. That is, the decision-maker or his support staff developing recommendations utilizes the expected value and standard deviation of a probability distribution as the only information required in the decision.

14.4 Value of Information of Moderate-Resolution Land Imagery or Agricultural Land Management to Reduce the Potential for Groundwater Contamination

14.4.1 Introduction

In this case study, the EO data are applied in a hypothetical land use adaptation policy to reduce the impact of agrochemical groundwater contamination in the Midwest United States. Recent U.S. Energy Policy has mandated increases in the use of biofuels for energy consumption. The U.S. Energy Policy Act of 2005 and the Energy Independence and the Energy Security Act of 2007 legislated biofuels supply requirement, including subsidies for farmers to increase the production of corn and other energy feedstocks, to reduce dependence on fossil fuels.

The case study is about how an increase in corn production (used as a feedstock in biofuel production) can change the quality of the groundwater where nitrogen is applied to crops nearby private and public drinking water wells. In certain soils and surficial geology, some of the nitrogen will leach into the soil and migrate into an aquifer. Once in the soil and depending on the surficial geology, the nitrogen may convert to nitrate compounds that can react with other chemicals to form carcinogenic compounds (Forney et al. 2012). The resulting nitrate accumulation in the groundwater has the potential to exceed the U.S. Environmental Protection Agency's (USEPA) health standard for drinking water, which cannot exceed 10.0 mg/L.

14.4.2 Landsat Imagery to Land Use Information

The Landsat archive was used to locate agricultural production from 2000 to 2010 for estimating the amount of nitrogen that could leach into the soil and to a subsurface aquifer. The EO was transformed from satellite imagery to geospatial information by classifying the data into crop type in a rectangular grid. The resulting classifications of land use were coupled with a statistical groundwater vulnerability model to determine the risk of future groundwater contamination. Landsat, the Moderate-Resolution Imaging Spectroradiometer (MODIS), and the Advanced Wide Field Sensor (AWiFS) archives were used in a spatiotemporal analysis of the impact of nitrates on groundwater resources (Forney et al. 2012).* The archive is critical because different crops leach more or less nitrogen and can be linked to groundwater pollution that accumulates from nonpoint sources over time. Both the parcel level and regional risks are related to the quality of information about crop production, variable production inputs, farm management practices, and parcel characteristics. Crop production is the amount of each crop produced on a parcel. EO provides excellent spatiotemporal information for environmental regulatory decisions, because the data for the study period provided the population of land uses. Forney et al. (2012) provided a summary of the accuracy of transformation from data to information for the study period.

The case study was an application of the Production Function Approach (see Chapter 5 for an explanation of the approach) that was described in Forney et al. (2012). The analytical challenge was to measure the incremental economic value of the EO to assess the impacts of nonpoint-source pollution from farm activities in northeastern Iowa (outlined in Figure 14.1). As background for the empirical analysis of the VOI, the economic model in Equation 14.1 was developed to solve the production and water quality interdependence problem for producers and regulators. In the model, economic agents used spatiotemporal information from markets and natural systems to aid in maximizing agricultural production and not decreasing groundwater quality. The State of Iowa was chosen because of the amount of corn produced for food and biofuel and the use of groundwater in private and municipal drinking water wells.

The study area outlined in red in Figure 14.1 includes the following:

- 5.4 million hectares in 35 counties in Northeast Iowa. A great majority of cropland in the region produces corn and soybeans.
- The region covers 603 watersheds (subbasins), with a median of 7910 hydrologic response units. Hydrologic response units are areas within a watershed that respond similarly to a given input, such as nitrogen.

---

* During the period 2006–2008, the Landsat sensor delivered MRLI that contained stripes that were not acceptable for use in the USDA Cropland Data Layer (CDL). The images that were used in the CDL were provided by AWiFS/MODIS. However, there was a reduction in the resolution and quality of the images for classification of land use.

FIGURE 14.1
Iowa study area. (From Forney, W. et al., An economic value of remote-sensing information—Application to agricultural production and maintaining groundwater quality, Professional Paper 1796, Reston, VA, U.S. Geological Survey, 2012. With permission.)

- The USDA Cropland Data Layer that is based on the corn and soybean production estimates for 2001–2010 provided by the EO data.
- The environmental impact of the relative distribution of corn and soybeans is a result of the amount of nitrogen applied, which is typically 114 lb/acre for corn and is 3 lb/acre for soybeans.
- Water quality was sampled from 32,000 wells, ranging from just below the surface to 1220 meters.
- 80% of Iowa drinking water is from groundwater.

14.4.3 Integrated Assessment Model

Miller et al. (2011) identified agriculture and environmental science in the federal sector as a large application of MRLI and, in particular, Landsat. Biofuel production from corn ethanol and other sources was incentivized, but the increase in renewable fuel production could not occur independently of constraining public health policies such as the Federal Clean Water Act of 1972 and Safe Drinking Water Act of 1974, state water quality codes, and groundwater protection acts. The U.S. Environmental Protection Agency (EPA) established the threshold Maximum Contamination Level (MCL) at 10 milligrams per liter (mg/L) nitrate (measured as nitrogen) for safe drinking water. Drinking water exceeding the MCL of nitrate causes human health impacts such as methemoglobinemia, which is also known as *blue baby syndrome* in infants. Nitrate is not a carcinogenic compound; however, it reacts with other chemicals to form carcinogenic compounds, such as nitrosamines and nitrosamides, that are associated with multiple different types of cancers (Mirvish 1995; Weyer et al. 2001; Ward et al. 2005). Hence, there is a physical interdependence of agrochemical application to enhance production and the impact that the chemicals have on groundwater quality.

There are adaptation and mitigation approaches to reduce the amount of nitrate that accumulates in the groundwater that can be implemented. Mitigation approaches to the problem are to impose command and control regulations or to tax farmers to limit nitrogen application. Alternatively, an adaptation policy would be to reallocate regional land use to preserve the drinking water quality of groundwater resources. To address any of these approaches to this issue in a decision framework, an integrated natural–economics science model would be useful. Here, MRLI is used in a policy analysis to evaluate land use adaptation decisions by the individual farmer and regulators (Forney et al. 2012).

The MRLI data relate agricultural production, environmental pollution, and the joint production of agricultural products (corn and soybeans) and groundwater contaminants (nitrates) (Bernknopf et al. 2012). Analysis was possible by adapting an integrated assessment approach (IAA) from Antle and Just (1991). A production decision produces the joint output of a marketable agricultural commodity (Antle and McGuckin 1993) and a nonmarket service of groundwater quality. Decisions at regional scale involve land uses and their impact on ecosystem services.

Archival and current MRLI observations of regional crop production and rotation can be linked to the current level and future accumulation of nitrates in the groundwater. Farmers and regulators can adapt to the environmental risk by using MRLI to inform a potential reallocation of regional land use to preserve the groundwater resources. The economic value of the MRLI is derived using Equation 14.1, in which both the farmer and the regulator seek to maximize agricultural production for any given location within the region, while avoiding an increase in groundwater pollution from those agricultural nitrogen sources.

The regional model incorporates both the producers' (an individual's perspective) and the regulators' (a regional perspective) priorities in accommodating the overall decision-making process. The regional economic model is based on an individual producer's objective to maximize profit, while constraining risks of a marketable crop in Equation 14.1. Given the regulations R, producers seek to maximize profit on each plot of land. The IAA developed for the case study was used to determine an efficient allocation of resources from a regulator's perspective (Bernknopf et al. 2012). As part of this approach, agricultural producers are expected to behave as profit maximizers under given regulatory constraints. Depending on the regulator's objectives and risk preference, a decision to regulate is made. Thus, regulators seek to maximize the value of agricultural output, while limiting the risk of resource damage. Given prevailing crop prices **P**, they choose regulations **R**:

$$\max_{R} PQ$$

$$s.t. \quad \text{risks} \leq \alpha$$

(14.1)

where **P** represents prices of relevant crops, **Q** represents aggregate production of those crops, and **α** represents the probability of exceeding a regulatory standard that causes damage to a resource, namely groundwater. Based on the economic model in Equation 14.1, a probabilistic estimate of cumulative pollution was predicted for an agricultural production portfolio. A forecast of the time to exceed a regulatory standard for resource consumption is calculated in a statistical survival function after the risk of contamination is determined. A statistical survival function uses a cumulative nitrate estimate to provide a conditional probability of exceeding a concentration level that represents a threshold of nitrate contamination that adversely affects humans. Survival (1—failure) analysis (Kalbfleisch and Prentice 1980; Lancaster 1990; Kleinbaum 1996) is applied by the regulator for a probability of exceedance α of a given economic loss. It is assumed that good regulatory policy reduces or eliminates the adverse health effects of nitrates on humans.

The case for the use of EO is made by linking land use information from the satellite archive with agricultural, geospatial, and hydrogeologic models. The MRLI archive is used to classify agricultural land into corn, soybean, or other crops. This baseline land use is linked to Iowa nitrogen application rates for acreage of each crop in a county.[*] A cumulative nitrate index was developed to accumulate the amount of nitrate that was migrating toward a drinking water well in a statistical regression analysis. Figure 14.2 represents the spatiotemporal history over a 10-year period of the zone of capture for a particular well overlaid on a land use map for one of the years of the production period. Depending on the crop planted in a specific year, the model is applied to obtain groundwater vulnerability from nitrogen application in a 10-year encapsulation for a particular drinking water well of a capture zone (shown as black polygons emanating backward in time from the well). Drinking water quality survivability is estimated in subbasins of the region retrospectively for 10 years and then projected for the next 10 years, as shown in Figure 14.3. Based on this risk map, the regulator can make decisions to avoid exceeding the USEPA drinking water quality standard for the period ending in 2020. Since this study was undertaken, the Hruby et al (2015) conducted a survey of Iowa groundwater and an evaluation of public wells' vulnerability to contaminants. The analysis

[*] Iowa State University 2007; http://www.ipm.iastate.edu/ipm/icm/2007/2-12/nitrogen.html; accessed November 17, 2016).

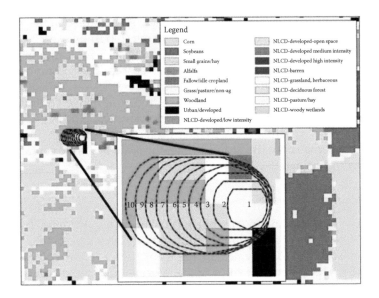

FIGURE 14.2
Well capture zone results on USDA Cropland data layer. Source: (From Forney, W. et al., An economic value of remote-sensing information—Application to agricultural production and maintaining groundwater quality, Professional Paper 1796, Reston, VA, U.S. Geological Survey, 2012.)

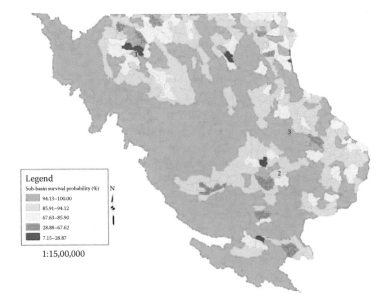

FIGURE 14.3
Groundwater failure and sub-basin probability of survival. Source: (From Forney, W. et al., An economic value of remote-sensing information—Application to agricultural production and maintaining groundwater quality, Professional Paper 1796, Reston, VA, U.S. Geological Survey, 2012.)

for nitrate + nitrite was detected in 26% of the wells sampled in the state, and 3% of these samples exceeded the MCL of 10 mg/L (Hruby et al. 2015).

Currently, the Iowa Department of Natural Resources maintains a state-wide source-water protection areas map of susceptible community water supplies. The determination of susceptibility identifies certain aquifers that are better protected than others, based on the cumulative confining layer thickness above the aquifer (http://www.iowadnr.gov/Environmental-Protection/Water-Quality/Source-Water-Protection). Although the State of Iowa has concerns about the groundwater, the issue of nitrate contamination has not been eliminated.

14.4.4 Results

The EO data provides an input to the decision to adapt agricultural land use to maximize the value of agricultural production and preserve potable ground-water resources. The VOI is calculated as the net present value of the benefits of the MRLI. The net present value of the information is the discounted mon-etized difference of the benefits of land use reallocation and water quality that can be attributed to the MRLI. Maximum estimated VOI for EO is an annual-ized $858 million ± $197 million per year (in 2010) and has a net present value of $38.1 billion ± $8.8 billion for northeastern Iowa (the equations for these monetary values are contained in the appendix to this part of Chapter 14). The VOI is a hypothetical estimate and most likely unattainable. More realistically, if it were possible to identify a 1% improvement by reallocating land use with EO, the VOI would be an annualized $43.0 million per year (in 2010) and has a net present value of $1.91 billion for northeastern Iowa.

14.4.5 Summary

The policy assessment was about an adaptation of land use to consider the economic impact of reallocating land for agricultural production, based on the future availability of potable groundwater in northeastern Iowa. Estimation of a dynamic nitrogen loading and transport model allowed the determination of the cumulative impact of nitrates in groundwater at specified distances from specific sites (wells) for 35 Iowa counties and two aquifers. Over a 10-year period, groundwater wells' probability of survival is so low that these wells could be threatened by nitrate contamination. These wells should be monitored, and exceedance of the nitrate standard should be avoided. Based on this statistical prediction, a statistical survivor function was specified and estimated to forecast the probability of preserv-ing drinking water quality in a private or municipal well. The probability is used by a regulatory agency for further evaluation of the water quality status relative to the chemical standard. The impact of incorporating EO was to demonstrate that it could be used to provide a stewardship type of land management by reallocating production of corn and soybeans to avoid the negative effects of nitrate on groundwater quality.

14.5 Estimate the Value of Information of the NASA Gravity Recovery and Climate Experiment (GRACE) Satellite Mission

14.5.1 Introduction

In this case study, a retrospective analysis[*] of USDA drought mitigation decision-making provides the context for evaluating the VOI of GRACE (Bernknopf et al. 2016). Drought-related decisions arise because the Secretary of Agriculture has the responsibility to declare a natural disaster that is due to severe and prolonged drought and to mitigate the financial impact of lost agricultural production. Specifically, the U.S. Agricultural Act of 2014 legislated payments to eligible livestock producers due to drought through the Livestock Forage Disaster Program. The USDA also promulgated regulations that assigned the Secretarial Disaster Designation Process (7 CFR Part 759) to the USDA Farm Service Agency. The process provides for emergency loans to eligible producers suffering losses due to drought. These USDA responsibilities necessitate the use of the U.S. Drought Monitor (USDM) to determine farmer eligibility for financial assistance. In addition, the USDM is used to inform several major drought management decisions, including eligibility for federal drought assistance programs and drought emergency declarations by state agencies.

Many government programs that allocate resources for drought assistance employ the USDM. The USDM utilizes a classification scheme that identifies general drought areas, labeling droughts by intensity, with Category D1 being the least intense and Category D4 being the most intense. Category D0 is used to indicate drought watch areas. The categorizations in any given USDM map are the result of a well-documented process (Svoboda et al. 2002) conducted by climatologists from the National Oceanic and Atmospheric Administration (NOAA), the USDA, and the National Drought Mitigation Center (NDMC).

The USDM is an ensemble of drought indicators and is a simplified version of the actual state of the environment (Bernknopf et al. 2015). Primary inputs for the USDM include the Palmer Drought Severity Index, NOAA Climate Prediction Center Soil Moisture Model percentiles, the U.S. Geological Survey (USGS) daily streamflow percentiles, precipitation data, NOAA National Climate Center Standardized Precipitation Index percentage of normal precipitation, and remotely sensed satellite vegetation health indices. Supplemental information includes evaporation-related indicators, reservoir, lake, and groundwater levels; field observations of soil moisture; Western Regional Climate Center Western Drought; Snow Telemetry; and GRACE products. The expert-based map weekly communicates the severity of droughts by county

---

[*] A retrospective study relates the outcome to risk and preventive factors present before start of the study (https://www.reference.com/world-view/difference-between-prospective-retrospective-studies-91a9e31f66eb4a55).

in the United States. A given USDM categorization can be estimated with an expected severity category and variance based on deliberations among the authors and variations in the physical indicators. The size of the variance can, in turn, affect the expected socioeconomic outcomes of management decisions that are made based on these categorizations. For example, high variance in USDM categorizations can result in misclassifications of eligibility for drought assistance. A modification to the USDM that can reduce the variance associated with a drought category can lead to reduced social losses.

Once the weekly USDM map is posted on the Internet by the USDA, it is applied as a screening instrument in agricultural regions to determine eligibility for financial assistance to farmers. Because, in some cases, the USDM is the sole criterion for disaster assistance eligibility, it is imperative that the USDM be accurate for cost-effective risk communication. The objective of the USDA is to identify counties that are classified as being in severe and exceptional drought to reduce agricultural losses. A drought severity misclassification is defined as an incorrect determination that the USDM selected a county to be in an exceptional or severe drought when it is not actually in either of those drought severity categories or a county that is actually in an exceptional or severe drought, which is not identified as being in those categories (Bernknopf et al. 2015).

14.5.2 GRACE Products for USDM Drought Severity Classification

The GRACE satellites measure variations in water stored at and below the land surface. The spatial (>150,000 km²) and temporal (monthly) resolutions of GRACE limit its direct applicability for drought monitoring. The GRACE Data Assimilation (GRACE-DA) ensemble model was developed, which integrates GRACE data with other ground- and space-based meteorological observations (precipitation, solar radiation, and so on) (Zaitchik et al. 2008; Houborg et al. 2012). The GRACE-DA model provides soil moisture and groundwater storage variations that are used to generate drought indicators, based on the cumulative distribution function of wetness conditions during 1948–2009. The three indicators are information produced as a surface soil moisture percentile, based on soil moisture anomalies in the top two centimeters of the column; a root zone soil moisture percentile, based on the top 100 centimeters; and a groundwater percentile, based on storage below the root zone. The surface soil moisture drought indicator is expected to vary rapidly in response to weather events, and the groundwater drought indicator is sensitive to meteorological conditions over longer time periods. The GRACE-DA drought indicators are provided as maps and raster datasets, with a resolution of approximately 25 km². The products are provided weekly to support production of the official USDM drought maps. In the case study, the GRACE-DA drought indicators are incorporated into the USDM as a core dataset rather than as supplementary data to demonstrate that they can provide improved inputs to determine the eligibility for drought disaster financial assistance.

14.5.3 Application of the Bayesian Decision Method

The objective of the analysis was to demonstrate how GRACE drought indicators could add value to the USDM. The GRACE products have a complementary role to datasets included in the USDM for soil moisture and other point data. The GRACE drought indicators have been used as supplementary input in the process of making the weekly USDM. The case study was to evaluate whether GRACE information can improve the correlation between incomes of farmers and drought severity classifications from the USDM (Bernknopf et al. 2016).

The analysis is an application of the Bayesian decision theory* and principles to assess the value added of the three GRACE drought indicators. The Bayesian model is used to evaluate if the GRACE drought indicators enhance the risk communication capability of the USDM and assist in improving regional economic outcomes. Specifically, it was designed to assess whether the GRACE data would improve the USDM classification of U.S. counties for the presence and severity of drought conditions (Bernknopf et al. 2016). The model assumes that the decision-maker seeks the best information as input for allocating farm relief during a drought disaster. Consequently, if additional information were made available that provides an improvement in the relationship between the signal, the USDM drought severity category, and the outcome, eligibility for government assistance or insurance, the decision maker would incorporate the enhanced information into the decision.

The Bayesian decision model requires two steps (Bernknopf et al. 2016). The first step is the description of how the decision-maker's information changes with the acquisition of new information. That is, how the decision-maker's probability density over an outcome of interest changes as a result of the new information. Before receiving new information, the decision-maker's belief regarding the probability of occurrence is referred to as the decision-maker's prior belief of probability density. On receipt of new information, the decision-maker observes a value for which the new observations provide an improvement in the prediction of the outcome. This expected outcome is referred to as the decision-maker's posterior belief regarding the probability of occurrence. The second step in the Bayesian decision model is to quantify the VOI by describing how the decision-maker's updated beliefs affect decisions and how these changes in the decisions affect the economic outcome relative to that which would have occurred had the decision-maker made the decisions based on prior beliefs.

---

* Bayesian decision theory refers to a decision theory, which is informed by Bayesian probability. It is a statistical system that tries to quantify the trade-off between various decisions, making use of probabilities and costs. An agent operating under such a decision theory uses the concepts of Bayesian statistics to estimate the expected value of its actions and update its expectations based on new information (https://wiki.lesswrong.com/wiki/Bayesian_decision_theory; accessed November 14, 2016).

14.5.4 Risk Assessment

In the Bayesian model, it is assumed that the decision-maker's decision is represented by the mean and standard deviation of a probability distribution (Sinn 1983). Further, in this example, it is assumed that two normal distributions are used in the eligibility decision (Bernknopf et al. 2016). The first normal distribution represents the intensity of drought in a county i during a particular week t, $S_{i,t}$. The decision-maker is uncertain about this value but has beliefs about it. The decision-maker believes that $S_{i,t}$ is a random variable that is normally distributed with mean $\mu_{S_{i,t}}$ and variance $\sigma^2_{S_{i,t}}$. The second distribution is the information from the USDM, which assigns a drought category to a county during a particular week. Based on USDM information from previous weeks, the decision-maker knows that the drought severity category is normally distributed. The decision-maker believes that $S_{i,t}$ and the drought severity category are correlated; that is, they have a statistical relationship that can be represented by a bivariate normal distribution. The bivariate normal distribution is the distribution for two jointly normal random variables[*] when their correlation coefficient is ρ.

In the model, the decision-maker observes that the USDM has assigned a drought category to county i for week t and updates the prior belief about the intensity of the drought (Bernknopf et al. 2016). The distribution of $S_{i,t}$ is conditional on observing the USDM severity category assignment to form the posterior distribution. If the correlation ρ, increases between a drought category assigned to a county by the USDM and the actual drought intensity in that county, there will be a reduction in the posterior distribution variance. It follows that the incorporation of GRACE products that provide a new set of drought indicators correlates better with drought intensity, and the decision-maker will face an uncertainty that is smaller than the current USDM drought categories without GRACE.

14.5.5 Risk Management and Value of Information

For VOI to be positive, the correlation must increase between the two distributions. The payoff of increased correlation to the decision-maker is represented by a change in the value of the losses to the agricultural sector that were avoided, given that a county experienced a drought of intensity $S_{i,t}$ and received drought assistance. Measurement of the relationship between the decision-maker's actions and the payoff to the decision-maker of undertaking those actions to assess the VOI is required. A decision-maker who makes an optimal decision based on the payoff function and believes that

---

[*] https://www.probabilitycourse.com/chapter5/5_3_2_bivariate_normal_dist.php, accessed November 15, 2016.

the distribution of $S_{i,t}$ and the drought severity category are bivariate normal will derive the following VOI:

$$VOI = \kappa_1 \left\{ \rho^2 \sigma_{S_{i,t}}^2 \right\} \tag{14.2}$$

where κ_1 is a constant. Value of information is proportional to the variance of drought and the square of the correlation coefficient. Thus, VOI increases with ρ^2.

The USDM information structure is the combination of input data that supports drought assistance eligibility decisions. Each variation in the input information can contain a variety of different indicators, depending on the location of county i in week t. Using Equation 14.2, alternative versions of the USDM can be indexed according to their relative informativeness (Lawrence 1999). By being able to index the various combinations of indicators and other input data, it is possible to rank the alternative choices for application to the eligibility decision, according to their VOI.

14.5.6 Econometric Analysis

Ranking alternative USDM information structures is an empirical evaluation. An econometric analysis was conducted to estimate the value added by including GRACE drought indicators to the USDM. The econometric model was used to estimate the correlation between incomes of farmers and the USDM with and without the contribution of GRACE drought indicators (Bernknopf et al. 2016). The econometric model uses observed data to estimate the incremental economic effect of drought on agricultural incomes, while accounting for the fact that some of the determinants of the outcome (including some dimensions of drought) cannot be observed.

The econometric analysis in Equation 14.3 employs data for the lower 48 U.S. states from the USDM and GRACE drought indicators as key explanatory variables. County-year panel data* are based on USDM maps from the National Drought Mitigation Center's website and GRACE data provided by the University of Nebraska–Lincoln. To evaluate the relationships between these drought signals and realized economic agricultural outputs, farm income data of the U.S. Bureau of Economic Analysis were collected for each county and each year that is covered by the USDM and GRACE data (Bernknopf et al. 2016). The main agricultural economic indicator of interest is realized as net income plus the value of inventory

---

* Panel data is a dataset in which the behaviors of entities are observed across time (https://www.princeton.edu/~otorres/Panel101.pdf; accessed November 16, 2016).

change. The dataset covers the period from 2002 to 2013. The econometric model is as follows:

$$FarmY_{it} = a + \beta_0 USDM\_D0wks_{it} + \cdots + \beta_4 USDM\_D4wks_{it}$$

$$+ \beta_5 GRACE\_sfsm\_D0wks_{it} \cdots \beta_9 GRACE\_sfsm\_D4wks_{it}$$

$$+ \beta_{10} GRACE\_rtzm\_D0wks_{it} \cdots \beta_{14} GRACE\_rtzm\_D4wks_{it}$$

$$+ \beta_{15} GRACE\_GW\_D0wks_{it} \cdots \beta_{19} GRACE\_GW\_D0wks_{it} + \lambda_t + \varphi_i + \varepsilon_{it}$$

(14.3)

where $FarmY_{it}$ represents realized net farm income in county i in year t, $USDM\_D0wks_{it}$ represents the number of weeks in year t that county i was designated as being in drought category D0, $USDM\_D4wks_{it}$ represents the number of weeks in year t that county i was designated as being in drought category D4, $GRACE\_sfsm\_D0wks_{it}$ represents the number of weeks in year t that county i was designated as being in drought category D0 by the GRACE surface soil moisture indicator, $GRACE\_sfsm\_D4wks_{it}$ represents the number of weeks in year t that county i was designated as being in drought category D4 by the GRACE surface soil moisture indicator, $GRACE\_rtzm\_D0wks_{it}$ represents the number of weeks in year t that county i was designated as being in drought category D0 by the GRACE root zone moisture indicator, $GRACE\_rtzm\_D4wks_{it}$ represents the number of weeks in year t that county i was designated as being in drought category D4 by the GRACE root zone moisture indicator, $GRACE\_GW\_D0wks_{it}$ represents the number of weeks in year t that county i was designated as being in drought category D0 by the GRACE groundwater indicator, and $GRACE\_GW\_D4wks_{it}$ represents the number of weeks in year t that county i was designated as being in drought category D1 by the GRACE groundwater indicator. The remaining variables are as follows: λ_t, which controls for unobserved, time-varying determinants of farm income that are equivalent for all counties; these effects can include changes in crop or livestock prices at the national level or changes in the availability of modern seed varieties and other improved agricultural production technologies; φ_i, which represents county fixed effects, for obtaining unbiased parameter estimates in the presence of unobserved, county-specific characteristics that do not vary over time; and ε_{it}, which is an error term. All categories of drought severity are present in each regression equation. The alternative combinations of the USDM and GRACE indicators, regressions, and statistical inferences can be found in Bernknopf et al. (2015).

The analysis results shown in Table 14.1 reveal that the USDM, with and without GRACE, exhibits a low correlation between realized net farm income and drought severity categories. However, standard measures of statistical inference with big datasets may not be useful, and other measures are necessary (Varian 2014). The econometric analysis in this example used statistical information criteria metrics to test whether adding GRACE drought indicators showed an improvement relative to the current USDM (Bernknopf et al. 2016).

TABLE 14.1

Statistics and F-Tests for Assessing the Goodness of Fit of Net Farm Income Models with and without GRACE-DA Indicators for the Lower 48 States (N = 36,624)

	No GRACE Indicators	Root Zone Soil Moisture Only	Surface Soil Moisture Only	Groundwater Only	Root Zone and Surface Soil Moisture	Root Zone Soil Moisture and Groundwater	Surface Soil Moisture and Groundwater	All GRACE Indicators
Adjusted R squared	0.0808	0.0817	0.0819	0.0838	0.0826	0.0845	0.0840	**0.0850**
Akaike information criterion	851,268	851,236	851,228	851,152	851,207	851,131	851,150	**851,117**
Bayesian information criterion	851,404	851,415	851,406	**851,331**	851,428	851,352	851,371	851,381
p-Values for F-test (all coefficients for GRACE variables = 0)	N/A	0.0000	0.0000	0.0000	0.0000	0.0000	0.0000	0.0000

Source: Bernknopf, R. et al., The value of remotely sensed information: The case of GRACE-enhanced drought severity index, Working Paper, Department of Economics, University of New Mexico, 2016.

Bold type represents the most preferred value for the goodness of fit statistic and information criterion.

Models are compared in Table 14.1 by calculating four statistics that are used to describe the goodness of fit: adjusted R-squared assumes that larger values are preferred, whereas the Akaike information criterion and the Bayesian information criterion assume that smaller values are preferred. Improvements in these statistics for models that include the GRACE drought indicators suggested that their inclusion in a model describing the relationship between drought and farm income improves the goodness of fit of the model. In addition, an F-test for the joint significance of the GRACE variables showed that their inclusion in models of drought and farm income is statistically warranted and that an improvement occurred in the prediction of the impact of drought on farm income. Adding GRACE-DA information to the USDM indicators in certain regions improved the prediction of losses in farm income in the presence of drought. The statistics and F-test generally show that the best goodness of fit is achieved in models in which all three GRACE drought indicators are present, in addition to the USDM indicators. This is true for all the lower 48 U.S. states as well as regionally for the Midwest and South. On the other hand, the cases in which the model is based on current USDM indicators in the Northeast, Southeast, and West regions have the best goodness of fit for the Bayesian information criterion, which heavily favors model simplicity.

Expressing improvements in goodness of fit in terms of statistics and information criteria can make it difficult to assess whether the improvements are economically significant. To compare the impact of the prediction of the error for models that included USDM drought indicators only and models that included both USDM and GRACE drought indicators, the economic value of misclassification for eligibility was estimated. The analysis demonstrated that the societal cost of misclassifying farmer eligibility for financial assistance is reduced by about $13.3 billion for 2002–2013 or about $1.1 billion per year with the addition of GRACE drought indicators to the USDM (Bernknopf et al. 2016).

14.5.7 A Policy Implication: Application to the Eligibility for the Livestock Assistance Grant Program

The USDA Livestock Assistance Grant Program is a state-by-state block fund for recovering forage production losses due to a 2006 summer drought. The U.S. Congress allocated $50 million for eligible counties. A county is eligible if it has experienced exceptional drought or extreme drought in the period from March 7, 2006, to August 31, 2006. Eligible counties would have been affected if more GRACE product information were taken into account (Bernknopf et al. 2016). To evaluate and contrast how decision-making might be affected, GRACE-DA drought indicators were compared with the USDM to identify where eligibility decisions overlapped and those counties where the decisions were different.

About $16 million of the $50 million distributed would have been allocated to different states. Counties that had significant differences in drought status between the USDM and the GRACE drought indicators would have been the most likely to switch eligibility status, had the GRACE information influenced the production of the USDM in 2006. Counties that were deemed eligible for assistance under the USDM (severe and exceptional drought) but had no indications of drought according to the GRACE groundwater indicators were clustered near the Ogallala Aquifer in the Midwest United States. Counties that were not in drought according to the USDM but were with the three GRACE drought indicators were clustered in the Pacific Northwest, Nevada, Utah, Michigan, and New England. Understanding the location of these counties is important, because these are the counties that are most likely to have received aid when it was not necessary or not received aid when it was indeed very necessary.

14.5.8 Summary

Based on the Bayesian decision model and its application with econometric modeling techniques, GRACE-DA has the potential to lower the uncertainty associated with understanding drought severity and that this improved understanding has the potential to change policy decisions that lead to tangible societal benefits. Econometric modeling showed an improvement in the goodness of fit of statistical models that account for the impact of drought on farm income when GRACE drought indicators were added to the USDM drought severity categories to improve the prediction of losses in farm income. With the addition of the GRACE drought indicators, the number of county misclassifications decreased, thus reducing the uncertainty of the information delivered by the USDM. The outcome suggests that the improvement in classification would increase the efficiency of management decisions.

Suggestive evidence of the societal benefits to incorporate GRACE-DA into the USDM is shown in a hypothetical case study of the USDA Livestock Assistance Grant Program. The program criteria determine county eligibility for assistance, based on the USDM. Using 2006 data, counties that had significant differences in drought status between the USDM and GRACE-DA were identified. These counties would have been the most likely to switch eligibility status, had GRACE-DA information influenced the production of the USDM in 2006. One caveat regarding these hypothetical changes in eligibility is that they assume that the eligibility decisions would be made entirely based on a GRACE-DA indicator, which is unlikely to occur in practice. In reality, decision-makers will use GRACE-DA as one of the several information sources, so that not every eligibility change that was actually simulated would occur.

14.5.9 Conclusion

New EO technologies provide better information, for example, population statistics rather than sample statistics to reduce decision risk. The agricultural sector studies summarized above utilized quantitative models and analysis to value EO in specific decisions. The transformation of EO data into geospatial information in the two examples could lead to improvements in government and business operations.

The VOI case studies affirm the efficiency of the technology investment choices. The studies demonstrated that connecting the information products to operational applications provides an economic value to the EO. Results showed that (a) EO is likely to lower the uncertainty associated with the understanding of factors that affect a decision and (b) this improved understanding has the potential to change policy decisions that lead to tangible societal benefits.

The case study research included evidence, relied on multiple sources of quantitative evidence, and estimated benefits of the application. In both examples, a form of Bayesian model was used to evaluate whether the MRLI or the GRACE-DA enhances the reliability of the risk communication to the decision-maker. The economic analysis suggests that including the scientific data and subsequent geospatial information products can support decisions and programs that could have resulted in smaller economic loss. Each example case study was an evaluation of a decision within its real-life context, that is, a value in use. However, the simulations are only illustrative in that they show the possibility of positive return on investment and the potential for successful implementation of the technology.

What remains are the necessary trials for testing the geospatial information in use cases. A use case is the interaction of various stakeholders with a decision process. The objectives would be to design implementation experiments of a system from the user's and contributor's perspectives and to communicate system behavior in their terms. A use case requires communication of system requirements, how the system operates and may be used, the roles that all participants play, and the value that the customer or user will receive from the system.[*] For example, USDA decision-makers should evaluate the benefits of integrating GRACE-DA and other types of EO to improve the statistical robustness of the USDM.

---

[*] https://en.wikipedia.org/wiki/Use-case_analysis.

Appendix

This appendix is a brief summary of the calculations by Forney et al. (2012) for estimating the net present value for MRLI benefits.

The economic value $\mathbf{P \Delta Q}$ is the difference in the benefits, with and without MRLI, respectively, which is the *VOI*:

$$VOI_{\omega(1)} = P[Q_{R(\omega(1)\alpha)} - Q_{R(\omega(0)\alpha)}] \qquad (14.A1)$$

where the regulations with the additional information from MRLI ($\omega(1)$) would be $R(\omega(1), \alpha)$ and without additional information ($\omega(0)$) would be $R(\omega(0), \alpha)$ for the probability of exceeding the regulatory standard for resource damage, α. The additional information may allow regulations to be better targeted, so that the crop production will be different $\mathbf{Q}_{R(\omega(1), \alpha)}$ with the information than without information, $\mathbf{Q}_{R(\omega(0), \alpha)}$.

The present discounted value of $\mathbf{PQ}$ is calculated by summing the quantities of corn and soybeans produced in each land unit for each year $\mathbf{Q}$, that is multiplied by the present discounted vector of real prices $\mathbf{P}$, that prevailed during the period of analysis. A possible combination of cropping choices across the study region is eliminated if the environmental constraint is exceeded, and among those choices not eliminated, the optimization algorithm steps through cropping choices until a maximum $\mathbf{PQ}$ is identified and annualized. The present discounted value of the difference between optimal (with MRLI and associated modeling data) and baseline (without MRLI data) is $\mathbf{P\Delta Q}$. The VOI expressed as an equivalent annual income (EAI) is:

$$EAI = P\Delta Q \frac{r(1+r)^t}{(1+r)^t - 1} \qquad (14.A2)$$

where r is the discount rate. Assuming a similar flow of benefits into the indefinite future because of the continuation of the availability of MRLI, for this region, the net present value (NPV) is calculated as follows:

$$NPV = EAI \frac{(1+r)}{r} \qquad (14.A3)$$

This net present value is an estimate of the value of using MRLI-based information for managing the corn/soybean crop patterns and groundwater resources in the study region into the indefinite future.

References

Antle, J. and R. Just. 1991. Effects of commodity program structure on resource use and the environment, Chapter 5. In Just, R. and N. Bockstael (Eds.), *Agricultural Management and Economics: Commodity and Resource Policies in Agricultural Systems* (pp. 97–128). Berlin, Germany: Springer-Verlag.

Antle, J. and T. McGuckin. 1993. Technological innovation, agricultural productivity, and environmental quality. In Carlson, G., D. Zilberman, and J. Miranowski (Eds.), *Agricultural and Environmental Resource Economics* (p. 528). New York: Oxford University Press.

Bernknopf, R., D. Brookshire, Y. Kuwayama, M. Macauley, M. Rodell, B. Zaitchik, A. Thompson, and P. Vail. 2015. The value of information from a GRACE-enhanced drought severity index, Final Report to the Applied Sciences Program, NASA.

Bernknopf, R., D. Brookshire, Y. Kuwayama, M. Macauley, M. Rodell, B. Zaitchik, A. Thompson, and P. Vail. 2016. The value of remotely sensed information: The case of GRACE-enhanced drought severity index, Working Paper, Department of Economics, University of New Mexico.

Bernknopf, R., D. Brookshire, M. Macauley, G. Jakeman, Y. Kuwayama, H. Miller, L. Richardson, and A. Smart. 2017. Societal benefits: Methods and examples for estimating the value of remote sensing information, Chapter 10. In Morain, S., M. Renslow, and A. Budge (Eds.), *Manual of Remote Sensing* (4th ed.). Bethesda, MD: American Society for Photogrammetry and Remote Sensing.

Bernknopf, R., W. Forney, R. Raunikar, and S. Mishra. 2012. Estimating the benefits of land imagery in environmental applications: A case study in nonpoint source pollution of groundwater, Chapter 10. In Laxminarayan, R. and M. Macauley (Eds.), *The Value of Information: Methodological Frontiers and New Applications in Environment and Health* (pp. 257–300). Dordrecht, the Netherlands: Springer.

Bouma, J.A., H.J. van der Woerd, and O.J. Kuik. 2009. Assessing the value of information for water quality management in the North Sea. *Journal of Environmental Management*, 90(2):1280–1288.

Forney, W., R. Raunikar, R. Bernknopf, and S. Mishra. 2012. An economic value of remote-sensing information—Application to agricultural production and maintaining groundwater quality. Professional Paper 1796. U.S. Geological Survey, Reston, VA.

Houborg, R., M. Rodell, B. Li, R. Reichle, and B. Zaitchik. 2012. Drought indicators based on model-assimilated Gravity Recovery and Climate Experiment (GRACE) terrestrial water storage observations. *Water Resources Research*, 48(W07525): 1–17. doi:10.1029/2011WR011291.

Hrubya, C., R. Libra, C. Fields, D. Kolpinb, L. Hubbard, M. Borchardt, S. Spencer et al. 2015. 2013 Survey of Iowa groundwater and evaluation of public well vulnerability classifications for contaminants of emerging concern. Iowa Geological and Water Survey Technical Information Series 57. Iowa Department of Natural Resources.

Isik, M., D. Hudson, and K.H. Coble. 2005. The value of site-specific information and the environment: Technology adoption and pesticide use under uncertainty. *Journal of Environmental Management*, 76: 245–254.

Kalbfleisch, J. and R. Prentice. 1980. *The Statistical Analysis of Failure Time Data* (321 p). New York: John Wiley & Sons.

Kalluri, S., P. Gilruth, and R. Bergman 2003. The potential of remote sensing data for decision makers at the state, local and tribal level: Experiences from NASA's Synergy program. *Environmental Science & Policy*, 6: 487–500.

Kleinbaum, D. 1996. *Survival Analysis: A Self-Learning Text* (324 p). New York: Springer-Verlag.

Lancaster, T. 1990. *The Econometric Analysis of Transition Data* (324 p). Cambridge, UK: Cambridge University Press.

Lawrence, D. 1999. *The Economic Value of Information*. New York: Springer-Verlag.

Macauley, M. 2006. The value of information: Measuring the contribution of space-derived earth science data to national resource management. *Space Policy*, 22: 274–282.

Macauley, M. and D. Diner. 2007. Ascribing societal benefit to applied remote sensing data products: An examination of methodologies based on the multi-angle imaging Spectro-radiometer experience. *Journal of Applied Remote Sensing*, 1(1): 1–27.

Macauley, M. and R. Laxminarayan. 2010. The value of information: Methodological frontiers and new applications for space policy. *Journal of Space Policy*, 26(4): 249–251.

Macauley, M., J. Maher and J.S. Shih. 2010. From science to applications: Determinants of diffusion in the use of earth observations. *The Journal of Terrestrial Observation*, 2(1): 11–20.

Miller, H.M., N.R. Sexton, L. Koontz, J. Loomis, S.R. Koontz, and C. Hermans. 2011. The users, uses, and value of Landsat and other moderate-resolution satellite imagery in the United States—Executive Report. USGS Open File Report 2011-1031. Reston, VA.

Mirvish, S.S. 1995. Role of N-nitroso compounds (NOC) and N-nitrosation in etiology of gastric, esophageal, nasopharyngeal and bladder cancer and contribution to cancer of known exposures to NOC. *Cancer Letters*, 97(2): 271.

Sinn, H.-W. 1983. *Economic Decisions Under Uncertainty*. Amsterdam, the Netherlands: North-Holland.

Svoboda, M., D. LeComte, M. Hayes, R. Heim, K. Gleason, J. Angel, B. Rippey et al. 2002. The drought monitor. *Bulletin of the American Meteorological Society*, 83(8): 1181–1190. doi:10.1175/1520-0477(2002)083<1181:TDM>2.3.CO;2.

Varian, H. 2014. Big data: New tricks for econometrics. *The Journal of Economic Perspectives*, 28: 3–27. Retrieved November 11, 2016 23:12 UTC from http://www.jstor.org/stable/23723482.

Ward, M.H., T. deKok, P. Levallois, J. Brender, G. Gulis, B.T. Nolan, and J. VanDerslice. 2005. Drinking water nitrate and health—Recent findings and research needs. *Environmental Health Perspectives*, 115: 1607–1614.

Weyer, P.J., J.R. Cerhan, B.C. Kross, G.R. Hallberg, J. Kantamneni, G. Breuer, M.P. Jones, W. Zheng, and C.F. Lynch. 2001. Municipal drinking water nitrate level and cancer risks in older women—The Iowa Women's Health Study. *Epidemiology*, 11(3): 327–338.

Williamson, R.A., H.R. Hertzfeld, J. Cordes, and J.M. Logsdon. 2002. The socioeconomic benefits of earth science and applications research: Reducing the risks and costs of natural disasters in the USA. *Space Policy*, 18: 57–65.

Zaitchik, B.F., M. Rodell, and R.H. Reichle. 2008. Assimilation of GRACE terrestrial water storage data into a land surface model: Results for the Mississippi River basin. *Journal of Hydrometeorology*, 9: 535–548.

15

Affordability of Natural Catastrophe Insurance: Game Theoretic Analysis and Geospatially Explicit Case Study

**Kecheng Xu, Linda Nozick, Jamie Brown Kruse,
Rachel Davidson, and Joe Trainor**

CONTENTS

15.1 Introduction

This chapter examines criteria for insurance affordability and their effect on uninsured losses within the context of an integrated model and a case study that includes homeowners, primary insurers, reinsurers, and government. The foundation of the analysis and case study is a loss model that relies on simulated hurricanes and their resulting wind fields and storm surge, as well as geocoded building inventories that include location, structural characteristics, and building value. Although the geospatial information is an essential component of the analysis, how the database is used and integrated using the insight from several disciplines determines its contribution and ability to inform. This study is offered as an example of what can be accomplished when starting with a sound foundation of geospatial information about extreme weather events and landscape characteristics.

In recent decades, the cost of natural disasters such as hurricanes has increased dramatically with the substantial growth in coastal populations (Kunreuther 1998). The two most important and effective ways to manage regional catastrophe risk is natural catastrophe loss insurance and mitigation. However, studies have shown that homeowners do not invest in sufficient preevent mitigation and often do not fully insure their properties to reduce losses (Kreisel and Landry 2004; Dixon et al. 2006; Kunreuther 2006). As a result, when a damaging hurricane occurs, government financial aid must come into the affected area to provide relief and fuel community recovery. These large and unanticipated expenditures strain local and state government budgets and represent an additional tax burden for society (Kunreuther and Pauly 2004). In order to encourage flood insurance adoption, in the past, the U.S. National Flood Insurance Program (NFIP) has provided flood insurance at highly subsidized rates that do not reflect the true actuarial risk. Offering insurance at rates that do not reflect the true risk is not financially sustainable, and consequently, the NFIP had a $24 billion deficit as of 2013 (Atreya et al. 2015). Similarly, many state wind catastrophe pools are also at risk of insolvency (Baker and Moss 2009).

To place the NFIP on sounder financial footing, the Biggert-Waters Flood Insurance Reform Act of 2012 (BW-12) was passed by Congress and called for flood insurance premiums that more accurately reflected the actual risk to properties from flooding. The move to implement risk-based rates drew a public outcry from property owners that faced insurance rates that could double or more. Hence, public pressure led Congress to reconsider the notion of the full implementation of risk-based premiums, passing the Homeowner Flood Insurance Affordability Act of 2014 (HFIAA 2014). The HFIAA 2014 suggests a standard for affordability of 1% of the value of the home. That is, premiums that exceed the 1% threshold are assumed to be unaffordable (NRC 2015). Both the BW-12 and HFIAA called on the Federal Emergency Management Agency to provide a framework for considering affordability. For discussions on affordability, see government accountability office (GAO 2016) and national research council (NRC 2015). Kousky and Kunreuther (2014) and Zhao, Kunreuther, and Czajkowski (2016) examined the affordability of coverage in Ocean City, New Jersey, and Charleston County, South Carolina, respectively. Both studies concentrate on resolving the challenge to implement risk-based premiums that are also sensitive to affordability issues. The question of insurance affordability is not unique to flood insurance. We examine the effect of affordability criteria on insurance for wind and flood damage caused by hurricanes in terms of insurance prices determined by a Cournot–Nash game theoretic model, the number of households uninsured due to the affordability threshold, the uninsured losses due to the affordability threshold, and the level of government support that would be needed to bring the cost of insurance within the affordability threshold. This chapter utilizes a framework developed in Gao et al. (2016) to explore the affordability issue in a case study of eastern North Carolina, United States.

Section 15.2 summarizes the modeling framework developed in Gao et al. (2016) and exercised in this context. Section 15.3 presents the results of a case study conducted in North Carolina. Finally, the conclusions are presented in Section 15.4.

15.2 Nested Modeling Framework

This section gives a brief description of the modeling framework used in this study. For a more complete description, see Gao et al. (2016). Figure 15.1 shows the structure of the interacting models. We consider homeowners, primary insurers, reinsurers, and the government as the key stakeholders in the natural catastrophe insurance market. We use a Cournot–Nash game theoretic model to represent the strategic interactions among insurers that compete in a regional insurance market. An expected utility-based home-owner decision model is used to predict homeowner's insurance purchase behavior (based on the cost of the policy, expected losses, and homeowner's risk attitudes). A stochastic optimization model is used to optimize primary insurers' pricing decisions and how much risk the insurer will retain or transfer to reinsurers. A regional catastrophe loss model that couples the frequency of damaging hurricanes with location and building resistance is used to estimate the loss to each homeowner and primary insurer. We include the impact of the government in the primary insurers' stochastic optimization model by placing requirements on the cash reserves to be held by the primary insurer proportional to the magnitude of his or her liability. The reinsurer is assumed to offer reinsurance at a formula-based price that depends on the expected loss of the primary insurer and the standard deviation of the loss. The loss model estimates the financial losses in each hurricane event, and based on the cost of insurance and the resultant take-up rate for insurance, these losses are allocated to homeowners, insurers, and reinsurers.

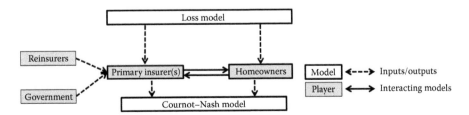

FIGURE 15.1
Structure of interacting models in nested framework.

15.2.1 Loss Model

There are two parts in the loss estimation model. One is the hazards simulation, and the other is loss estimation for each residential building for each
hazard. As input to the loss model, we use a set of 97 probabilistic hurricane
scenarios developed in Apivatanagul et al. (2011), using the optimization-
based probabilistic scenario method. In the hazards simulation, for each
scenario, open terrain 3-second peak gust wind speeds and surge depths
were computed throughout the study region by using the storm surge model
advanced circulation model (ADCIRC) (a hydrodynamic model of coastal
ocean, inlets, rivers and floodplains) (Westerink et al. 2008). In order to
include the financial implications of the number and severity of hurricanes
experienced over time, we create 2000 scenarios of a 30-year period developed in Peng et al. (2014). The loss estimation is a component-based building loss model that relates probabilistic resistances of building components
to wind speeds and flood depths, considering the effects of wind pressure
and missiles as well as the internal pressure gradient resulting from a breach
of the building envelop. The residential buildings are classified into different groups by their location (i), building category (m), resistance level (c),
and risk regions ($v \in [H, L]$), where H denotes high-risk region and L denotes
a lower-risk region. The number of buildings in each group is identified,
which is denoted as X_{imcv}, and the loss for a specific hurricane hazard event
h for a specific building of type i, m, c, v is estimated by the model developed
in Peng et al. (2013), which is denoted as L_{imcv}^h.

15.2.2 Homeowner Model

We assume that all the homeowners in the catastrophe insurance market are
risk-averse rational decision-makers that maximize expected utility. Their
level of risk aversion differs with their location by region, $v \in [H, L]$, and with
the expenditures, from either catastrophe damage or insurance payments.
In the event of a damaging hurricane, the portion of loss that homeowners
are assumed to pay, B_{imcv}^h, is the minimum of the loss experienced and the
deductible level (d). We also assume that each homeowner has an affordability threshold for insurance (and that threshold can vary based on the
home's location in either the high- or low-risk region), where that threshold
is a percentage, κ_v, of their home values, V_m.

15.2.3 Insurer Model

Each insurer is assumed to maximize profit by using a stochastic optimization
procedure that prices insurance coverage, given actuarial risk, homeowner
demand, the number of competitors, and the cost of reinsurance. Insurers
select prices $p_v, v \in [H, L]$ per dollar of expected loss to be covered in each
risk region. The price per dollar of coverage is the sum of 1 plus a specified

administrative loading factor (τ) and a profit loading factor (λ_v). If the premium is less than a minimum annual cost associated with servicing a policy, the insurer will not offer the policy. The minimum specified value is r. For the price offer (p_v), homeowners choose whether or not to buy the insurance. The premium collected is the price (p_v) multiplied by the expected insured loss, (Q_v). In the event of a hurricane (h), the loss to an insured building (L^h) is covered by several parties. Homeowners pay the first portion of the loss up to deductible (d), denoted as B^h. Reinsurers pay a specified coparticipation percentage ($\beta\%$) of the loss between the attachment point (A) and the maximum limit (M). Reinsurers require an annual premium (r^{sy}), including a base premium (b) and a reinstatement payment. Primary insurers pay the remaining part of the loss. The primary insurer uses an optimization model to choose the price of a policy tailored to each homeowner (p_v), the attachment point (A), and the maximum limit (M) for the reinsurance policy. The government requires that primary insurers have cash reserves in order to limit the risk of insolvency. We assume that the primary insurers will start their business with an initial investment (C^{s0}) that equals a specified constant (k) multiplied by the annual premiums received in all risk regions. In each year (y), they will reallocate the amount of their accumulated surplus (C^{sy}) larger than C^{s0} into other lines of business. If the accumulated surplus becomes zero or less, the primary insurer becomes insolvent.

15.2.4 Cournot–Nash Model

A Cournot–Nash game theoretic equilibrium model is used to capture the competition among primary insurers. All carriers have the same knowledge of the risk and only provide full-coverage insurance to homeowners. In other words, they face the same cost structure. Gao et al. (2016) discusses the Cournot–Nash game and a collusive joint profit maximization framework to describe the range of possible outcomes of a dynamic game in this context. The homeowner model is used to derive a demand function for each risk region, $Q_v = D_v(p_v), v \in [H, L]$, with its inverse: $p_v = P_v(Q_v) = D_v^{-1}(Q_v), v \in [H, L]$, where Q_v is the total insured loss covered by primary insurers or reinsurers in the entire region (v) and p_v is the price for that region. If there are n primary insurers (carriers) in the market, by symmetry, we can rewrite the inverse demand function as: $p_v = P_v(Q_v) = D_v^{-1}(Q_v) = D_v^{-1}(nq_v), v \in [H, L]$, where q_v is the expected loss insured by one primary insurer in the region v. From the stochastic optimization insurer model, we can derive a cost function for each primary insurer in terms of the expected loss insured: cost $= C(q_H, q_L)$. Therefore, we can derive the net profit for each primary insurer as follows:

$$\pi(q_H, q_L) = \sum_v q_v P_v(nq_v) - C(q_H, q_L) \forall v = [H, L]$$

q_{vj}^* is the coverage offered by insurer j that maximizes profit in the region v, given that there are n competitors. This model takes into consideration the optimal reaction of competitors to every q_{vj} chosen by insurer j.

15.3 Case Study

15.3.1 Inputs

The case study uses 503 census tracts covering eastern North Carolina, as shown in Figure 15.2. The case study area includes low-lying coastal counties and extends westward to include half of Raleigh, the state capital. On average, a tropical storm or hurricane is expected to make landfall on the North Carolina coast every 4 years (state climate office of north Carolina (SCONC) 2010). Recent hurricanes affecting North Carolina include Floyd (1999), Isabel (2003), Irene (2011), and Mathew (2016). We classify this region into two risk (H and L) zones by distance from the coastline, which yields 731 geographic zones. We define eight categories of buildings based on the number of stories, garage, and roof type. For each category, there are 68 building resistance

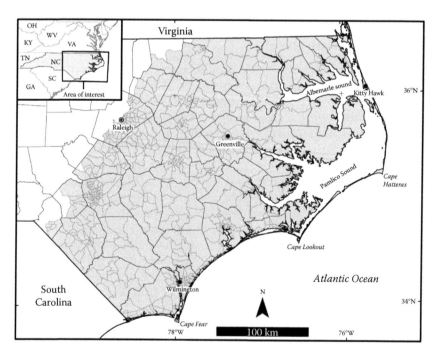

FIGURE 15.2
Study area.

levels. The total inventory reaches 931,902 in two risk zones. We used the component-based loss simulation model to estimate both wind and flood damages for each type of building at each location. The loss calculation process includes 2000 scenarios of (probabilistic event-based, with a set of 97 events) 30-year (with 20 time steps per year) hazard simulation and a joint probability distribution estimation of annual loss for each type of building under each possible hurricane event. We embed the loss model into individual homeowner and insurer models to derive the insurer's optimal cost function. We repeat the loss estimation with a varying portfolio of insured buildings (as determined by the homeowner decision model) and conduct a stochastic optimization for managing the risk for the book of business. In the insurer model, the deductible level $d = \$5000$, coparticipation factor $\beta\% = 95\%$, administrative loading factor $\tau = 0.35$, and factor defining allowed surplus $k = 3$. We set the minimum premium required by the insurer to offer insurance at $\rho = \$100$.

15.3.2 Results

In this section, we give an overview of the households and total expected losses for homes whose expected loss exceeds 1% of the home value. This overview does not consider the requirements for a profitably sustainable insurance industry. We then present the results of the nested model and consider 1% and 2% affordability thresholds. Lastly, we examine how much in public funds would be necessary to bring the cost of insurance down to the affordability threshold.

First, strictly in terms of expected loss, are there homes with expected loss that exceeds 1% of home value? The shaded areas in Tables 15.1 and 15.2 identify homes with expected loss of more than 1% of the home value. Table 15.1 shows the count of households that fit categories of expected loss as a percentage of home value that ranges from less than 0.5% to 6.0%. In the

TABLE 15.1

Number and Proportion of Households as a Function of the Expected Loss Expressed as a Percentage of the Home Value for Low- and High-Risk Regions

Region		Low Risk		High Risk	
Number of households		649,012	100.00%	282,890	100.00%
Expected loss/home value	[0,0.5%]	584,335	90.03%	159,847	56.51%
	[0.5%, 1%]	60,069	9.26%	41,466	14.66%
	[1%, 2%]	3,794	0.58%	34,484	12.19%
	[2%,3%]	83	0.01%	26,154	9.25%
	[3%,4%]	731	0.11%	15,721	5.56%
	[4%,5%]	0	0.00%	4,589	1.62%
	[5%,6%]	0	0.00%	629	0.22%
	[6%,+∞]	0	0.00%	0	0.00%

TABLE 15.2

Proportion of Total Expected Loss Assigned to Homes that Exceed 1% of Home Value for Low- and High-Risk Regions

Region		Low Risk		High Risk	
Total expected loss ($)		148,610,960	100.00%	445,686,656	100.00%
Expected	[0,0.5%]	112,736,200	75.86%	63,886,828	14.33%
loss/home	[0.5%, 1%]	22,838,570	15.37%	43,772,532	9.82%
value	[1%, 2%]	8,127,765	5.47%	88,921,568	19.95%
	[2%,3%]	332,270	0.22%	117,121,040	26.28%
	[3%,4%]	4,576,575	3.08%	91,312,312	20.49%
	[4%,5%]	0	0.00%	35,565,316	7.98%
	[5%,6%]	0	0.00%	5,106,162	1.15%
	[6%,+∞]	0	0.00%	0	0.00%

low-risk region, there are 4,608 (0.7%) of the 649,012 homes in the risk region whose expected loss exceeds 1% of the home value. In the high-risk region, 81,577 (28.8%) of the 282,890 homes in the area within 2 miles of the coast have expected losses that exceed 1% of the home value. Table 15.2 provides the total expected loss associated with each category. When we examine the total expected losses for each region that fall past the 1% threshold, we find expected loss of $13,036,610 and $338,026,398 for the low- and high-risk regions, respectively. Using information from the loss model alone, we find that 9.2% of the homes account for roughly 59% of the total expected loss. Clearly, a relatively small proportion of homes accounts for a relatively large proportion of expected losses.

The remainder of our analysis moves beyond the loss model to incorporate the effect of the interaction of all stakeholders in the nested model. The system equilibrium price of insurance and hence the cost of insurance to homeowners depends on the expected value of the loss, homeowner's risk attitudes, the number of insurance carriers in the market, the cost of reinsurance, and government-imposed reserve requirements. However, all insurers offer insurance in both markets, potentially at different prices per dollar of the expected loss. We will focus on affordability thresholds of the cost of insurance as percentages of home value of 1% and 2%. If a home falls below the affordability threshold, it will not be insured.

The equilibrium prices for one through five insurers for both the low- and high-risk regions and 1% and 2% thresholds are shown in Figure 15.3. As expected, when we step away from a single monopoly insurer, insurance prices in equilibrium decline as the number of competitors increases. The affordability threshold takes potential buyers out of the market when the price of insurance exceeds either 1% or 2% of the home value. The buyers

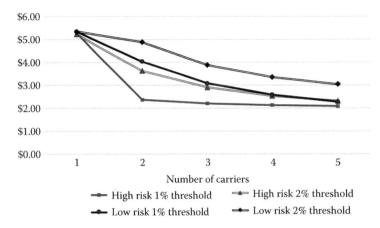

FIGURE 15.3
Equilibrium prices for 1 through 5 carriers for each affordability threshold.

left out due to affordability have either lower home values or the higher risk, or both. The extreme case would be the monopolist that would find it most profitable to *cherry pick* and serve only the lower-risk/highest-home-value clients in both the high-risk and low-risk regions.

The effect of the 1% and 2% affordability thresholds on the number of uninsured households and the proportion of total expected loss covered are summarized in the four-panel Figures 15.4 and 15.5 for the low- and high-risk regions, respectively. In Figure 15.4a and b, the solid black portion represents the proportion of the 649,012 homes in the low-risk region that would exceed the affordability threshold. The left portion of each bar represents the homes that would choose to insure, and the right portion of each bar represents homes that would be uninsured due to reasons other than the affordability constraint. Other reasons include that the expected utility-optimizing homeowner's choice of not insuring or the minimum premium required by the insurer to offer insurance is not met. Under fairly competitive conditions, with five firms in the market, roughly one half of the homeowners are predicted to insure. Panels c and d of Figure 15.4 present the proportion of $148,610,488 total expected loss that would be covered by insurance. Using the five-firm case, focusing on the homes uninsured due to the affordability threshold, although they represent around 2%–4% of the number of homes in the low-risk region, they potentially account for approximately 18%–27% of the total expected losses for the two affordability thresholds.

Moving to the high-risk region summarized in Figure 15.5, results indicate an even more dramatic impact of the affordability thresholds. Of the 282,890 homes in the high-risk region, 31%–43% would be uninsured due to

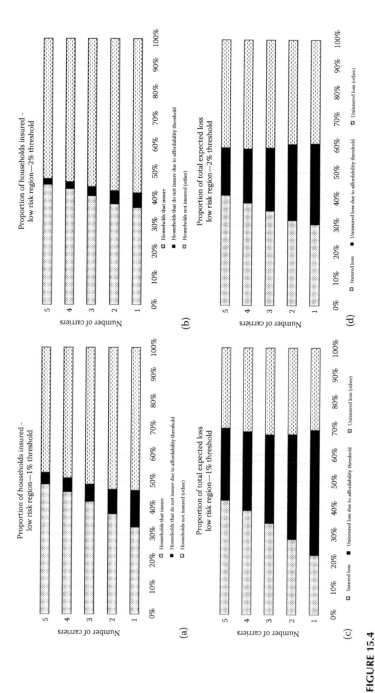

FIGURE 15.4

(a, b) Proportion of households that do and do not insure (c, d). Proportion of total insured and uninsured expected losses in low-risk region.

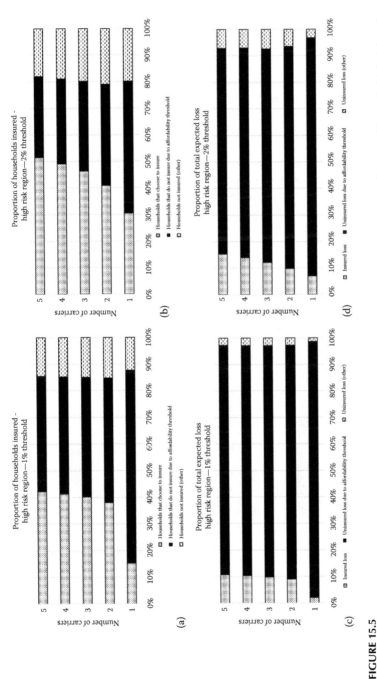

FIGURE 15.5
(a, b) Proportion of households that do and do not insure (c, d). Proportion of total insured and uninsured expected losses in high-risk region.

the affordability threshold for the five-firm case. At the 1% (2%) threshold, 86% (78%) of the total expected losses would be uninsured due to the affordability constraint. The affordability constraint affects a considerable proportion of the expected losses in both risk regions.

Figure 15.6 provides a summary of the total insurance premiums that would be collected and the total primary insurance industry profit predicted for the case study area. Raising the threshold from 1% to 2% qualifies more homes to purchase insurance and thus increases the number of homes insured and the total premiums collected. The five-firm and 2% affordability scenario at the more competitive range of the industry would collect a total of $165,032,330 in premiums and produce $62,927,880 in profit, divided between five firms.

Finally, we ask the question, "How much subsidy would be required to make insurance affordable under the 1% and 2% threshold rule?" Figure 15.7 illustrates the total cost of a subsidy for the high- and low-risk regions. The total budget necessary to fully subsidize insurance premiums to meet 1% or 2% threshold is high. For the five-firm case and 2% affordability threshold, it would still require a $417,520,192 expenditure for the high-risk region and $9,919,247 for the low-risk region, with all other cases requiring an even larger expenditure of public funds.

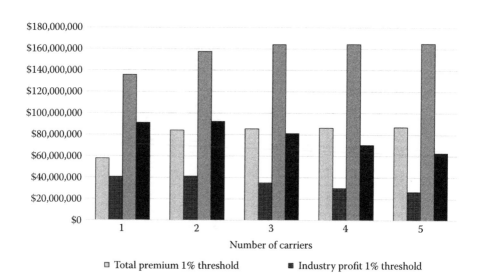

FIGURE 15.6
Total premium collected and industry profit at 1% and 2% thresholds.

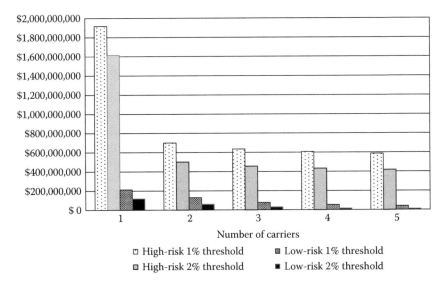

FIGURE 15.7
Total subsidy required to bring cost of insurance to 1% and 2% affordability thresholds.

15.4 Conclusion

This study explores the effect of two levels of affordability criteria in a nested model that takes account of the interaction of homeowners, primary insurers, reinsurance, and the government in a region that is vulnerable to flooding and wind damage due to hurricanes. This study focuses on the current stocks of homes in a study area of eastern North Carolina and their level of resistance to hurricane damage. Resistance depends on location of the buildings and their structural characteristics. Insurance decisions are assumed to be made by risk-averse expected utility-maximizing homeowners. Homeowners face insurance prices determined by an insurance industry composed of one to five carriers that interact within a noncooperative Cournot–Nash game theoretic framework. All stakeholders have knowledge of the loss distribution due to hurricane damage. We find that a relatively low percentage of homes in the region accounts for a considerable proportion of the expected losses. In case of the affordability criterion that cost of insurance cannot exceed 1% or 2% of the home value, we find that about 80% of the total expected losses would represent homes that fail to meet this criterion in a high-risk region within two miles of the coastline.

This study does not take into account any retrofit strategies to mitigate the risk and to make homes more resistant to hurricane damage. We estimate the total public fund necessary to subsidize insurance for the homes that do not meet the affordability thresholds. Without coupling the subsidy with strategies to encourage homeowners to mitigate the risk, the required subsidy is excessively relative to the total expected losses for the study area. Our conclusions, based on this study, are similar to those given by Zhao, Kunreuther, and Czajkowski (2016), which recommends against subsidized premiums. Rather, accurately priced insurance premiums that communicate the true risk of hazardous locations should be coupled with assistance for low-income homeowners, and approaches that encourage individual and community-wide hazard mitigation will be more effective policy instruments for mitigating risk.

References

Apivatanagul, P., Davidson, R., Blanton, B., and Nozick, L. (2011). Long-term regional hurricane hazard analysis for wind and storm surge. *Coastal Engineering*, 58(6):499–509.

Atreya, A., Ferreira, S., and Michel-Kerjan, E. (2015). What drives households to buy flood insurance? New evidence from Georgia. *Ecological Economics*, 117:153–161.

Baker, T. and Moss, D. (2009). Chapter 4 government as risk manager. In Cisternino, J. and D. Moss (Eds.) *New Perspectives in Regulation*. Cambridge, MA: The Tobin Project.

Dixon, L., Clancy, N., Seabury, S., and Overton, A. (2006). *The National Flood Insurance Program's Market Penetration Rate: Estimates and Policy Implications*. Santa Monica, CA: RAND Corporation.

Gao, Y., Nozick, L., Kruse, J., and Davidson, R. (2016). Modeling competition in the natural catastrophe loss insurance market. *The Journal of Insurance Issues*, 39(1):38–68.

GAO (2016). National Flood Insurance Program: Options for Providing Affordability Assistance GAO-16-190. United States Government Accountability Office.

Kousky, C. and Kunreuther, H. (2014). Addressing affordability in the national flood insurance program. *Journal of Extreme Events*, 1(1):1450001.

Kriesel, W. and Landry, C. (2004). Participation in the National Flood Insurance Program: An empirical analysis for coastal properties. *Journal of Risk and Insurance*, 71(3):405–420.

Kunreuther, H. (1998). *Insurability Conditions and the Supply of Coverage*. Washington, DC: Joseph Henry Press.

Kunreuther, H. and Pauly, M. (2004). Neglecting disaster: Why don't people insure against large losses? *Journal of Risk and Uncertainty*, 28(1):5–21.

Kunreuther, H. (2006). Disaster mitigation and insurance: Learning from Katrina. *The Annals of the American Academy of Political and social Sciences*, 604(1):208–227.

National Research Council (2015). Affordability of National Flood Insurance Program Premiums: Report 1. Washington, DC: The National Academies Press.

Peng, J., Shan, X., Davidson, R., Nozick, L., Kesete, Y., and Gao, Y. (2013). Hurricane loss modeling to support regional retrofit policymaking: A North Carolina case study. *11th International Conference on Structural Safety and Reliability—ICOSSAR'13*. New York, June 16–20.

Peng, J., Shan, X., Gao, Y., Kesete, Y., Davidson, R., Nozick, L., and Kruse J. (2014). Modeling the integrated roles of insurance and retrofit in managing natural disaster risk: A multi-stakeholder perspective. *Natural Hazards*, 74:1043–1068.

Westerink, J., Luettich, R., Feyen, J., Atkinson, J., Dawson, C., Powell, M., Dunion, J., Roberts, H., Kubatko, E., and Pourtaheri, H. (2008). A basin-to-channel-scale unstructured grid hurricane storm surge model as implemented for Southern Louisiana. *Monthly Weather Review*, 136(3):833–864.

Zhao, W., Kunreuther, H., and Czajkowski, J. (2015). Affordability of the national flood insurance program: Application to Charleston County, South Carolina. *Natural Hazards Review*, 17(1):04015020.

16

Socioeconomic Value of Hydrometeorological Information in Austria

Nikolay Khabarov, Andrey Krasovskii, Alexander Schwartz,
Ian McCallum, and Michael Obersteiner

CONTENTS

16.1 Introduction

Since the 1960s, assessments of economic benefit are on the World Meteorological Organization's (WMO) agenda (WMO 1968). These estimates are still of academic and practical interest, as they help analyze the efficiency of funding operation, expansion, and maintenance of national hydrometeorological services (NHMSs). The importance of keeping their efficiency high is due to the large total value of global infrastructure maintained and operated by NMHSs, which is estimated to be more than US$10 billion (WMO 2015). An overview of the history of research on the economic value of meteorological information from early works up to the present day, along with a rich set of bibliographical references, can be found in the WMO 2015 report (WMO 2015).

This section presents a rather pragmatic quantitative assessment of social benefits generated by the use of weather information in Austria. The analysis of costs associated with the provision of the hydrometeorological information is beyond the scope of this work. The goal of the case study is to develop a country-specific estimate of the value of weather information for Austria. The assessment has been carried out in two mutually supplemental tracks.

The first track deals with transferring the willingness to pay (WTP)-based estimates to Austria. The focus of the analysis is on studies related to the WTP estimates, as assessed generally for households at a country scale (Anaman and Lellyett 1996; Lazo and Chestnut 2002; World Bank 2008) and WTP estimates for agriculture (Kenkel and Norris 1995, 1997).

The second track is devoted to the transfer of other foreign sectoral-based benefit estimates. It is based on studies performed by the Technical Research Centre of Finland and VTT Finnish Meteorological Institute (Leviäkangas et al. 2007; Hautala et al. 2008) and focused on Finland, Croatia, and Southeastern Europe, as well as on a study in Switzerland (Frei 2010). Foreign estimates were adjusted to Austria, employing the Austrian statistical data (Austrian Economic Chambers [WKO] 2015).

16.2 Willingness-to-Pay-Based Estimates

The focus of the analysis is on studies related to the WTP and direct elicitation of values from individuals through surveys (Arrow et al. 1993). These include estimates for households at a country scale (Anaman and Lellyett 1996; Lazo and Chestnut 2002; World Bank 2008) and sector-specific estimates for agriculture (Kenkel and Norris 1995, 1997). In general, the value of meteorological information for a household or an individual is in knowing what clothes to wear and what actions to take (e.g., take an umbrella, close the windows, and so on). The potential damage from not knowing the weather forecast is relatively small. The respective damage in agriculture has a potentially higher value. Starting irrigation, spraying pesticides, or fertilizing the soil just before it rains could have considerable economic and environmental effects.

16.2.1 Agricultural Value

The assessments of WTP in agriculture are based on a U.S. study undertaken in 1994 for Oklahoma (Kenkel and Norris 1995). This study states that the majority of agricultural producers are actually paying US$30 per month per farmer (current quality of weather forecasts). Another study highlighted by Kenkel and Norris (1995) was carried out in Texas, United States, in 1984 (Vining et al. 1984). It estimated the WTP value to be US$40 per month per farmer (current quality of weather forecasts). Two methodological approaches are employed for a WTP transfer, which are as follows:

- Gross domestic product (GDP) per capita based on purchasing power parity (World Bank, International Comparison Program database 2016). It is further referred to as *GDP cap^{-1} PPP* and denoted as $P_{\text{country,year}}$, as shown in Equation 16.1.
- Gross domestic product at market prices[*] (World Bank national accounts data, and the Organisation for Economic Co-operation and Development [OECD] national accounts data files 2016), together with the share of agricultural production within GDP (U.S. Department of Commerce, Bureau of Economic Analysis 2015), denoted as $G_{\text{country,year}}$ and $S_{\text{country,year}}$, respectively, as shown in Equation 16.2.

$$W_{A,y1} = W_{C,y0} \cdot \left(\frac{P_{A,y1}}{P_{C,y0}}\right) \cdot \left[\frac{\left(D_A \cdot G_{A,y1}^{-1}\right)}{\left(D_C \cdot G_{C,y1}^{-1}\right)}\right] \cdot N_{A,y1} \cdot M \qquad (16.1)$$

$$W_{A,y1} = W_{C,y0} \cdot \left[\frac{\left(G_{A,y1} \cdot S_{A,y1} \cdot N_{A,y1}^{-1}\right)}{\left(G_{C,y0} \cdot S_{C,y0} \cdot N_{C,y0}^{-1}\right)}\right] \cdot \left[\frac{\left(D_A \cdot G_{A,y1}^{-1}\right)}{\left(D_C \cdot G_{C,y1}^{-1}\right)}\right] \cdot N_{A,y1} \cdot M \qquad (16.2)$$

Here, brackets are optional and are used intentionally for better visibility of scaling coefficients. Units are both current U.S. dollar per year (USD year^{-1}). Here, the following notation is used:

$W_{A,y1}$ is the derived estimate of WTP for agricultural sector in Austria for the year y_1 (current) and its unit of measurement is USD year^{-1}.

$W_{C,y0}$ is the reported estimate (in literature) of WTP for agricultural sector in a country (e.g., the United States) in year y_0 and its unit of measurement is USD farm^{-1} month^{-1}.

$\left(P_{A,y1}/P_{C,y0}\right)$ represents the ratio between PPP in Austria in year y_1 (*now*) and PPP in a country year y_0 (*back then*)—employed to scale for countries' differences in PPP *back then* and *now*.

$\left(D_C \cdot G_{C,y1}^{-1}\right)$ represents the ratio between D_C, the annual average hydrometeorological disasters' damage in a country over the period 1990–2014 in billion U.S. dollars (Guha-Sapir et al. 2016), and country's GDP in the year y_1 (current)—employed to scale for severity of disasters' impact on a country, as compared with another country (letter C replaced with letter A in this formula results in a respective term for Austria).

---

[*] World Bank's indicator code NY.GDP.MKTP.CD.

TABLE 16.1

Agricultural Willingness to Pay Value Derived for Austria, Based on Several Literature Sources

WTP Value, M Euro Year^{-1}	Source Study
33.6	U.S. agriculture 1984 Texas, method (1)
21.7	U.S. agriculture 1994 Oklahoma, method (1)
16.7	U.S. agriculture 1984 Texas, method (2)
7.7	U.S. agriculture 1994 Oklahoma, method (2)

$N_{A,y1}$ is the number of farms in Austria (Lowder et al. 2014) in the year y_1; replacing A with C and y_1 with y_0 implies another country and year.

M is the number of months in a year when the hydrometeorological information is important for a farmer (average growing season length of 6 months).

$\left(G_{A,y1} \cdot S_{A,y1} \cdot N_{A,y1}^{-1}\right)$ is the added value per farm per year (for year y_1) in Austria, where replacing A with C and y_1 with y_0 implies another country and year—employed to scale for countries' differences at per-farm level *back then* and *now*.

An application of these two methods described above to each of the two U.S.-based WTP estimates (Oklahoma, 1994 and Texas, 1984) has led to Austrian WTP values presented in Table 16.1. The officially published U.S. Federal Reserve exchange rates (Board of Governors of the Federal Reserve System 2016) were used to convert the values to Euro.

16.2.2 Households

The assessment of the Austrian households' WTP value is based on the following studies: in the United States, the respective value was estimated at US\$286 and US\$109 per household per year in 2009 and 2002, respectively (Lazo and Chestnut 2002; Lazo et al. 2009). In Australia, in 1996, it was estimated at the level of 24 Australian dollars per household per year (Anaman and Lellyett 1996), and in Azerbaijan and Serbia, the respective estimates were US\$1.55 and US\$4.37 per household per year in 2008 (World Bank 2008). In all cases, the methodology based on Equation 16.1 was employed, assuming that all farm-relevant input variables (units: per month) have the analogous household meaning (units: per year). To distinct household values, a *hat* diacritic has been placed on top of formerly used variables, leading to the WTP representation:

$$\hat{W}_{A,y1} = \hat{W}_{C,y0} \cdot \left(\frac{P_{A,y1}}{P_{C,y0}}\right) \cdot \left[\frac{\left(D_A \cdot G_{A,y1}^{-1}\right)}{\left(D_C \cdot G_{C,y1}^{-1}\right)}\right] \cdot \hat{N}_{A,y1} \cdot \hat{M} \qquad (16.3)$$

TABLE 16.2

Household Willingness to Pay Value Derived for Austria, Based on Several Literature Sources

WTP Value, M Euro Year^{-1}	Source Study
390.0	U.S. 2009 (Lazo et al. 2009)
183.0	U.S. 2002 (Lazo and Chestnut 2002)
106.5	Azerbaijan 2008 (World Bank 2008)
80.9	Australia 1996 (Anaman and Lellyett 1996)
19.5	Serbia 2008 (World Bank 2008)

The value of $\hat{M}$ is equal to 1 year, as households are interested in and do receive weather information throughout the whole year. The number of households in Austria in 2015, denoted as $\hat{N}_{A,2015}$, equals to 3,816,766 (Statistik Austria 2016). The respective results are presented in Table 16.2. For currency conversion of these estimates to Euro, the official exchange rates published by the U.S. Federal Reserve were used (Board of Governors of the Federal Reserve System 2016).

16.3 Sectoral-Based Benefit Estimates

16.3.1 Road Transport

Traffic accidents represent a very significant annual cost for the Austrian society, estimated at around 10 billion Euro, which is about 3.4% of GDP (OECD/ITF 2015). Weather is an important factor that explains the spatiotemporal pattern of road crashes, as reported in statistics on traffic accidents in Austria (ASFiNAG 2010; OECD/ITF 2015). The estimates of benefits in terms of avoided damage on the roads were transferred from Finland to Austria. According to an expert interview, information and warning services targeted to road users in Finland reduce the number of accidents on public roads involving personal injury or death by 1%–2% (Leviäkangas et al. 2007). In Finland, 21.4% of deaths and 22.3% of injuries in road accidents happen when the road surface is snowy, slushy, or icy (Statistics Finland 2007). In Austria, the majority of accidents occur on dry road surfaces, whereas wet road surfaces have been observed in about one-fourth of accidents (i.e., about 25%) and cause 20% of deaths (ASFiNAG 2010). Under such conditions, the road surface parameters (grip and lane grooves) may play a role in accidents. Multiple collisions due to fog happen less frequently but result in high accident severities, and they consistently take place at least once a year (ASFiNAG 2010).

The Finnish values of 1%–2% reduction in number of accidents due to the use of weather information were scaled to Austrian conditions. The reduction in fatal road accidents in Austria can be calculated according to the formula used to carry out a similar scaling from Finnish to Croatian numbers (Leviäkangas et al. 2007):

$$r_{aut} = r_{fi} \times \frac{a_{aut}}{a_{fi}}$$

where a_{aut} is the share of fatal road accidents happening in adverse weather conditions in Austria and a_{fi} is the share of fatal road accidents happening in adverse weather conditions in Finland. The symbol r_{fi} denotes the reduction of fatal road accidents on public roads in Finland (relative to accidents under all road conditions) because of the provided information services (that alleviate accidents only under adverse road conditions). All the values, r_{fi}, a_{aut}, and a_{fi}, are calculated in percentage.

The reduction in the number of fatal road accidents on public roads in a year was calculated using the equation (Leviäkangas et al. 2007):

$$x = \frac{1}{1 - r_{aut}} \times c_{fatal} - c_{fatal}$$

where c_{fatal} is the total number of fatal road accidents on public roads in Austria and x is the number of fatal road accidents avoided because of the present information services. Similarly, the reduction in injury accidents on public roads, further referred to as y, was calculated. Meteorological information and warning services also reduce the number of accidents that cause property damage. According to an expert interview (Leviäkangas et al. 2007), the effect of meteorological information and warning services targeted to road users is based on both lower speeds and increased vigilance of road users. According to the studies carried out by (Nilsson 2004; Leviäkangas et al. 2007), the reduction of the number of property damage only accidents, in percentage p, can be estimated based on the derived above number of road accidents (injuries and fatal) before and after the weather information/warning has been received by the road users. Then, the number of avoided accidents involving only property damage is calculated by using the formula similar to that for x.

In calculating the socioeconomic benefit, the numbers of avoided accidents are multiplied by relevant unit cost values. Austrian unit cost values for a fatal road accident (u_f), for a road accident involving personal injury (u_i), and for an accident involving only property damage (u_p) were used. The data for injury costs in Austria are available for two categories (OECD/ITF 2015): hospitalized people and slight injuries. Thus, the following modification in the calculation of the socioeconomic benefit (B) in a year, as compared with (Leviäkangas et al. 2007), was introduced:

$$B = x \times u_f + y \times \left(\alpha u_i^1 + \beta u_i^2 \right) + z \times u_d$$

Here, x, y, and z are respective reductions in the number of fatal accidents, injuries only, and property damage only road accidents; coefficient $\alpha \geq 0$ is the ratio of hospitalized people, $\beta \geq 0$ is the ratio of slightly injured people, and $\alpha + \beta = 1$. Here we have denoted the unit cost for hospitalized people as u_i^1, and unit cost for light injuries as u_i^2. Ratios $\alpha = 0.185$ and $\beta = 0.815$ were calculated for Austria using publicly available data on the number of road accidents, number of injuries (separately light and serious), and number of fatalities reported for 2014 (Austrian Road Safety Fund 2014). Cost estimates are available for the year 2011. The average unit costs were projected to 2014 by applying a price level coefficient $PLC_{\text{Austria}}^{2011,2014} = 1.08$ to the unit costs (Euro) of road crashes in Austria in 2011 (Sedlacek et al. 2012; OECD/ITF 2015). The $PLC_{\text{country}}^{\text{year1,year2}}$ is defined as the ratio between the *past* (year 1) and the *more recent* (year 2) PPP-adjusted GDP per capita in a country in current international dollars, as published by the World Bank (World Bank, International Comparison Program database 2016). In this study, PLC was applied to transfer prices or costs from a *past* to a *more recent* point in time, unless there was a justification for not using it. The road users' benefits from the provision of the weather information are summarized in Table 16.3.

TABLE 16.3

Benefits of the Weather Information Targeted to Road Users (Present Information Services)

	Low Estimate (1% Avoided Damage)	High Estimate (2% Avoided Damage)
Accidents involving personal injury on public roads (2014)	47,670	
Fatal road accidents on public roads (2014)	430	
Accidents involving property damage only on public roads (2014)	37,957	
The effect of information on the reduction of the number of accidents causing property damage only	0.85%	1.71%
The effect of information on the reduction of number of fatal accidents	0.94%	1.9%
The effect of information on the reduction of number of injury accidents	1.13%	2.29%
Injury accidents without present information services	48,210	48,763
Accidents involving property damage only without present information services	38,279	38,606
Fatal road accidents without present information services	434	438
Injury accidents avoided with present information services	540	1,093
Fatal road accidents avoided with present information services	4	8
Accidents involving property damage only avoided with present information services	322	649
Socioeconomic benefit of the information service (million Euro year^{-1})	80	142

The estimates obtained for Austria are compatible with Swiss estimates. It was shown (Frei et al. 2014) that the use of meteorological information in the road transportation sector in Switzerland generates an economic benefit to the national economy of 65.7–79.77 million Swiss Francs (1 Swiss Franc ~0.90 Euro as of 2011 and 2014). Using $PLC_{Switzerland}^{2011,2014} = 1.09$ (World Bank, International Comparison Program database 2016), the Swiss estimate can be translated to 2014: 64.51–78.32 million Euro. The fatality rate in road accidents (representing by far the largest part of damage) in Austria is approximately 1.6 times higher than that in Switzerland (OECD/ITF 2015). Taking the average estimates for 2014, 110.88 million Euro (Austria) and 71.42 million Euro (Switzerland), leads to approximately the same ratio of 1.55 $\approx$ 1.6.

16.3.2 Railway Transport

The railway transportation system of the Alpine country, Austria, plays an important role in the European transit of passengers and goods (Kellermann et al. 2016). There is an evidence of positive impacts of weather information system on benefits of railway system in Austria (Doll et al. 2014). The focus of the European Union's (EU) FP7 WEATHER project (Przyluski et al. 2012) was on adaptation strategies for transport infrastructure in Europe. In this subsection, their estimates were used to assess the value of current meteorological information in Austria, based on the methodology of transferring foreign estimates (Leviäkangas et al. 2007).

The costs in the European transport sector from 2010 (Przyluski et al. 2012) were projected to 2014 by applying PLC for the EU (World Bank, International Comparison Program database 2016): $PLC_{EU}^{2010,2014} = 1.085$. The major losses in the transportation system, which are around 73%, were allocated to road traffic (Przyluski et al. 2012). Losses due to weather events in rail and air transports were comparable and constituted around 12% and 15%, respectively, and floods were the most important cause for damages in rail transport (Przyluski et al. 2012).

The passenger-kilometer (further p-km) equivalents were used in this study. They are calculated as p-km plus ton-kilometer (further t-km) times a mode-specific ton-passenger equivalency factor (Przyluski et al. 2012). Considering the average occupancy (passengers) and load (tons) per vehicle as well as the vehicle-specific capacity use, this factor is set to 0.3 for road and rail transports and to 1.0 for air transport. The cost estimates for the Alpine region (Przyluski et al. 2012) were applied to the Austrian data. The majority of losses in rail transport were caused by rain and flood, whereas the second level of losses was due to ice, snow, and storm. Those factors impact the railway infrastructure, fleet operation, and user time. The total average cost for the Alpine region in 2010 was 0.89 Euro 1,000^{-1} p-km^{-1} (Przyluski et al. 2012). In order to project the cost to 2014, $PLC_{Austria}^{2010,2014} = 1.096$ is employed (World Bank, International Comparison Program database 2016). According to the data reported for the rail transportation in Austria in 2014 (Eurostat 2014), the transport flow comprised 11,345 million p-km and 98,281 million t-km of goods. Using the t-km to p-km conversion factor of 0.3, suggested in (Przyluski et al. 2012), results in 40,829.3 million p-km equivalent.

The average annual cost of weather events in rail transport in Austria in 2014 Euro value was estimated as follows:

$$C_{RT} = \frac{40,829.3 \times 0.89 \times 1.096}{1000} = 39.83 \text{ million Euro}$$

Assuming that the present weather information delivers 10%–20% of the reduction in the damage in railway transport (Leviäkangas et al. 2007; Doll et al. 2014), the following socioeconomic benefit range was derived: $B_{RT} \in [3.98, 7.96]$ million Euro annually.

16.3.3 Air Transport

Here, the average cost for air transport, 0.73 Euro 1000^{-1} p-km^{-1}, derived in the WEATHER project for the Alpine region in 2010 (Przyluski et al. 2012) was applied. The most affected aspects of air transport in Alpine region were user time, fleet operation, and user safety. The most losses were caused by ice, snow, flood, and storm (Przyluski et al. 2012).

According to the data reported for the air transportation in Austria in 2014 (The World Bank 2014a, 2014b), the transport flow comprised 15,210,488 passengers and 370 million t-km of freight. In order to calculate the p-km equivalent, the average flight distance in Europe equal to 2270 km (German Aerospace Center 2008) and the vehicle-specific capacity factor equal to 1.0 were applied. The total p-km equivalent was calculated as 34,897.81 million p-km.

The annual weather-extremes-related cost for air transportation in Austria in 2014 was calculated using the formula:

$$C_{AT} = \frac{34897.81 \times 0.73 \times 1.096}{1000} = 27.92 \text{ million Euro}$$

The losses caused by weather to airlines can be derived, given the existing meteorological services (Mason 1966). Further, estimates of what the losses might be if there were no such services can be obtained. The total weather losses suffered by British European Airways (BEA) were estimated at about 1.5 million British Pounds, and the value of the present meteorological services, on a most conservative basis, was estimated at 2.4 million British Pounds (Mason 1966). Thus, the ratio between the value of present information and weather losses was equal to 1.6. Departing from this estimate and assuming that the value of information increases with time, a factor range from 1.5 to 2.5 (expert opinion) was applied in our study, leading to the following range for socioeconomic benefit in air transport: $B_{AT} \in [41.88, 69.8]$ million Euro.

TABLE 16.4

Benefits of the Weather Information to the Transport
Sector in Austria 2014, in million Euros

Sector/Estimate	Road	Rail	Air	Total
Upper estimate	141.68	7.96	69.80	219.44
Lower estimate	80.07	3.98	41.88	125.93

There are only a few similar estimates available in the literature for air transport. The total economic benefits of terminal aerodrome forecasts to Switzerland's domestic airlines at both main Swiss airports (Zurich and Geneva) add up to somewhere between 13 million and 21 million Swiss francs per year (von Gruenigen et al. 2014). In Europe, domestic flights constitute around 18% of all flights (Eurostat 2016). Thus, by converting this estimate to all flights, the values of 72–117 million Swiss francs, or 65–94.5 million Euro, can be derived. Taking into account that air traffic in Switzerland is around 1.7 times higher than that in Austria in terms of the number of transported passengers (Eurostat 2016), these estimates are comparable with those obtained in this chapter (a respective range for Austria would be 65×1.7^{-1} and 94.5×1.7^{-1}, i.e., 38–56 million Euro). Let us note that Finnish estimate in 2007 was 66.5 million Euro (Hautala and Leviäkangas 2007), and taking into account that air traffic in Finland is around 1.5 times lower than that in Austria in terms of the number of transported passengers (Eurostat 2016), the estimate derived in this chapter (upper bound is 69.8 million Euro) seems to be reasonable.

The benefit estimates of weather information to the transport sector in Austria are summarized in Table 16.4.

The transfer of the sector-based estimates to Austria, as applied to transport sector and presented in Table 16.4, delivers the highest benefit in the road transport: the lower estimate of the socioeconomic benefit of the information service is 80.07 million Euro, and the higher estimate is 141.68 million Euro (both estimates are copied from Table 16.3 for convenience). These values correspond to approximately 0.8%–1.42% of the total annual cost of traffic crashes, which is 10 billion Euro in Austria, according to the OECD (OECD/ITF 2015). The benefits in the rail transport are in the range of 3.98–7.96 million Euro and those in the air transport are in the range of 41.88–69.8 million Euro.

16.3.4 Construction and Building

Here, the approach for the assessment of the value of weather information in construction and building sectors of the economy (Hautala et al. 2008) was employed. The percentage of construction with weather risk was taken at the level of 33.3%, which is a climate-related value that was previously uniformly applied for a range of Southeast European countries (Hautala et al. 2008). The estimate of weather risk equal to 3% (Hautala et al. 2008) was applied to

Austria, describing the potential losses/additional construction costs in case of unfavorable weather conditions. In 2014, the construction sector contributed 18.51 billion Euro to the Austrian GDP (Austrian Economic Chambers [WKO] 2015). Based on the estimate that weather services can yield about 25%–50% savings in additional costs (Hautala et al. 2008), the following lower (higher) estimate was obtained:

$$S = 18.51 \times 33.3\% \times 3\% \times 25\% \, (50\%) = 46.2 \, (92.5) \text{ million Euro}$$

The range for the socioeconomic benefit due to weather information in the construction sector thus obtained is $B_{CS} \in [46.2, 92.5]$ million Euro.

For a simple cross-verification of the above assessment, the weather services benefit estimate obtained for Finland in 2006, that is, 15 million Euro (Hautala and Leviäkangas 2007), was employed. It can be used as a current estimate, because the construction sector in Finland is rather stable over the last decade (delivering annually approximately 2.5 billion Euro share in Finnish GDP [Trading Economics 2016]). Therefore, the direct construction GDP-based scaling from Finland to Austria leads to a higher Austrian benefit estimate of $15 \times 2.5^{-1} \times 18.51 = 111$ million Euro. From this point of view, the range of estimates derived by applying the methodology (Hautala et al. 2008), originally targeted at South European countries, seems to be reasonably conservative.

16.4 Conclusion

A condensed summary of the obtained benefit estimates is presented in Table 16.5. There is a considerable uncertainty associated with the available source estimates. The study (Lazo et al. 2009) cited by WMO (2015) induces a rather high estimate presented in Table 16.2 and might represent a biased WTP value. Another remark that should be taken into account regarding the other WTP estimates is that the WTP might be actually lower than the received benefits (Kenkel and Norris 1997). Although the direct users' feedback on the WTP helps us understand their perceived value of weather forecasts, the main challenge pertinent to this method still consists of obtaining an unbiased estimate. The sector-specific assessments presented

TABLE 16.5

Summary of the Weather Information Benefit Estimates in Austria 2014, in million Euros

Sector/Estimate	Households	Agriculture	Transport	Construction	Total
Upper estimate	390.0	33.6	219.4	92.5	735.5
Lower estimate	19.5	7.7	125.9	46.2	199.3

in the literature, and particularly those employed in this study, often rely on expert opinions and therefore might be subjective and hard to verify. This aspect limits the reliability of the obtained estimates and calls for a careful consideration when using those for decision-making.

A conceptual limitation in benefit assessment (that IS DUE TO the LIMITED scope of this particular study) RELATES TO the difficulty in capturing the extreme nature of rare hydrometeorological events by routinely employing an average annual avoided loss as the benefit indicator. A low-probability high-magnitude flood might have a disproportionally large impact when no protection infrastructure is in place or no timely forecast is provided for preparing the infrastructure. From this point of view, the average annual avoided damage might not be able to catch the true benefit of a timely weather information for the respective risk reduction. This consideration is especially relevant to Austria, where extreme floods have caused excessive damage in recent years: about 3 billion Euro in summer 2002, about 700 million Euro in summer 2005 (Stiefelmeyer et al. 2006), and about 1 billion Euro in 2013 (Guha-Sapir et al. 2016). From this standpoint, the estimate derived in this case study should be considered rather conservative, as it is not explicitly focused on risk.

Acknowledgments

This work was supported by Zentralanstalt für Meteorologie und Geodynamik (ZAMG, Austria) and by the EU Framework Program for Research and Innovation (SC5-18a-2014-H2020, grant agreement no. 641538 ConnectinGEO). The authors are grateful to Klaus Stadlbacher, Alexander Beck, Michaela Rossini, and Matthew Cantele for useful discussions and help.

References

Anaman, K.A. and Lellyett, S.C. 1996. Contingent valuation study of the public weather service in the Sydney metropolitan area. *Econ. Pap. J. Appl. Econ. Policy* 15:64–77.
Arrow, K., Solow, R., Portney, P.R., Leamer, E.E., Radner, R., and Schuman, H. 1993. Report of the NOAA panel on contingent valuation. *Fed. Regist.* 58:4601–4614.
ASFiNAG. 2010. Road safety program 2020. Vienna, Austria. Retrieved from https://www.asfinag.at/documents/10180/12493/VSB+2020+Verkehrssicherheitsprogramm+englisch/4dbb3a3f-9ef6-41d6-b091-8731c89f1904.
Austrian Economic Chambers. (WKO). 2015. Statistical yearbook 2015. Vienna, Austria.
Austrian Road Safety Fund. 2014. Road safety in Austria. Annual report 2014. Retrieved July 24, 2017 from https://www.bmvit.gv.at/en/service/publications/transport/downloads/roadsafety_report2014.pdf.

Board of Governors of the Federal Reserve System. 2016. FRB: G.5A release—foreign exchange rates—January 4, 2016. Retrieved June 2, 2016 from http://www.federalreserve.gov/releases/g5a/current/default.htm.

Doll, C., Trinks, C., Sedlacek, N., Pelikan, V., Comes, T., and Schultmann, F. 2014. Adapting rail and road networks to weather extremes: case studies for southern Germany and Austria. *Nat. Hazards* 72:63–85.

Eurostat. 2014. Transport. Retrieved from http://ec.europa.eu/eurostat/web/transport/data/main-tables.

Eurostat. 2016. Air transport statistics. Retrieved from http://ec.europa.eu/eurostat/statistics-explained/index.php/Air_transport_statistics.

Frei, T. 2010. Economic and social benefits of meteorology and climatology in Switzerland. *Meteorol. Appl.* 17:39–44.

Frei, T., von Grünigen, S., and Willemse, S. 2014. Economic benefit of meteorology in the Swiss road transportation sector. *Meteorol. Appl.* 21:294–300.

German Aerospace Center. 2008. Analyses of the European air transport market. Annual report 2008. Retrieved July 24, 2017 from http://citeseerx.ist.psu.edu/viewdoc/download;jsessionid=0BA8F6BAC785E0E3106D671EA43B557E?doi=10.1.1.149.1442&rep=rep1&type=pdf.

von Gruenigen, S., Willemse, S., and Frei, T. 2014. Economic value of meteorological services to Switzerland's airlines: The case of TAF at Zurich airport. *Weather Clim. Soc.* 6:264–272.

Guha-Sapir, D., Below, R., and Hoyois, P. 2016. EM-DAT: The international disasters database–Université Catholique de Louvain–Brussels–Belgium. Retrieved June 2, 2016 from http://www.emdat.be/.

Hautala, R. and Leviäkangas, P. 2007. Effectiveness of Finnish Meteorological Institute (FMI) services. Espoo, Finland.

Hautala, R., Leviäkangas, P., Räsänen, J., Oorni, R., Sonninen, S., Vahanne, P., Hekkanen, M. et al. 2008. Benefits of meteorological services in South Eastern Europe. Finland: VTT Technical Research Centre of Finland.

Kellermann, P., Bubeck, P., Kundela, G., Dosio, A., and Thieken, A. 2016. Frequency analysis of critical meteorological conditions in a changing climate—Assessing future implications for railway transportation in Austria. *Climate* 4:25.

Kenkel, P.L. and Norris, P.E. 1995. Agricultural producers' willingness to pay for real-time mesoscale weather information. *J. Agric. Resour. Econ.* 356–372.

Kenkel, P.L. and Norris, P.E. 1997. Agricultural producers' willingness to pay for real-time mesoscale weather information: A response. *J. Agric. Resour. Econ.* 22:382–386.

Lazo, J.K. and Chestnut, L.G. 2002. United States, National Oceanic and Atmospheric Administration, Office of Program Planning and Integration 2002. Economic value of current an improved weather forecasts in the U.S. household sector. Retrieved from http://www.economics.noaa.gov/library/documents/benefits_of_weather_and_climate_forecasts/econ_value-weather_forecasts-households.pdf.

Lazo, J.K., Morss, R.E., and Demuth, J.L. 2009. 300 billion served: Sources, perceptions, uses, and values of weather forecasts. *Bull. Am. Meteorol. Soc.* 90:785–798.

Leviäkangas, P., Hautala, R., Räsänen, J., Sonninen, S., Hekkanen, M., Ohlström, M., Venäläinen, A., and Saku, S. 2007. Benefits of meteorological services in Croatia. VTT Technical Research Centre of Finland.

Lowder, S.K., Skoet, J., and Singh, S. 2014. What do we really know about the number and distribution of farms and family farms in the world? *Backgr. Pap. State Food Agric.* 8. ESA Working Paper No 14-02 Food and Agriculture Organization of the United Nations.

Mason, B. 1966. The role of meteorology in the national economy. *Weather* 21:382–393.

Nilsson, G. 2004. Traffic safety dimensions and the power model to describe the effect of speed on safety. Bulletin 221 Thesis. Lund, Sweden: Lund Institute of Technology.

OECD/ITF. 2015. Road safety annual report 2015. Paris, France: OECD Publishing.

Przyluski, V., Hallegatte, S., Nadaud, F., Tomozeiu, R., Cacciamani, C., Pavan, V., and Doll, C. 2012. Weather trends and economy-wide impacts‖ Deliverable 1 within the research project WEATHER (Weather Extremes: Impacts on Transport Systems and Hazards for European Regions) European Commission, 7th framework programme. Project co-ordinator: Fraunhofer-ISI. Karlsr. Paris Bologna.

Sedlacek, N., Herry, M., Pumberger, A., Schwaighofer, P., Kummer, S., and Riebesmeier, B. 2012. Unfallkostenrechnung Straße 2012. Vienna, Austria: Österreichischer Verkehrssicherheitsfonds Bundesministerium für Verkehr, Innovation und Technologie.

Statistics Finland. 2007. Road traffic accidents 2006. Helsinki, Finland.

Statistik Austria 2016. Privathaushalte 1985–2015. Retrieved June 3, 2016 from http://statistik.gv.at/web_de/statistiken/menschen_und_gesellschaft/bevoelkerung/haushalte_familien_lebensformen/haushalte/index.html.

Stiefelmeyer, H., Hanten, K.-P., and Pleschko, D. 2006. Hochwasserschutz in Österreich. Bundesministerium für Land- und Forstwirtschaft, Umwelt und Wasserwirtschaft.

The World Bank. 2014a. Air transport, freight (million ton-km). Retrieved from http://data.worldbank.org/indicator/IS.AIR.GOOD.MT.K1.

The World Bank. 2014b. Air transport. Retrieved from http://data.worldbank.org/indicator/IS.AIR.PSGR.

Trading Economics. 2016. Finland GDP from construction 1990–2016. Retrieved from http://www.tradingeconomics.com/finland/gdp-from-construction.

U.S. Department of Commerce, Bureau of Economic Analysis. 2015. BEA: Gross-domestic-product-(GDP)-by-industry data, value added 1947–2015. Retrieved June 2, 2016 from http://www.bea.gov/industry/gdpbyind_data.htm.

Vining, K.C., Pope, C.A., and Dugas, W.A. 1984. Usefulness of weather information to Texas agricultural producers. *Bull. Am. Meteorol. Soc.* 65:1316–1319.

World Bank. 2008. *Weather and Climate Services in Europe and Central Asia: A Regional Review*. Washington, DC: The World Bank.

World Bank, International Comparison Program database. 2016. GDP per capita, PPP (current international $). Retrieved June 2, 2016 from http://data.worldbank.org/indicator/NY.GDP.PCAP.PP.CD/countries?display=default.

World Bank national accounts data, and OECD National Accounts data files. 2016. GDP at market prices (current US$). Retrieved June 2, 2016 from http://data.worldbank.org/indicator/NY.GDP.MKTP.CD/countries?display=default.

World Meteorological Organization. 1968. The economic benefits of national meteorological service. Geneva.

World Meteorological Organization. 2015. Valuing weather and climate: Economic assessment of meteorological and hydrological services. WMO-No. 1153.

17

Performance Measurement of Location Enabled e-Government Processes: A Case Study on Traffic Safety Monitoring

Danny Vandenbroucke, Glenn Vancauwenberghe,
Anuja Dangol, and Francesco Pignatelli

CONTENTS

17.1 Introduction

Over the past 10 years, important efforts have been made to improve the access and sharing of location information, for example, through the Infrastructure for Spatial Information in the European Community (INSPIRE) directive (European Commission 2007) and the Global Monitoring for Environment and Security (GMES)/Copernicus initiative (European Commission 2016a) at European level and the development of Spatial Data Infrastructures (SDI) at national and regional levels (Masser and Crompvoets 2014). However, it is expected that European Union (EU) institutions and member state public administrations could benefit more from the potential of a consistent and integrated use of location information in e-government processes.

17.1.1 From Spatial Data Infrastructures to Location-Enabled e-Government Processes

Throughout the world, many countries and regions have developed a (national) SDI to facilitate the access, sharing, and use of spatial data. In Europe, the INSPIRE directive aims at establishing an infrastructure for spatial information in Europe to support community environmental policies and policies or activities that may have an impact on the environment. The focus of INSPIRE is on the sharing and harmonization of the many spatial data resources available in the member states, in order to allow their cross-border use. The data are made accessible through network services, which allow discovering, viewing, and downloading them through the web. Member states already make many location information and services available. For example, at the end of 2016, the European geoportal provided access to 135,571 spatial datasets and 59.364 network services (European Commission 2016b). However, the uptake and integration of this information and the services in e-government processes remain relatively weak (Pignatelli et al. 2014; Vandenbroucke et al. 2014).

In order to improve the uptake and integration of location information, the European Interoperability Solutions for European Public Administrations (ISA) program has set up a specific action, called the European Union Location Framework (EULF). The EULF aims to identify barriers and possible solutions for a consistent and interoperable use of location information and services, while promoting the reuse of INSPIRE where possible and feasible. In more general terms, the ISA program supports interoperability solutions as well as sharing and reuse of frameworks, architectures, and reusable components among European public administrations to enable more cost-effective e-government services and support cross-border applications. The EULF aims to deliver the location framework that underpins this broader vision (Pignatelli et al. 2014).

In the public sector, the implementation of policies mainly takes place through processes, in which policy is translated into a sequence of activities. A public-sector process can be defined as a set of related activities that transform a certain input of resources (e.g., a [spatial] dataset, a register, or statistical data) into an output of products or services (e.g., a decision, a permit, or an answer), which are delivered to citizens, businesses, or other administrations. Usually, the transformation requires the processing of the input data and information to generate the required output (Johansson et al. 1993). In the context of each policy domain, many processes are running. Moreover,

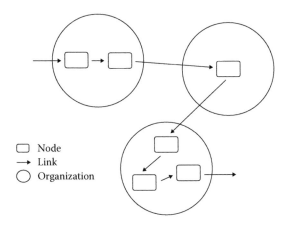

FIGURE 17.1
The process as a chain of activities within and between organizations. (From Dessers, E. et al., SDI at work, The spatial zoning plans case, KU Leuven, Leuven, 2011. With permission.)

many of the processes are interlinked. This means that, for example, a process might need the output of another process as input. For example, the initiation of a building permit might depend on the result of the control of the location of the cadastral parcel against flood risk areas.

Typical for the public sector is that processes are often similar in structure, as their outcome is typically determined by law. Moreover, most processes involve different organizations (Figure 17.1), at different administrative levels and/or in different thematic areas (Dessers et al. 2011; Gereffi et al. 2005).

Most processes consist of different—intra- and interorganizational—process steps and involve several interactions between stakeholders. These interactions can be divided into government-to-citizens (G2C), government-to-business (G2B), and government-to-government (G2G) interactions. Each of these interactions can take place in different phases of the process: at the start, at the end, or during the process. Many government processes often start with a G2C or G2B interaction, for example, a request from a citizen or a company, and end with a G2C or G2B interaction, for example, the delivery of a product or permission to a citizen or a company. However, these G2C and G2B interactions can also take place during the process (e.g., public consultations), whereas other processes exist without any G2C or G2B interactions. The latter often correspond with the so-called back-office processes (Pignatelli et al. 2016).

17.1.2 Location-Enabled e-Government Services

In order to support the interactions in processes, e-government services are developed. In the context of e-government, the term *service* is defined as the execution or result of an activity by a public administration that serves the citizen, businesses, or another public agency (Federal Ministry of Interior of

TABLE 17.1

Different Type of e-Government Services

Type of Service	Description	Interaction Types
Information services	Disclosure of government information or other information that is relevant for their citizens and businesses, for instance, the possibility to view or download reports, data, brochures, and other official documents.	One-way interactions (G2C and G2B)
Contact services	Possibility for citizens and businesses to contact public administration, for instance, to ask questions to civil servants and politicians about the application of certain rules and programs or to make a complaint.	One-way interactions (C2G and B2G)
Transaction services	The electronic intake and handling of requests and applications of rights, benefits, and obligations, such as digital tax assessments and the granting of permits, licenses, and subsidies.	Two-way interactions (G2C and G2B, as well as C2G and B2G)
Participation services	Allowing citizens and businesses to get involved in the formulation and evaluation of policy programs, by informing them about different policy options and allowing them to provide input and participate in discussions on these options.	Two-way interactions (G2C and G2B, as well as C2G and B2G)
Data transfer services	They refer to the exchange and sharing of (basic and standard) information between public (and private) organizations, including the exchange of information between different processes.	Two-way interactions (G2G and G2B/B2G)

Source: Bekkers, V., *Information Polity*, 12, 103–107, 2007.

Germany 2008). Nowadays, many e-government services take the form of web applications, apps on mobile devices, and so on; however, other channels might also be used (e.g., e-mail). A distinction can be made between five main types of e-government services (Bekkers 2007), as shown in Table 17.1.

The EU e-Government Benchmark reports define 20 basic e-government services, including 12 services related to citizens and 8 services related to businesses (Lörincz et al., 2010). Citizen services are, for example, income taxes, personal documents, car registration, application for a building permit, address change, and so on. Services for businesses are, for example, corporate tax, registration of a new company, environment-related permits, public procurement, and so on. Many of these services require location information; hence, the related processes can potentially be location-enabled. The concept of *location enablement* or *spatial enablement* of processes refers to the access, integration, and use of location information in different process steps, especially in the various interactions that take place. Location-enabled e-government services are services provided by public authorities and are supported by—digital—location data. The use and integration of location data allow us to make the

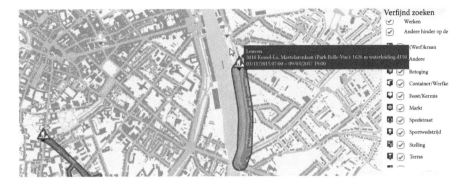

FIGURE 17.2
GIPOD, an example of location-enabled transaction service. (Agentschap voor Geografische Informatie Vlaanderen, http://www.geopunt.be/, 2016. With permission.)

existing services more efficient and effective or to provide new and innovative services. An examination of the same 20 basic e-government services shows that there are several services for which the use of location information is essential, such as the application for building permits, the granting of environment-related permits, and citizens informing municipalities of moving to another address. Figure 17.2 provides an example of a location-enabled transaction e-government service (Agentschap voor Geografische Informatie Vlaanderen 2016).

Generiek Informatieplatform Openbaar Domein (GIPOD) is an e-government service that allows to register different types of events (spanning a specific period of time) such as public works, public demonstrations, cycling events, rock concerts, and so on. The service allows to define the exact location (point or address, road segment, and so on) and describe the characteristics (type of event and period). The back office can then process the request, accept or reject it, and prepare the necessary permit(s).

Although location enablement of e-government processes has been a subject of research, the impact on process performance is less studied. This chapter aims to help fill this gap. Therefore, the objective of this chapter is to define and test a method to describe and estimate the impact of location-enabled e-government processes on the performance of those processes. First, process performance is defined in the context of e-government. Three indicators are proposed based on the literature: time, cost, and quality. The method is applied to one e-government process, that is, the Traffic Safety Monitoring processes. The process is described in terms of data flows and how G2G, G2B, and G2C interactions could be location-enabled. Finally, the impact of this location enablement on process performance is described in qualitative terms for the three indicators defined. The chapter ends with a discussion of the results and some conclusions and ideas for further work.

17.2 Method to Estimate Process Performance

The impact of location enablement of e-government processes on the process performance can be measured (or estimated) in different ways. Ideally, it is measured before adding location information components (ex-ante) and after the process is location-enabled (ex-post). However, this requires collecting performance information at least at two moments in time, which is not always possible when analyzing the existing e-government processes. This section first describes what process performance means, and then, it proposes a method and a series of indicators for measuring process performance.

17.2.1 Process Performance

Bouckaert and Halligan (2008) make a difference between the *span of performance* and *depth of performance*. The depth of performance refers to the level at which performance can be measured and managed. A distinction can be made between micro, meso, and macro levels. Performance measurement at the micro level deals with the performance of individual organizations or even at the level of a specific e-government service, whereas performance at the meso level deals with the performance of an entire e-government process or policy chain. Finally, macro performance is at a government-wide level or at the level of a government program (e.g., the impact of INSPIRE). The span of performance relates to the extent of the performance measurement: Is the performance measurement only dealing with inputs against outputs (looking merely to efficiency), or is it also looking at the outcomes (efficiency and effectiveness) and eventually even at the broader impact (e.g., on society)? Looking at effectiveness means that the span of performance is broadened beyond the borders of a single organization or process, and also, the effects outside the organization or process are taken into account.

Analyzing public-sector performance considers both span and depth of performance, that is, measuring input, activities, outputs, effects/outcomes, and even trust at different levels. Heckl and Moormann (2010) provide an overview of the existing methods for performance measurement, comparing the main objectives and characteristics of each method. Well-known methods, among others, are the following: balanced scorecards, return on investment studies, economic value added, and process performance. Figure 17.3 compares the different methods in terms of the focus and scope of the performance that they address. The figure clearly shows the key characteristics of the process performance (systems) method, where focus is on business processes and performance, is defined in the broader sense (efficiency and effectiveness).

Focus on		
	... the entire business or an organizational unit	... a single business process
Performance in a broad sense (efficiency and effectiveness)	Balanced scorecard Self-assessment	Process performance measurement systems
Performance in a narrow sense (primarily measuring efficiency)	Traditional controlling (e.g., return on investment, economic value added)	Activity-based costing Workflow-based monitoring Statistical process control

FIGURE 17.3
Comparison of the existing performance measurement approaches. (From Heckl, D. and Moormann, J., Process performance management, In J. vom Brocke and M. Rosemann, (Eds.) *Handbook on Business Process Management 2*, pp. 115–135, Springer-Verlag, Berlin Heidelberg, 2010. With permission.)

In the context of this study, the focus will be on the analysis of one particular case, that is, one well-defined e-government process. Therefore, the analysis is made at the meso level, taking into account the full process cycle, including the input, throughput, output, and outcomes in terms of efficiency and effectiveness (and with potential impacts on other processes). Therefore, the performance is analyzed at the level of a full process, taking into account the potential impact on other processes and hence performance in a broader sense, according to Figure 17.3. The analysis will be performed in two steps: First, the process is modeled to understand the required input, the process steps (throughput), and the output created. It allows to analyze the data and information flows and the existing interactions between the different actors. In a second step, the performance information for all the process steps is collected according to a series of indicators defined.

17.2.2 Indicators for Process Performance

Measuring the performance of a process is a necessary first step for assessing and improving the performance. Process performance measurement requires collection of quantitative and qualitative information about the process. Van der Aelst (2011) distinguishes between three main dimensions of performance, for which different indicators can be defined:

Time can be looked at from different angles and can be measured in different ways. An indicator that is often used is the lead time or flow time, that is, the total time from the start of a process to the completion of a process. Other indicators are the actual processing time, which is the working time for the process steps, and the waiting time (or *dead time*), which is the time for which a process step is waiting for a resource to become available or a previous process step to be finalized.

Cost can also be measured in several ways. Distinction is made between different cost factors for creating time indicators: labor costs, IT costs, production costs, product costs, service costs, and so on. Moreover, the distinction between fixed and variable costs is often used for designing cost indicators.

Quality mainly focuses on the product or service that is delivered at the end of the process to the customer. Measurements of user satisfaction are often used to measure the *quality of a process*; however, for measuring the quality of a process, several other indicators can be used (e.g., number of complaints).

Table 17.2 provides an overview of the applicability for location-enabled processes of each of these three key indicators.

The ultimate goal of performance analysis and management is to improve processes with respect to time, cost, and/or quality. This means that the process modeled in the first step of the analysis will be extended with performance data regarding flow time, cost, and quality. Measuring the performance of a process is not an end in itself but should provide a basis for actions for improving the performance and optimizing the process. Process optimization means that changes or adjustments to a process are made in order to get better results. The goal of developing and implementing spatially enabled processes is to improve the quality of the process and to reduce the process flow time and total cost of the process through the integration of spatial data and services in one or more process steps.

TABLE 17.2

Three Key Indicators and Their Applicability for Location-Enabled Processes

Key Indicators	Applicability in Different Processes
Flow time	*Building permits*: Period of time between the application and the issuance of a building permit
	Address registration: Period of time between the request of a new address and the inclusion of this new address into the national address register
Cost	*Flood mapping*: Resources needed to create a map of recently flooded areas
	Animal transport: Resources needed for spatiotemporal monitoring of a single animal transport
Quality	*Business register*: Correctness, completeness, and up-to-dateness of the national register of private companies
	Creation of spatial development plans: The degree to which new spatial development plans are considered of high quality by the different stakeholders of the spatial planning process and by related policy domains

17.3 The Traffic Safety Monitoring Case

In this subsection, we describe the traffic safety monitoring process as an example of an e-government process in general terms and the mechanisms to manage it, that is, through the Traffic Safety Policy Research Centre as key driver for policy preparation, monitoring, and evaluation. We also describe the results of the application of the method for estimating the process performance. The case was selected among a series of potential cases, including *monitoring animal transport, monitoring land use, managing flood areas*, and several others. Each potential case must meet certain criteria: (1) it must correspond to an existing e-government process; (2) the process should cover multiple organizations and preferably multiple levels of government; (3) it must show a potential for making it more location-enabled (already using location information, but not to its full potential); (4) it should be possible to involve the process stakeholders (owners and managers of the process); and (5) it should be feasible to work within the process without major developments. The traffic safety monitoring case was finally chosen because it has a huge potential for improvement and location enablement and also because the exercise fitted well in ongoing activities of the Traffic Safety Policy Research Centre.

17.3.1 The Traffic Safety Monitoring Process

The Traffic Safety Policy Research Centre was established in 2001 by the Flemish Government, with the aim to provide a scientific basis for policymaking in the field of transport and mobility, and, in particular, traffic safety. The goal of the Traffic Safety Policy Research Centre was to understand the anticipated social developments and challenges in order to take proactive measures to reduce the number of accidents and victims, which have a huge impact not only on individuals and their families but also on society and its economy as a whole. The specific goals assigned for traffic safety were as follows: data collection, short-term research on various policy matters, fundamental scientific research focusing on the traffic safety of its citizens, and the provision of policy support (Steenberghen et al. 2015). In order to attain tangible results, there was a subdivision into five work items relating to the following:

- Development of a traffic safety monitor based on data and indicators as a basis for traffic safety assessment: Analysis and dissemination of data and indicators concerning traffic safety and its underlying factors through a road safety monitoring system and an annual road safety report.

- Risk analysis: Computation of the relative safety level at various locations, based on analysis of registered data. Network safety management and analyzing road crash patterns by using collision diagrams.
- Human behavior in relation to system components and vehicle environment: Evaluation of the road system (on three dimensions: driver, environment, and vehicle) as a whole, with regard to road accidents and exploration of innovative solutions for each of the three dimensions of the road system.
- Development of road safety measures: Focus on the usefulness of education and engineering strategies that might intrinsically motivate drivers to behave safely. Investigation of influence on behavior by using three approaches: simulator-based training, in-vehicle technology, and road design and infrastructure.
- Ranking and evaluation of measures: Evaluation of traffic safety policies and provision of a ranking for publicly acceptable measures to meet the traffic safety targets, ensuring best possible use of available resources.

The traffic safety monitoring process consists of several subprocesses: (1) the registration of traffic accidents, (2) the assessment of traffic safety at national level, and (3) the assessment of traffic safety at European level (Vandenbroucke et al. 2016). There also exist related processes such as the traffic accidents insurance and the traffic accidents legal processes (in case of dead and/or heavy wounded). Owing to time constraints and because the exercise was conducted in collaboration with the Traffic Safety Policy Research Centre of Flanders, the performance assessment was analyzed only for the first two subprocesses.

17.3.1.1 Registration of Traffic Accidents (Operational Level)

The basic subprocess is the process of registering all the relevant traffic accidents that occur somewhere on the road network in Flanders. This is an operational process that focuses on the collection of all the relevant information in the field, each time an accident occurs (see start of the subprocess in Figure 17.4—lighter circle). It means that it is an event-driven process; however, only important accidents are registered. Information is collected on the location of the accident (X/Y coordinate, address, and linear referencing) and on the accident characteristics. This information is further processes by the local or the federal police and integrated and consolidated in a federal information system (called integrated system for local police (ISLP) and feeding information system (FEEDIS)). In a subsequent step, the information is aggregated, in order to generate statistics (algemene directie statistiek en economische informatie (ADSEI)—the National Statistical Institute). Finally, the information goes to the mobiliteit en openbare werken (MOW) (and the provinces) to prepare the information for assessing causes and trends in

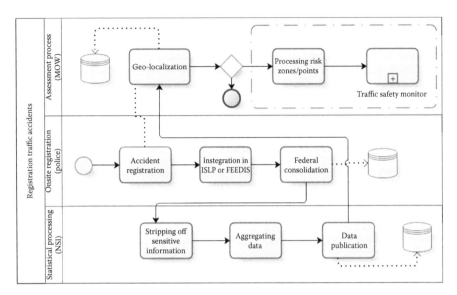

FIGURE 17.4
Overview of the traffic accident registration (sub-)processes. (Based on Vanhaverbeke, L. et al., SDI at work: The traffic accidents registration case, KU Leuven, Leuven, 2011.)

view of policy measures (see top of Figure 17.4). Therefore, it is this process that then feeds the traffic safety assessment subprocess. Figure 17.4 provides an overview of this subprocess (which ends at the darker circle) and also provides a link to the other subprocess, the traffic safety assessment process, which makes use of the *traffic safety monitor* platform.

17.3.1.2 Traffic Safety Assessment (Policy Level, National)

The core policy process is the traffic safety assessment subprocess in which an assessment is made of all the traffic accidents as they occurred over a certain period of time, in order to better understand them (e.g., causes) and to formulate policy measures to reduce their number and the negative impact that they have, for example, by reducing the number of deaths and heavy injuries. In this subprocess, the information collected in the registration process is used as a starting point. The information is being prepared for use in the so-called *traffic monitor,* an information platform that is based on a series of indicators. The indicators are defined and regularly revised, and mapping is prepared to evaluate patterns, for example, by preparing black point maps.

17.3.2 Performance of the Traffic Safety Monitoring Process

In order to understand the different steps in the traffic safety monitoring process, the process and its subprocesses are modeled. This is done using the BPMN standard (Open Management Group 2011). After modeling the process and its subprocesses, the objective is to evaluate the use of location information and location-enabled web services in the traffic safety monitoring process and subprocesses and to identify where in the subprocesses the newly added value e-government services might be developed. The assessment was done for the two subprocesses separately, based on the information collected through interviews and observations. For each subprocess, the impact of the spatial enablement of some process steps on the (sub-)process performance was estimated.

17.3.2.1 Registration of Traffic Accidents

Before discussing where the traffic accident subprocess could be better spatially enabled and how process performance could be enhanced, some general considerations should be made with regard to the organization of the traffic accident registration subprocess and how this might impact the process performance.

1. The overall registration process (from begin to end) is taking much time. The throughput or flow time for the subprocess is minimum 2 years, but often, it is taking 3 years. This makes the data available for policy purposes to be out of date when they reach the policy stakeholders involved. However, the registration itself, in the field, including the geolocalizing and entering into the systems of the police, is going relatively fast.

2. The *slowdown* of the process is happening at specific activities (steps) in the subprocess, which makes the traffic accident data unavailable for a long period of time. This happens, for example, at the time the data are delivered by the federal police to the statistical office (*dead time* or waiting time): traffic accident information is collected over the time span of 1 year. The data are then grouped before being delivered to and processed by the statistical office: the single accidents are not processed, but this is done in a batch process. The same happens when the data arrive at the Department of Mobility of the Flemish Government and the provinces. The process to *fix* the data takes around 6 months annually.

3. The Belgian privacy laws and rules slow down the subprocess as well: not only are exact locations stripped from the original database, but the data are also aggregated to *hide* what is considered *personal* information.

4. The order in which the data are processed seems not entirely logic. In fact, the statistical office might be considered as one of the (key) end users of the data at the end of the chain rather than in the middle of it. The Department of Mobility and the provinces could get the data earlier in the process (in parallel), directly from the federal police, and could then process the data according to their needs with respect to the privacy laws and rules.

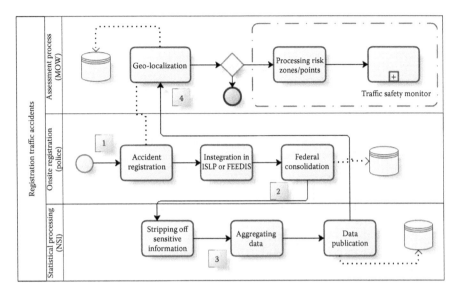

FIGURE 17.5
Places in the accident registration subprocess where the integration of location information could be improved.

Figure 17.5 provides the same BPMN scheme of the traffic accidents registration subprocess as Figure 17.4, but this time, it has four areas indicated where location information could be better integrated and used. The four areas were identified based on observations of the process (flows) and were confirmed by process owners and managers.

1. Accident registration app: A drastic change would be to automate the collection of information on the accidents in the field. Police officers could make use of tablets or other mobile devices to collect information about the accident by making use of the global positioning system (or, in the future, Galileo technology), such as to locate the precise location of the accident, to sketch and annotate, to fill in forms and directly write the information into the central database of the police, and so on.

2. Harmonized address register: Along the process, several address databases are used. This leads to mismatching and difficulties in the localization of certain accidents. The use of one central address register would avoid this problem. Work is on its way to harmonize and streamline addresses, but the current situation is still very mixed.

3. Use of OGC web services: For the processing and aggregation of the data at the statistical office, OGC web services could be used for reference data such as administrative boundaries, the road network,

TABLE 17.3

Expected Impacts of Further Location Enablement of the Traffic Safety
Registration Subprocess on Process Performance

	Flow time	Cost	Quality
1. Accident registration app	Very high	High	Medium to high
2. Harmonized address register	Medium	Medium	High
3. Use of OGC web services	Low to medium	Very low	Medium
4. Geolocalization service	Medium	Medium	Low

and so on. Currently, local geospatial datasets are used, which
require regular updates.

   4. Geolocalization service: In the phase of geolocalizing the accident
      data by the Department of Mobility and the provinces, more OGC
      web services could be used, which will allow to harmonize the way
      the localization is done (e.g., same underground reference data such
      as the large-scale reference database of Flanders). Moreover, an
      option would be to offer geolocalization as an e-government service
      that could be reused in other processes.

Table 17.3 provides an overview of the estimated impact on the process performance in terms of flow time, cost, and quality. The impact is expressed in
terms of major categories but could of course be refined.

   Since the possible (required) actions were relatively complex and time-
consuming and also required agreement with many stakeholders, none of
the described improvements were tested for this subprocess during the case
study (but might be tested in the future).

17.3.2.2 Assessing Traffic Safety

For the traffic safety assessment subprocess, the impact of some location-
enablement measures on the process performance were also estimated
in terms of flow time, cost, and quality, and some of them were tested.
They are summarized in Table 17.4, whereas Figure 17.6 provides the
Business Process Model and Notation (BPMN) schema of the traffic
safety assessment subprocess, indicating the five areas where location
enablement could improve the process.

   1. Quality assurance service: During the preprocessing activity, after the
      traffic accident data are received from the statistical office, the data are
      investigated using different kind of tools, including geospatial and
      statistical tools and additional spatial data layers. The activity corre-
      sponds to a quality control assurance procedure, which could even-
      tually be automated and offered as a back-office G2G service, which

TABLE 17.4

Expected Impacts of Further Location Enablement of the Accident Monitor Subprocess on Process Performance

	Time Flow	Cost	Quality
1. Quality Assurance service	High	Medium	High
2. Use of OGC web services	Medium	Very low	Medium
3. Service for generating indicators	High	Medium	High
4. Publication of OGC web services	Low	Low	High
5. Use of linked data	Unknown	High	High

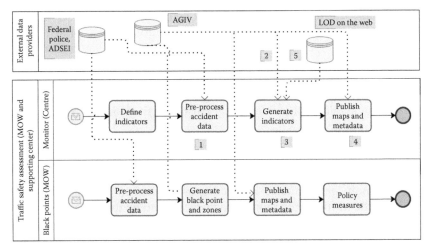

FIGURE 17.6
Places in the traffic safety assessment subprocess where the integration of location information could be improved.

could also inform the relevant stakeholders of the traffic accidents registration process of the shortcomings/errors. This was not tested.

2. Use of Open Geospatial Consortium (OGC) web services: For generating the indicators, several geospatial datasets are used from the Agency for Information in Flanders (AIV). This could be done by accessing them via OGC web services. This was tested, and the expected impact was confirmed: efficiency gains are in the order of magnitude of 1 week to a few weeks, because there is no need any more to prepare local copies of the data.

3. Service for generating indicators: Supporting the generation of the indicators is a fixed process for each indicator. Once the indicators are defined and once the calculation method is fixed, they can be generated automatically. In the longer term, this activity could be offered as a web processing service. This was not tested either.

4. Publication of OGC web services: The publication of the traffic safety information happened in the past in a static manner through statistics and static maps. In this use case, the resulting indicators were published as web map services. Unfortunately, downloading of the information in the form of a web feature service is not allowed for the same privacy rules and regulations mentioned previously. Moreover, metadata on the indicators are published along with the services. The impact of this measure is not only important for the policy-makers (instant access to all the information) but also for other users or processes that make use of the services.

5. Use of linked data: Much information on traffic and traffic safety is available through the web. Different communities are active in this regard (e.g., cycling organizations). Therefore, the use of semantic web technologies such as linked data might be a way to link the base information from the traffic safety monitoring to this information. This was tested, and it was estimated that especially the quality of the information was improving in terms of new insights in the already available information by enriching the basic data delivered in the monitor.

17.3.3 Discussion

The three main performance indicators, as defined in Section 17.2.2, were applied on the traffic safety monitoring process and on the two subprocesses, in particular, that is, the registration of traffic accidents and the assessment of traffic safety. With the information available, the indicator *time* was assessed from the perspective of the *flow time* or throughput time. The *cost* indicator was assessed based on the required resources to create and analyze the traffic accident database, respectively. Moreover, within each of the two subprocesses, the indicators could be applied on individual process steps, in order to assess the performance of each individual step. This especially allowed us to identify bottlenecks in the process, that is, process steps in which the performance of the process is particularly low or in which the process is even temporarily stopped (*dead time*). An example is given in Figure 17.7. The major reason for having a long flow time (can be more than 1 month) at the

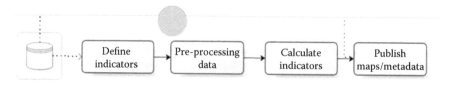

FIGURE 17.7
A bottleneck or slowdown in the early stage of the traffic safety assessment subprocess.

TABLE 17.5

Comparison of the Current Performance of the Traffic Accident Registration Process and the Optimal Performance

Indicator	Optimal Process	Process Flanders
Flow time	After registration on the spot by the local police zone, a traffic accident is immediately uploaded in the central database of traffic accidents	In some cases, it takes more than 3 years before a traffic accident is included into the central traffic accident database
Cost	No further processing activities are needed after the registration of the traffic accident by the local police	After registration of the accident by the local police, four more public organizations spend resources on the processing of the individual accidents into one single traffic accident database
Quality	A complete, correct, and up-to-date traffic accident database is available to all stakeholders (with respect for privacy regulations)	The central traffic accident database is outdated, does not include all traffic accidents, and is not available to the public

beginning of the preprocessing of the data is the need to harmonize the data because of the changes in the data model in the course of the registration subprocess.

An interesting way of analyzing, assessing, and describing the performance of the existing processes (or subprocesses) is by comparing them with the performance of the optimal process. To illustrate this, Table 17.5 provides a summary of the performance of the current traffic accident registration process in Flanders and how a more performant traffic accident process could look like. The table shows how time, cost, and quality are useful indicators for describing the performance of processes.

The total flow time of the process is generally considered as one of the main weaknesses of the registration process. It can take up to 3 years between the occurrence and registration of an accident and its inclusion in the consolidated central database. This is mainly due to the fact that the process is organized in a strongly sequential manner; that is, some process steps are initiated only after the previous step is fully completed for several cases. This makes the amount of waiting time for some process steps significantly higher.

Even without detailed information on the cost associated with the creation of a central traffic accident database, some evidence is available on the amount of resources spent on the creation of the database. Besides the local police who is in charge of the registration of traffic accidents on local roads, four more public authorities—at three different administrative levels—are involved in the creation of a traffic accident database: the federal policy, the Directorate-General for Statistics and Economic Information (ADSEI), the provinces, and the Flemish Department for Mobility and Public Works. Each of these organizations is responsible for a particular process step.

With regard to the quality of the process, and of its outputs, in particular (database of accidents), some important observations can also be made. An important problem in the traffic accident registration is the underregistration of accidents. Only accidents with injuries or death are included in the database; data on accidents with material damage are missing. Moreover, the database of traffic accidents in Flanders is not online accessible for the general public. A third quality issue—or weakness—is also related to the time dimension. The most recent data on traffic accidents in Flanders are the data related to accidents in 2013 (at the time of writing).

The goal of developing and implementing spatially enabled processes is to improve the quality of the process and to reduce the process flow time and total cost of the process through the integration of spatial data and services in one or more process steps, as was demonstrated in the previous sections. Several examples can be given of how the use of spatial data and services has improved the time, cost, and/or quality of the traffic accidents registration process:

- Availability of an up-to-date detailed, large-scale reference map improves the quality of the *on-the-spot* registration (*drawing*) of an individual accident, as data on the surrounding of the accident (road characteristics, traffic signs, and so on) should not be registered manually.

- Availability of a reference address database means that the majority of all traffic accidents in Flanders could be localized automatically, without intervention of an operator (time and cost savings).

- Direct exchange of traffic accidents data, including information on the location of the accident, between the local police zones and the Flemish Department of Mobility could dramatically decrease the flow time of the process.

17.4 Conclusion and Further Work

In this chapter, a case study approach has been defined in order to assess the impact of location enablement of e-government processes. The assessment is done in two steps: first by mapping the process and subprocesses by means of the BPMN standard, and second, by the assessment of the process in terms of where in the process location information could play a more prominent role, as well as in terms of estimating the impact on the process performance. For estimating the process performance, three indicators were defined: flow time, cost, and quality. One process was selected to test the approach, based on a series of criteria, that is, the traffic safety monitoring process.

This process consists of several subprocesses and related processes. The most important ones were analyzed in more detail: the traffic accidents registration and the traffic safety assessment subprocess. Both subprocesses were described in detail by using the BPMN standard in terms of actors, input, process activities, and outputs, with particular attention to the use of location information and the data flows throughout the different process steps. Subsequently, the process was assessed in terms of where in the process the use of location information and location-based services could be improved and the impact that such location enablement could have on process performance.

The integration of the existing OGC web services in the traffic safety monitor proved to be an easy step to make the use of the geospatial data in the process more efficient and effective. In addition, the publication of the indicator maps as OGC web services was very straightforward and provided added value, mainly for other stakeholders that can reuse them in their own processes (e.g., provinces). Several potential e-government services were also identified. Owing to time and resource constraints, these could not be implemented as part of the case study. Finally, the importance of focusing on a harmonized approach for data gathering, in this case the traffic accidents information cannot be overestimated. Harmonized data have a major impact on the performance further in the process, for example, because of the possibility to avoid the need for extensive reengineering of data structures. In the context of the traffic safety case, some initial work started to enrich the information of the traffic safety monitor with other transport and mobility data found on the web by using linked data technologies. Owing to the same time and resource constraints, this topic could only be handled partially but will be further explored in the future.

References

Agentschap voor Geografische Informatie Vlaanderen (2016). http://www.geopunt.be/.
Bekkers, V. (2007). Modernization, public innovation and information and communication Technologies: The emperor's new clothes? *Information Polity*, 12:103–107.
Bouckaert, G. and Halligan, J. (2008). *Managing Performance: International Comparisons*. London, UK: Routledge.
Dessers, E., Crompvoets, J., Janssen, K., Vancauwenberghe, G., Vandenbroucke, D. and Vanhaverbeke, L. (2011). SDI at work. The spatial zoning plans case. Leuven: KU Leuven.
European Commission (2016a). Retrieved November 12, 2016 from http://www.copernicus.eu/.
European Commission (2016b). Retrieved November 12, 2016 from http://inspire-geoportal.ec.europa.eu/.
European Commission (2007). Directive 2007/2/EC of the European Parliament and of the Council of March 14, 2007 establishing an Infrastructure for Spatial Information in the European Community (INSPIRE). Retrieved January 1, 2017 from http://eur-lex.europa.eu/legal-content/EN/ALL/?uri=CELEX:32007L0002.

Federal Ministry of the Interior of Germany (2008). SAGA–Standards and architectures for eGovernment applications, Version 4.0. Federal Ministry of Interior: Berlin, Germany.

Gereffi, G., Humphrey, J. and Sturgeon, T. (2005). The governance of global value chains. *Review of International Political Economy*, 12(1), 78–104.

Heckl, D. and Moormann, J. (2010). Process performance management. In J. vom Brocke and M. Rosemann (Eds.) *Handbook on Business Process Management 2* (pp. 115–135). Strategic Alignment, Governance, People and Culture. Berlin, Germany: Springer-Verlag.

Johansson, H.J., McHugh, P., Pendleburg, A., and Wheeler, W.A. (1993). *Business Process Reengineering: Breakpoint Strategies for Market Dominance*. Chichester, UK: John Wiley & Sons.

Lörincz, B., Tinholt, D., van der Linden, N., Colclough, G., Cave, J., Schindler, R., Cattaneo, G., Lifonti, R., Jacquet, L. and Millard, J. (2010). Digitizing public services in Europe: Putting ambition into action. 9th benchmark measurement. Brussels, Belgium: Capgemini, IDC, Rand Europe, Sogeti and DTi.

Masser, I. and Crompvoets, J. (2014). *Building European Spatial Data Infrastructures*. Redlands, CA: ESRI Press.Open Management Group (2011). Business process model and notation (BPMN), Version 2.0. Retrieved November 12, 2016 from http://www.omg.org/spec/BPMN/2.0/.

Pignatelli, F., Smits, P., Vandenbroucke, D., Vancauwenberghe, G., Boguslawski, R. and Borzachiello, M. (2016). *European Union Location Framework: Design of Location-enabled e-Government Services*. Ispra, Italy: EC JRC.

Pignatelli, F., Smits, P., Boguslawski, R., Vowles, G., Borzacchiello, M., Vandenbroucke, D., Vancauwnberghe, G. and Crompvoets, J. (2014). *European Union Location Framework (EULF): Strategic Vision*. Ispra, Italy: EC JRC.

Steenberghen, T., Dangol, A., Tirry, D. and Dewaelheyns, V. (2015). Handleiding instrumentarium verkeersveiligheidsmonitor. Steunppunt Verkeersveiligheid 2012–2015: Diepenbeek, Belgium.

Vandenbroucke, D., Vancauwneberghe, G., Dangol, A., Boguslawski, R. and Pignatelli, F. (2016). *European Union Location Framework– Use Case on Traffic Safety Monitoring*. Ispra, Italy: EC JRC.

Vandenbroucke, D., Vancauwenberghe, G., Crompvoets, J., Pignatelli, F., Smits, P., Boguslawski, R., Vowles, G. and Borzacchiello, M.T. (2014). *Assessment of the Conditions for a European Union Location Framework*. Luxembourg: Publications Office of the European Union.

van der Aalst, W.M.P. (2011). *Process Mining: Discovery, Conformance and Enhancement of Business Processes*. Berlin, Germany: Springer.

Vanhaverbeke, L., Crompvoets, J., Dessers, E., Janssen, K., Vancauwenberghe, G. and Vandenbroucke, D. (2011). SDI at work: The traffic accidents registration case. Leuven: KU Leuven.

18

The Value of Geospatial Information, As We Look Forward—A Summing Up

Joep Crompvoets, Jamie Brown Kruse, and Françoise Pearlman

The value of geospatial information is widely recognized throughout the world. The 2030 Agenda for Sustainable Development was established by the United Nations (UN), including 17 Sustainable Development Goals (SDGs) and 169 associated targets. The SDGs were adopted and launched by the UN General Assembly in September 2015 and will frame the global development agenda through to 2030 (United Nations 2015). In order for the goals and targets to be implemented and achieved, strengthening geospatial data production and the use of better geospatial data in policy-making and monitoring have been recognized as critical. The United Nations had already set the foundation almost a decade ago in creating the UN Spatial Data Infrastructure (UNGIWG 2008) and established the UN Committee of Experts on Global Geospatial Information Management in 2011 (UN-GGIM 2017). More recently, as per one of the main recommendations contained in the report titled *A World That Counts* (United Nations 2014), presented in November 2014 by the UN Secretary-General's Independent Expert and Advisory Group on Data Revolution for Sustainable Development, the Statistical Commission agreed that a UN World Data Forum on Sustainable Development Data (UN World Data Forum) would be a suitable platform for intensifying cooperation with various professional groups, such as information technology, geospatial information managers, data scientists, and users, as well as with civil society stakeholders (UN World Data Forum 2017).

Many of the SDG targets are thematically based and geographic in nature. This provides an ideal opportunity for the global geospatial community to ensure that the value that geospatial information has in the development of the targets and indicators is realized. All the SDG developments occur in a location-based environment, and geospatial information provides a fundamental baseline for the global indicator framework, as well as for measuring and monitoring the SDGs. Many of the sustainable development challenges are cross-cutting in nature and are characterized by complex interlinkages that will benefit from using location as a common reference framework. To effectively measure, monitor, and mitigate challenges, we need to link demographic, statistical, and environmental data together with the one thing that they have in common—geospatial data (UN Committee of Experts on Global Geospatial Information Management 2016).

While the United Nations and the SDGs are considered a global effort, decisions that are made at national, regional, and local levels significantly benefit from geospatial information. The paradigm of geospatial information is changing; it is no longer used strictly for mapping and visualization but is also used for integrating with other data sources, data analytics, and modeling. Once the geospatial data are created, they can be used numerous times to support a wide range of different applications and services. Geospatial information has become an essential component in decision-making processes at many levels. Using location as the unifying unit of analysis, knowing where people and things are, and their relationship with each other are essential for informed decision-making. Not only is real-time information needed to prepare for, and respond to, natural disasters and political crises, but location-based services are also helping to develop strategic priorities, make decisions, and measure and monitor outcomes. The point of *services* is that the geospatial data are moving beyond data, to information and knowledge as key services.

Our ability to create data is still, on the whole, ahead of our ability to solve complex challenges of using the data. As big data gets bigger, there remains no doubt that there are untapped and undiscovered benefits that are still to be gained from the knowledge that can be derived from the data. The growth in the amount of data collected not only brings with it a growing requirement to be able to find the right information at the right time but also presents challenges of how to store, maintain, and use the data and information that are created.

Despite the societal imperatives and intrinsic value of data, it is surprising that the valuation of geospatial information is still in its infancy in the sense that it is still difficult to exactly quantify the value of geospatial information and examine how its societal benefits develop and accrue over time. In any case, GEOValue approaches require time and effort for *building a track record* (Broom 1995). What we can see is that the valuation approaches likely will evolve to be more extensive, comprehensive, user-oriented, demand-driven, diverse, and more closely tied to explicit targets.

As previous chapters have shown, there is a number of important issues related to the value of geospatial information from conceptual, technical, sociotechnical, political, legal, institutional, and financial perspectives. Behind the saying "You can't manage what you don't measure," attributed to W. Edwards Deming and Peter Drucker (McAfee and Brynjolfsson 2012), the process of using appropriate and rigorous valuation methods for geospatial information will serve its dual capacity as a performance metric. In this book, some of the chapters have been devoted to geospatial information as a performance metric for measuring changes due to management actions applied to environmental resources and the built environment. In other chapters, the valuation exercise produces a performance metric that pertains to the management of geospatial information. A body of knowledge and experiences relating to performance assessment activities gives us some insight (Neely 2005; Bouckaert 2006; Bouckaert and Halligan 2007; Van Dooren et al. 2015). From this, it appears that certain temporal patterns or sequences are quite common and might also be applicable for the GEOValue community. It is vital to understand how strategies in different countries and different application domains are evolving. According to Ingraham et al. (2000), monitoring and understanding "performance must become a way of life and a critical part of the culture" of the public and private sectors in the future.

If we consider geo-valuation (i.e., valuation of geospatial information) as a new example of performance management (Pollitt 2007), we can expect the following:

- The GEOValue community will become bigger.
- The growing GEOValue community will include a wide range of disciplinary approaches. The community of discourse will contain inter alia spatial scientists, data scientists, political scientists, management specialists, sociologists, economists, statisticians, public administration scientists, and domain experts in the application areas of agriculture, energy supply, insurance, spatial planning, regional development, transport, and so on.
- The foci of the specific interest will broaden. Geo-valuation will range from high-level studies that address frameworks of valuation criteria to detailed contextual use cases.
- Information and communication technologies (ICTs) will support and sometimes shape the geovaluation. For example, there are studies of how new ICTs may facilitate the collection and dissemination of geospatial valuation (see, e.g., different INSPIRE Member States reports published in 2016).*
- The current geovaluation studies will be complemented with case studies of how the valuation approaches are used by practitioners.
- The community will advance the science of valuation of geospatial information by discovering and promoting better approaches that are valid and reliable, so that valuation studies are consistent, repeatable, and as unbiased as possible.

---

* According to Article 21 of EU INSPIRE Directive (European Parliament and Council, 2007), EU Member States shall send a report about the implementation progress of the directive every 3 years, starting from 2010. At INSPIRE Website "INSPIRE in your country," an overview of the reports is presented, including the reports of 2016 (see http://inspire.ec.europa.eu/inspire-your-country-table-view/51764).

So far, this final chapter looked at how the GEOValue community is expected to develop over time. The development over time will be accompanied by evolution of the principles and criteria. How are the GEOValue approaches expected to evolve? On the basis of literature (e.g., Neely 2005; Pollitt 2007; Van Dooren et al. 2015) and experiences relating performance management activities in the public and private sectors, we might expect the following:

- *Culturally shaped* paths: The societal culture may shape how valuations are used or, indeed, whether they are used at all.
- *Steady, incremental development*: This refers to gradual assessment shifts of focus on inputs and processes to outputs and finally outcomes aligned with the trend to focus more on the demands of users (Van Dooren et al. 2015).
- *Patterned alternations*: This mainly refers to regular changes within the approach to fit changing times and priorities. Sometimes, the changes are small (the technical definition of a criterion is changed slightly) or bigger (criteria are dropped altogether and new ones are introduced).

It will be no surprise that future geovaluation studies will change their criteria, since there is a number of reasons why criteria may be altered. One good reason is that the experts learn from the process and want to replace an existing criterion with the one that it is simpler or more comprehensive. Another reason is that shifts in public and political attention may require that new criteria be added. A third reason is that new procedures or technologies are adopted that require new measures for valuation (no one was valuing web services in 1990). A fourth reason is that a value criterion just becomes irrelevant, and it is altered or dropped. A fifth reason is that there may be a tendency to cycle between many criteria and few criteria. As valuation approaches become increasingly sophisticated in an attempt to capture the full contribution of geospatial information, the pendulum may swing to move away from complex approaches to simpler key criteria.

But what does the book contribute to our understanding of the societal value of geospatial information? We believe this contribution is highlighted by three themes that are reflected in the book. These are (1) the discussions surrounding the concept of GEOValue, (2) the detailed exploration of the component pieces of the multidimensional value chain linking observations to outcomes, and (3) existing research that links theory to geovaluation in practice.

The book adopted a broad scope of GEOValue, recognizing many perspectives and proving that geovaluation is evolving. It identified potential drivers and dilemmas that are currently influencing the development of geovaluation and explored in depth the importance of a user focus and recognition of what happens in practice.

It appears that what is valid for performance management in other domains is likely applicable for geovaluation as well. It is also likely to be shaped by prevailing cultures and by accidents and other foreseen events. Geovaluation will likely have developmental trajectories over time. Behaviors will adapt to the presence of particular valuation regimes. Some approaches and their value criteria will become obsolete. We assert that geovaluation methods will not stand still. They are subject to endogenous and exogenous pressures, which lead to change—sometimes incremental and at other times transformational.

It should be noted that the objectives of the geospatial information management and the purpose and usage of geovaluation are keys to the selection and refinement of the GEOValue methodology. The goal is to design an overall framework for geovaluation that takes into account the different purposes, views, approaches, developmental trajectories, and value regime. Designing such an *evolutionary* framework will be a challenge for the GEOValue community, as it is very likely that applications continue to be more comprehensive, realistic, and critical where resource limitations must be addressed.

References

Bouckaert, G. (2006). The public sector in the 21st century: Renewing public sector performance measurement. *Köz-Gazdaság*, 1: 63–79.

Bouckaert, G. and J. Halligan (2007). *Managing Performance: International Comparisons*. London: Routledge/Taylor & Francis Group.

Broom, C.A. (1995). Performance-based government models: Building a track record. *Public Budgeting & Finance*, 15(4): 3–17.

European Parliament and Council (2007). Directive 2007/2/EC of the European Parliament and of the Council of 14 March 2007—Establishing an Infrastructure for Spatial Information in the European Community (INSPIRE). *Official Journal of the European Union*, L108: 1–14.

Ingraham, P., Coleman, S.S., and D.P. Moynihan (2000). People and performance: Challenges for the future public service, the report from the Wye River Conference. *Public Administration Review*, 60: 54–60.

McAfee, A. and E. Brynjolfsson (2012). Big data: The management revolution. *Harvard Business Review*, 90: 62–68.

Neely, A. (2005). The evolution of performance measurement research: Developments in the last decade and a research agenda for the next. *International Journal of Operations & Production Management*, 25(12): 1264–1266.

Pollitt, C. (2007). Who are we, what are we doing, where are we going? A perspective on the academic performance management community. *Köz-Gazdaság*, 1: 73–82.

United Nations (2014). A world that counts–Mobilising the data revolution for sustainable development. Report prepared at the request of the United Nations Secretary-General, by the Independent Expert Advisory Group on a Data Revolution for Sustainable Development, New York.

United Nations (2015). Transforming our world: The 2030 agenda for sustainable development. Resolution adopted by the General Assembly on September 25, 2015, New York.

United Nations Committee of Experts on Global Geospatial Information Management (2016). *Future Trends in Geospatial Information Management: The Five to Ten Year Vision* (2nd ed).

United Nations Committee of Experts on Global Geospatial Information Management (2017). About UN-GGIM. Retrieved May 17, 2017, from http://ggim.un.org/about.html.

UNGIWG (2008). United national spatial data infrastructure–UNSDI. United Nations Geographic Information Working Group. Retrieved May 17, 2017, from http://www.ungiwg.org/content/united-nations-spatial-data-infrastructure-unsdi.

UN World Data Forum (2017). About. Retrieved May 17, 2017, from https://undataforum.org/WorldDataForum/about/.

Van Dooren, W., Bouckaert, G., and Halligan (2015). *Performance Management in the Public Sector* (2nd ed.). London, UK: Routledge.

Index

Note: Page numbers followed by f and t refer to figures and tables respectively.

For Product Safety Concerns and Information please contact our EU representative GPSR@taylorandfrancis.com Taylor & Francis Verlag GmbH, Kaufingerstraße 24, 80331 München, Germany

T - #0200 - 160425 - C356 - 234/156/17 - PB - 9780367878894 - Gloss Lamination